Electronic Tongues

Fundamentals and recent advances

IOP Series in Sensors and Sensor Systems

The IOP Series in Sensors and Sensor Systems includes books on all aspects of the science and technology of sensors and sensor systems. Spanning fundamentals, fabrication, applications and processing, the series aims to provide a library for instrument and measurement scientists, engineers and technologists in universities and industry.

The series seeks (but is not restricted to) publications in the following topics:

- Advanced materials for sensing
- Biosensors
- Chemical sensors
- Industrial applications
- Internet of Things
- Lab-on-a-chip
- Localization and object tracking
- Manufacturing and packaging
- Mechanisms, modelling and simulations
- MEMS/NEMS
- Micro and nanosensors
- Non-destructive testing
- Optoelectronic and photonic sensors
- Optomechanical sensors
- Physical sensors
- Remote sensors
- Sensing for health, safety and security
- Sensing principles
- Sensing systems
- Sensor arrays
- Sensor devices
- Sensor networks
- Sensor technology and applications
- Signal processing and data analysis
- Smart sensors and monitoring
- Telemetry

Authors are encouraged to take advantage of electronic publication through the use of colour, animations, video and interactive elements to enhance the reader experience.

Electronic Tongues

Fundamentals and recent advances

Edited by
Flavio M Shimizu, Maria L Braunger and Antonio Riul Jr
*'Gleb Wataghin' Institute of Physics, University of Campinas (UNICAMP),
São Paulo State, Brazil*

IOP Publishing, Bristol, UK

ISBN 978-0-7503-3687-1 (ebook)
ISBN 978-0-7503-3685-7 (print)
ISBN 978-0-7503-3688-8 (myPrint)
ISBN 978-0-7503-3686-4 (mobi)

DOI 10.1088/978-0-7503-3687-1

Version: 20211101

IOP ebooks

British Library Cataloguing-in-Publication Data: A catalogue record for this book is available from the British Library.

Published by IOP Publishing, wholly owned by The Institute of Physics, London

IOP Publishing, Temple Circus, Temple Way, Bristol, BS1 6HG, UK

US Office: IOP Publishing, Inc., 190 North Independence Mall West, Suite 601, Philadelphia, PA 19106, USA

Dedicated to all those who directly or indirectly contributed to getting us this far on this road.

Contents

Preface

This book brings fundamentals and theoretical aspects about detection techniques applied for electronic tongue systems (e-tongue) that will be interesting to all materials scientists and researchers from universities and national laboratories working in the field of multisensory systems. Our primary motivation was to bring the reader basic knowledge and fundamental principles involved in e-tongue sensing, avoiding redundancy with previous publications. At the same time, to provide an overview of developments in different systems and look towards the future. The final spice behind the scenes was to group the key players in a single edition during planning. The book features most of the main researchers responsible for the development of this research line, such as Kioshi Toko, Corrado Di Natale, Roberto Paolesse, Manel del Valle, Larisa Lvova, Osvaldo N. Oliveira Jr, and Antonio Riul Jr. We invite the reader to immerse themselves in the basic knowledge, passing through developments including the integration of emerging technologies in this type of sensing, closing with a fundamental roadmap upon the use of computational resources in data analysis.

In Chapter 1, a brief introduction describing pioneering works, bringing the reader common definitions and terminologies used in the area till the IUPAC definition. We will call the readers' attention to different materials and techniques applied in e-tongue systems (that will be presented in the following chapters), closing it by showing how the area has evolved, being now spread worldwide. Chapter 2 covers the fundamentals of potentiometric e-tongue using lipid/polymer membrane, i.e., taste sensor. This taste sensor can respond to each taste quality independently and hence can quantify and classify the taste of foods into six kinds of taste qualities composed of saltiness, sourness, sweetness, bitterness, umami, and astringency. Chapter 3 presents a step-by-step procedure of the development of potentiometric e-tongues, exemplifying with interesting experimental setups and a short review of different applications of these sensors. The idea of chapter 4 is to present the fundamentals of voltammetric e-tongues and their use with enzyme-modified electrodes. Chapter 5 explores the use of MIPs (molecularly imprinted polymers) to modify electrodes and produce the sensor array in the voltammetric e-tongues, showing results from the literature and their research group. Chapter 6 presents the fundamentals of impedimetric e-tongues, starting with a brief introduction to the principles of impedance spectroscopy and how this technique can be used in e-tongues for various applications. The authors will also elaborate upon the types of interdigitated electrodes (IDEs) employed in the sensing units of impedimetric e-tongues, their fabrication process, geometric characteristics, and experimental conditions that may affect sensing performance. In chapter 7 emergent technologies that are paving the way for innovation in the development of impedimetric e-tongues using microfluidics, 3D printing, paper-based devices among other approaches, aiming at low-cost implementation and/or portability of the systems are exploited. Chapter 8 depicts the sensoactive layer coating the sensing unit, which is the first element in the e-tongue system with which the analyte establishes contact.

The right choice for materials that will compose such a sensoactive layer is, therefore, a crucial aspect in the e-tongue design and fabrication. In chapter 9 the main aspects of optical and hybrid sensor arrays employed for liquid system analysis will be discussed. The functioning principles, advantages, and challenges of optical multisensory systems in respect to electroanalytical ones will be considered. Finally, chapter 10 presents a roadmap for e-tongue data analysis, discussing the advantages and disadvantages of existing approaches and also the future in combining multiple solutions to support more informative analysis. Computational solutions are proposed to make sense of e-tongue data, predominantly using data mining or machine learning techniques, and visualization strategies.

We wish a tasty reading of the book to all students and researchers in the field of e-tongue and multisensory systems.

Acknowledgments

We thank all universities, research institutes, and funding agencies that for decades have contributed to human resources formation and the development of research and innovation.

Editors' biographies

Flavio M Shimizu

Flavio M Shimizu is a collaborative researcher at the Department of Applied Physics, '*Gleb Wataghin*' Institute of Physics, University of Campinas (UNICAMP), São Paulo State, Brazil. He received his degree in Physics (2005), MSc (2008), and PhD (2012) in Materials Science and Technology (POSMat) from Universidade Estadual Paulista (UNESP—Presidente Prudente, Brazil), and conducted his post-doctoral research at the São Carlos Institute of Physics, São Paulo University (IFSC/USP São Carlos, Brazil) from 2012 to 2017, and Cranfield University (2014, UK). His main research areas are microelectrodes and nanostructured films for applications in sensing and biosensing, electronic tongues, and information visualization techniques.

Maria L Braunger

Maria L Braunger is a postdoctoral researcher at Faculty of Science and Technology, São Paulo State University (UNESP) and a collaborative researcher at the 'Gleb Wataghin' Institute of Physics (IFGW), University of Campinas (UNICAMP), São Paulo State, Brazil. She received her degree in Physics (2008) and MSc in Applied Physics (2011) from UNESP (Rio Claro-SP, Brazil) and PhD in Materials Science and Technology (2015) from UNESP (Presidente Prudente-SP, Brazil) and Chemistry from Université de Pau et des Pays de l'Adour (UPPA, Pau—France). She was a postdoctoral researcher at IFGW-UNICAMP from 2015 to 2020 and a visiting scholar (2017–2018) at School of Engineering and Applied Science, University of Pennsylvania (Penn, PA-USA) working in the development of microfluidic devices and electronic tongues.

Antonio Riul Jr

Antonio Riul Jr is currently an Associate Professor at the Department of Applied Physics, '*Gleb Wataghin*' Institute of Physics, University of Campinas (UNICAMP), São Paulo State, Brazil. He researches Materials Science using nanostructured films from self-assembled materials for the nano-engineering of interfaces using tunable materials. He also works with impedance measurements in e-tongue devices, nowadays focusing on multifunctional coatings formed by self-healing and conductive nanocomposites.

List of contributors

Rafaela S Andre
Embrapa Instrumentação
São Carlos
Brazil

Marta Bonet-San-Emeterio
Universitat Autònoma de Barcelona
Barcelona
Spain

Maria L Braunger
University of Campinas
Campinas
Brazil

Xavier Cetó
Universitat Autònoma de Barcelona
Barcelona
Spain

Daniel S Correa
Embrapa Instrumentação
São Carlos
Brazil

Manel del Valle
Universitat Autònoma de Barcelona
Barcelona
Spain

Corrado Di Natale
University of Rome 'Tor Vergata'
Rome
Italy

Murilo H M Facure
Embrapa Instrumentação
and
University of São Carlos
São Carlos
Brazil

Marcin S Filipiak
Polish Academy of Sciences
Warsaw
Poland

Lucas Fugikawa-Santos
São Paulo State University
Rio Claro
Brazil

Gabriel Gaál
University of Campinas
Campinas
Brazil

Masaaki Habara
Intelligent Sensor Technology
Japan

Anna Herrera-Chacon
Universitat Autònoma de Barcelona
Barcelona
Spain

Hidekazu Ikezaki
Intelligent Sensor Technology
Japan

Elżbieta Jarosińska
Polish Academy of Sciences
Warsaw
Poland

Larisa Lvova
University of Rome 'Tor Vergata'
Rome
Italy

Wojciech Mazurkiewicz
Polish Academy of Sciences
Warsaw
Poland

Luiza A Mercante
Federal University of Bahia
Salvador
Brazil

Emilia W Nery
Polish Academy of Sciences
and
Sensoliq Sp.z o.o
Warsaw
Poland

Osvaldo N Oliveira Jr
University of São Paulo
São Carlos
Brazil

Roberto Paolesse
University of Rome 'Tor Vergata'
Rome
Italy

Leonardo G Paterno
University of Brasília
Brasilia
Brazil

Fernando V Paulovich
Dalhousie University
Halifax
Canada
and
University of São Paulo
São Paulo
Brazil

Mario Popolin Neto
University of São Paulo
and
Federal Institute of São Paulo
São Paulo
Brazil

Antonio Riul Jr
University of Campinas
Campinas
Brazil

Flavio M Shimizu
University of Campinas
Campinas
Brazil

Yusuke Tahara
Shinshu University
Nagano
Japan

Kiyoshi Toko
Kyushu University
Fukuoka
Japan

Karthika K Valapil
Polish Academy of Sciences
Warsaw
Poland

IOP Publishing

Electronic Tongues

Fundamentals and recent advances

Flavio M Shimizu, Maria L Braunger and Antonio Riul Jr

Chapter 1

A first taste to electronic tongues

Antonio Riul Jr and Daniel S Correa

The seminal study prospected by Krishna Persaud and George Dodd [1] mimicking the mammalian olfactory system was essential for a new age of sensing devices without highly specialized detection. It was very keenly welcomed by the perfume, food, and odor industries and decisive for subsequent developments searching in close analogy for an electronic tongue. That was the necessary impulse to pursuit for rapid, reliable qualitative or quantitative evaluation of liquids, spurring the pivotal work done by Yamafuji and Toko [2, 3] using a multichannel sensor acting in liquid systems capable of identifying various taste substances. At that time, the most natural strategy would be mimicking the human gustatory system [4] that works with the highest known efficiency, resulting from thousands of years of evolution. Besides regulating our feeding behavior, the gustative system performs periodic exchange of receptor cells and taste buds, dispensing validation and calibration tests, while avoiding drifts in messages sent to the central nervous system, which would be deleterious to our edible food criteria. The gustative neuronal pathway is long known, with magnetic resonance imaging indicating state-dependent brain areas responding to tastes. Nevertheless, how taste neurons act individually and the different internal pathways that lead the brain to taste recognition is still poorly comprehended and a subject of vast scientific research [5–7].

Despite the complexity involved, the comprehension of biological mechanisms directly impacted the development of devices capable of analyzing liquids reliably. Briefly, the initial studies [2, 3] used multichannel sensors based on lipid membranes as transducers, deeply impacting future developments that were also inspired by multisensory systems [8–13]. Besides using multisensor systems, another contribution was the use of non-specific sensing units to form a fingerprint of measured samples from unspecific, overlapping signals. These concepts expanded rapidly, and the inclusion of new transducer materials and measurement techniques [14–18] contributed to the expansion of e-tongue or taste sensors that were initially focused on

doi:10.1088/978-0-7503-3687-1ch1

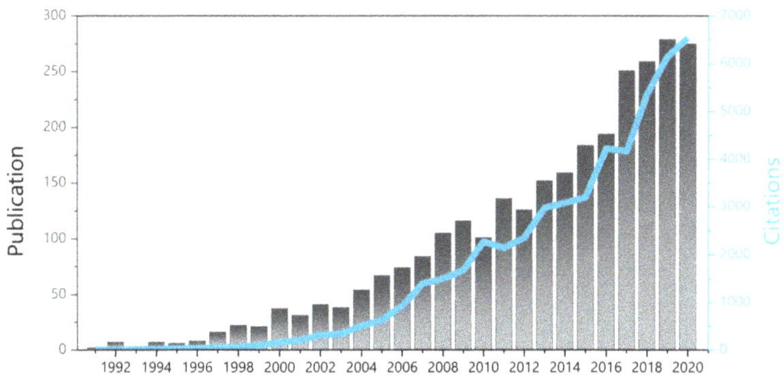

Figure 1.1. Number of publications reporting electronic tongues for the period 1990–2020, according to data retrieved from Web of Science [24].

taste analysis, food, and beverages to a gradual development for pharmaceuticals [19], environmental monitoring [20, 21] and cancer detection [22, 23]. Figure 1.1 illustrates such expansion.

Today this research field interacts with several areas, with bolder interfaces in chemistry, food science, engineering, instrumentation, electrochemistry, pharmacology, physics, and computer science. We also would like to draw the reader's attention to another growing and increasingly essential interface with sensory data analysis assisted by computational tools for clustering, classification, and data mining.

Nowadays, e-tongues are undoubtedly a powerful tool for analyzing complex liquid systems, despite the low number of commercial devices available. The main drawbacks related to commercialization are fouling formation on the sensing units with the continuous monitoring of aqueous systems, the calibration need of the sensor set or drift problems caused by the exchange of sensing units, difficulties in maintaining the same set of materials as transducers for diverse applications and, in some cases, the low repeatability and reproducibility. The literature is beginning to present papers concerned with handling, application, meaningful results, and clear protocols for sensitivity, selectivity, and detection limit for multisensory systems [25, 26], a daunting task considering distinct methodologies and materials used, which will undoubtedly lead to natural differences in sensitivity and detection limits of analytes.

Bearing this in mind, our aim here is to introduce the reader to the fascinating world of e-tongues, presenting basic concepts, terminologies, methodologies, materials employed as transducers, recent advances and applications, ending the book with a roadmap for sensorial data analysis. Chapter 2 brings a historical view and basic concepts on potentiometric lipid/polymer sensors. Chapter 3 extends those concepts for other potentiometric e-tongues and related data analysis tools. Chapter 4 introduces voltammetric e-tongues, and chapter 5 explores voltammetric e-tongues using molecularly imprinted polymers. Chapter 6 features basic concepts and applications on impedimetric (impedance measurements without a DC voltage offset) e-tongues. Chapter 7 focuses on novelties integrating emerging technologies (microfluidics and 3D-printing) with impedimetric e-tongues. Chapter 8 brings a broad range of materials used as sensing units in diverse e-tongues. Chapter 9 introduces the theoretical basis, trends, and applications in hybrid multisensory systems. Chapter 10 ends the book with a roadmap for sensorial data combining machine learning and data mining techniques to support informative analysis.

We wish a tasty reading of the book, and we are also very grateful to all contributors who made possible this collection of varied knowledge on e-tongues.

References

[1] Persaud K and Dodd G 1982 Analysis of discrimination mechanisms in the mammalian olfactory system using a model nose *Nature* **299** 352–5

[2] Hayashi K, Yamanaka M, Toko K and Yamafuji K 1990 Multichannel taste sensor using lipid membranes *Sens. Actuators* B **2** 205–13

[3] Toko K, Yamanaka M, Toko K and Yamafuji K 1994 Multichannel taste sensor using potential changes in lipid membranes *Biosens. Bioelectron.* **9** 359–64

[4] Spector A C and Travers S P 2021 Taste buds and gustatory transduction: a functional perspective *Oxford Research Encyclopedia of Neuroscience* (Oxford: Oxford University Press) pp 1–36

[5] Lemon C H 2021 Tasting temperature: neural and behavioral responses to thermal stimulation of oral mucosa *Curr. Opin. Physiol.* **20** 16–22

[6] Fu O, Minokodhi Y and Nakajima K 2021 Recent advances in neural circuits for taste perception in hunger *Front. Neural Circuits* **15** 609824–31

[7] Ahmad R and Daziel J E 2020 G protein-coupled receptors in taste physiology and pharmacology *Front Pharmacol.* **11** 587664–90

[8] Vlasov Y G, Legin A V, Rudnitskaya A M, Di Natale C and D'Amico A 1996 Multisensor system with an array of chemical sensors and artificial neural networks (electronic tongue) for quantitative analysis of multicomponent aqueous solutions *Russ. J. Appl. Chem.* **69** 848–53

[9] Sangodkar H, Sukeerthi S, Srinivasa R S, Lal R and Contractor A Q 1996 A biosensor array based on polyaniline *Anal. Chem.* **68** 779–83

[10] Legin A, Rudnitskaya A, Vlasov Y G, Di Natale C, Davide F and D'Amico A 1997 Tasting of beverages using an electronic tongue *Sens. Actuators B Chem.* **44** 291–6

[11] Vlasov Y, Legin A and Rudnitskaya A 1997 Cross-sensitivity evaluation of chemical sensors for electronic tongue: determination of heavy metal ions *Sens. Actuators B Chem.* **44** 532–7

[12] Winquist F, Wide P and Lundstro I 1997 An electronic tongue based on voltammetry *Anal. Chim. Acta* **357** 21–31

[13] Lavigne J J, Savoy S, Clevenger M B, Ritchie J E, Mcdoniel B and Yoo S *et al* 1998 Solution-based analysis of multiple analytes by a sensor array: toward the development of an 'Electronic Tongue' *J. Am. Chem. Soc.* **120** 6429–30

[14] Vlasov Y, Legin A and Rudnitskaya A 2002 Electronic tongues and their analytical application *Anal. Bioanal Chem.* **373** 136–46

[15] Riul A, Dantas C A R, Miyazaki C M and Oliveira O N 2010 Recent advances in electronic tongues *Analyst* **135** 2481–95

[16] del Valle M 2010 Electronic tongues employing electrochemical sensors *Electroanalysis* **22** 1539–55

[17] Ciosek P and Wróblewski W 2007 Sensor arrays for liquid sensing—electronic tongue systems *Analyst* **132** 963–78

[18] Kirsanov D, Correa D S, Gaal G, Riul A, Braunger M L and Shimizu F M *et al* 2019 Electronic tongues for inedible media *Sensors* **19** 5113–33

[19] Zheng J Y and Keeney M P 2006 Taste masking analysis in pharmaceutical formulation development using an electronic tongue *Int. J. Pharm.* **310** 118–24

[20] Oliveri P, Casolino M C and Forina M 2010 Chemometric brains for artificial tongues *Adv. Food Nutr. Res.* **61** 57–117

[21] Oleneva E, Khaydukova M, Ashina J, Yaroshenko I, Jahatspanian I and Legin A *et al* 2019 A simple procedure to assess limit of detection for multisensor systems *Sensors* **19** 1359–69

[22] Solovieva S, Karnaukh M, Panchuk V and Andreev E 2019 Chemical potentiometric multisensor system as a possible simple tool for non-invasive prostate cancer diagnostics through urine analysis *Sens. Actuators B Chem.* **289** 42–7

[23] Fitzgerald J and Fenniri H 2017 Cutting edge methods for non-invasive disease diagnosis using e-tongue and e-nose devices *Biosensors* **7** 59–98

[24] Web of science [Internet] 2021 https://webofscience.com/wos/woscc/citation-report/660d1839-7936-43a5-84ca-d03b09d1c900

[25] Legin A and Kirsanov D 2019 Avoiding nonsense in electronic taste sensing *Trends Anal. Chem.* **121** 115675–82

[26] Parastar H and Kirsanov D 2020 Analytical figures of merit for multisensor arrays *ACS Sensors* **5** 580–7

IOP Publishing

Electronic Tongues
Fundamentals and recent advances
Flavio M Shimizu, Maria L Braunger and Antonio Riul Jr

Chapter 2

Potentiometric electronic tongue using lipid/polymer membrane

Kiyoshi Toko, Yusuke Tahara, Masaaki Habara and Hidekazu Ikezaki

An electronic-tongue (e-tongue) using a lipid/polymer membrane, i.e. as a taste sensor, was first developed in 1990. This taste sensor could respond to each taste quality independently, and hence, could quantify and classify the taste of foods according to six types of taste qualities: saltiness, sourness, sweetness, bitterness, umami and astringency. The sensor transforms the taste information of chemical substances into electrical signals. Taste-sensing devices based on this mechanism have been commercialized and approximately 600 machines are now utilized in food and pharmaceutical companies. The tastes of foods such as coffee, tea, beer, wine, soup, fruits, vegetables, meat, and rice have been digitized and visualized using this e-tongue.

2.1 History of the development of taste sensor using lipid/polymer membrane

The taste sensor, otherwise known as the e-tongue, was patented in 1989 and the associated paper describing it was published in 1990 [1]. It has since been researched and developed so that at present there are different versions of the e-tongue worldwide. The taste sensor developed in Japan is said to have global selectivity [2, 3]. It uses a lipid membrane to receive the substance to be tasted; the measurement of taste is performed using the potentiometric method. In this chapter, we introduce this potentiometric e-tongue using lipid/polymer membranes (taste sensor). Its principles and application examples have been introduced in a number of reviews [4–11]; please refer to them as well.

Figure 2.1(a) shows a taste sensor (Taste Sensing System TS-5000Z) developed, produced and sold by Intelligent Sensor Technology, Inc. This technology was first developed in Japan and is currently used by food manufacturers, pharmaceutical manufacturers, universities, and public centres worldwide to guarantee taste and

doi:10.1088/978-0-7503-3687-1ch2

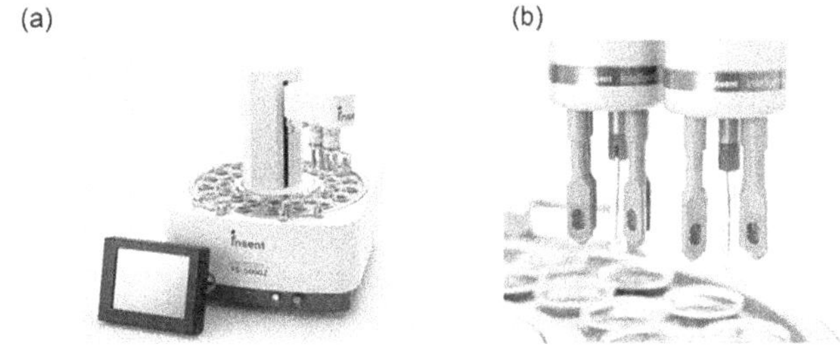

Figure 2.1. Taste sensor with lipid/polymer membranes (Taste sensing system TS-5000Z, Intelligent Sensor Technology, Inc.) (a) and the sensor electrode (b).

quality, create new desired flavors, and for marketing and research on the chemical properties of foods. It has been applied to food products (such as coffee, tea, beer, Japanese sake, shochu, wine, rice, bread, dashi stock, soup, meats, milk, mineral water, miso, soy sauce, seasoning, fruits and vegetables) and pharmaceutical products to successfully quantify their tastes. Figure 2.1(b) shows the sensor electrode that receives the substance to be tasted and converts its taste into a voltage signal. The lipid/polymer membrane (from now on referred to as lipid membrane) acts like the biological membrane that constitutes the taste cells of the human tongue. In other words, it converts chemical substances into the quality called 'taste'. This principle is almost identical to that in biological systems. In the sensor, the chemical substance interacts with the lipid membrane, which then converts the taste information into a voltage transmitted to a computer through a cable. As a result of improvements in lipid membranes, the current membranes respond well to sourness, saltiness, etc, with specific responses for each taste (global selectivity). Taste quality and intensity can be determined from the voltage output of multiple types of membranes (typically five) that respond to a specific taste.

The taste sensor is a biomimetic technology inspired by the structure and function of biological membranes. The lipid membrane generates a voltage that changes according to its interactions with the taste substance. The response characteristics of the membrane can be adjusted by changing the types of lipids and the plasticizer used as well as the ratio of multiple lipid types. This ability to modify response characteristics led to the successful development of membranes that selectively respond to specific tastes. For example, sensors responding to saltiness have an increased number of charged lipid molecules and increased surface hydrophilicity, making electrostatic interactions with ions more likely. By contrast, bitterness sensors have relatively low lipid content and high hydrophobicity. An advanced general-purpose taste sensor (TS-5000Z) was developed and sold in 2007.

Figure 2.2 shows a section of the sensor (1989) initially developed at Kyushu University. Eight membranes were attached to a single acrylic plate (probe). The sensor section had a structure with holes inside the acrylic plate through which Ag/AgCl wire and electrolyte solutions were inserted and attached with the membrane. The differences between this early sensor and the current sensor (figure 2.1(b)) are quite evident, but more important than differences in external appearance are the differences in selectivity. In the initial phases of development, the

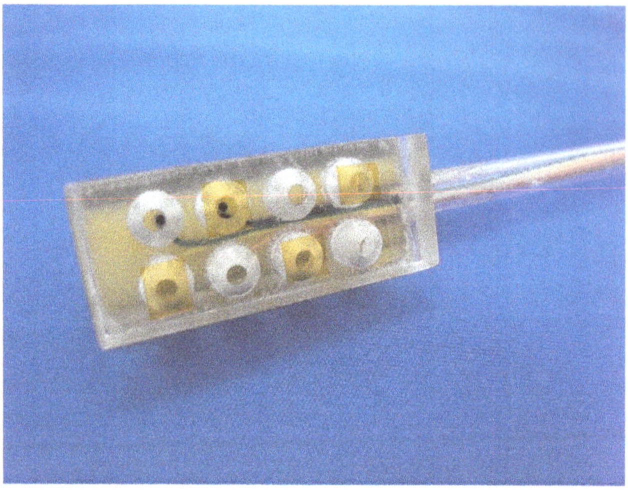

Figure 2.2. Multichannel taste sensor probe initially developed in the 1980s.

lipid membrane could not selectively respond to specific tastes, and had what could be referred to as 'fuzzy' overlapping responses characteristic of taste in each membrane. Thus, it was difficult to interpret the measurement data uniquely. This necessitated having a trained researcher look over the data and comparing them with previously recorded taste experiences, or alternatively, conducting additional experiments to quantify the taste.

The development of the lipid membrane in 2007 addressed the above-mentioned problem and led to the development of a taste sensor that can be used by anyone. This membrane embodies the global selectivity of biological systems. A single type of taste receptor in biological systems can receive multiple chemical substances. For example, the sweet taste receptor can receive substances such as monosaccharides (glucose) and disaccharides (sucrose). This is a major difference from the antigen-antibody interaction seen in the immune system, where one antibody (protein) specifically recognises one type of antigen (e.g., chemical substance). Enzymatic reactions used in glucose sensors also generally have highly selective specificities for particular chemical substances.

There are more than 30 types of taste receptors, of which 25 are for the bitter taste. For example, despite the presence of several hundred types of chemical substances in tea, there are only 30 types of receptors to respond to it. Biological systems receive several types of chemical substances and classify them into five tastes (sourness, bitterness, sweetness, saltiness, and umami). The taste receptor does not have a high selectivity or one-to-one correspondence with a particular chemical substance, but instead shows global selectivity, i.e. it accepts chemicals across a wide range, and then breaks them down by taste. The receiving membrane for taste sensors developed in recent years possesses the same properties.

As mentioned earlier, in the initial stages of the development of the taste sensor, taste was determined using a response pattern composed of the output voltages of multiple lipid membranes. It was possible to break down tastes for specific foods (measurement targets), However, this is not an ideal generalization method. This is because different food products have widely varying response patterns, so it is possible that the same result would not be obtained, i.e., the correct solution (or taste) would not be outputted, even if the same method was used. Relative values of taste for measurement targets could be obtained, but the absolute value would remain unknown. Food products could be identified, but tastes could not be quantified. This is a drawback of using statistical analysis, including principal component analysis. Like devices and tools for other quantities, such as length and weight, the taste sensor provides a scale for taste.

The patent application for the taste sensor was filed in 1989, and the first taste sensor (Taste Sensing System SA401) was prototyped and sold in 1993, indicating that knowledge from university research was put to practical use relatively quickly. This may, in part, be due to the fact that there had never been a 'taste measuring' device before and the taste sensor was a novelty.

2.2 Potentiometric measurement of two kinds of model membranes

In this section, we go back slightly in time to introduce very early research on taste sensors in the 1980s. Figure 2.3 illustrates a simple experimental setup from that time, when a membrane was sandwiched between two acrylic cells, and the potential (membrane potential) generated was measured through the membrane. Each acrylic cell contained 100 mM and 1 mM KCl solution. A salt bridge was used to measure the potential, and another cell was attached to both, containing a solution with equal concentration (3.0 M KCl) to counteract the difference in electrode potential due to the Ag/AgCl wire. The DC supply applied an external voltage. Synthetic dioleyl phosphate (DOPH) was used as the lipid in this study (figure 2.4). A membrane constructed from a porous filter (Millipore Corp, cellulose ester) with average pore

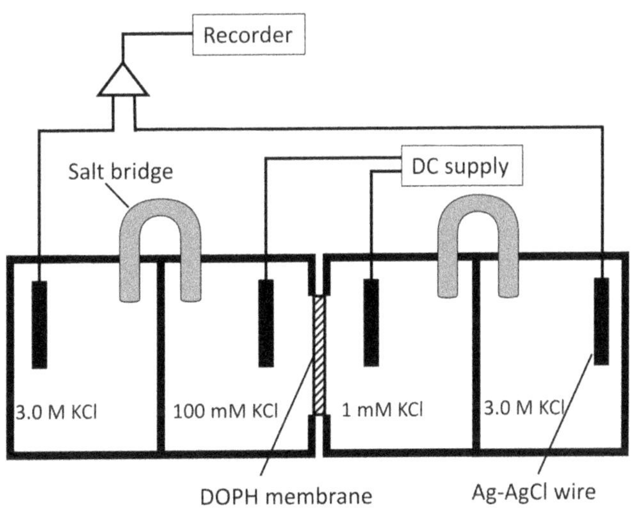

Figure 2.3. Experimental setup for measuring the electrical characteristics of the model membrane.

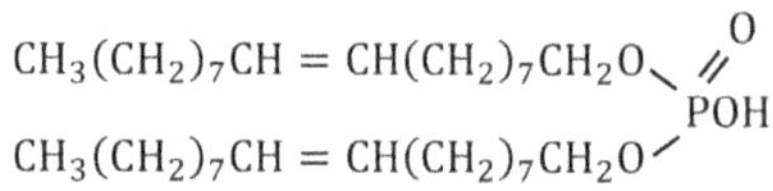

Figure 2.4. Chemical structure of DOPH.

size 5 to 8 μm was studied as a neural model membrane because its electrical characteristics could be varied by changing the external salt concentration and applying external pressure or voltage [12, 13]. The structure of DOPH, a yellow liquid oil at 20 °C, is a simplification of the typical phospholipid structure seen in biological systems. The electrical characteristics of the DOPH-adsorbed membrane, i.e., the membrane electrical resistance, membrane capacitance, and self-sustained oscillation, could change, and the mechanisms by which these occur have been elucidated [11, 14].

Research on the taste response of this lipid membrane initially began using the self-sustained oscillation of the DOPH-adsorbed membrane. Details are summarized in [11], but due to reproducibility problems with the experimental data, the measurement method was changed to one using the static state of the membrane potential instead of dynamic responses, such as oscillation, which was used till the present day. As shown in figure 2.3, this method does not use externally applied voltage. It adds the taste substance to a 1 mM KCl solution and observes the potential changes in the DOPH-adsorbed membrane. The reference of the electric potential was taken at 1 mM KCl. Results of this experiment are shown in figure 2.5 [11, 15–18]. The horizontal axis shows the taste substance concentration, and the vertical axis shows the relative value of the membrane potential. The initial value of the electric potential of DOPH-adsorbed membrane without the addition of taste substances changes across a range of −100 to −125 mV, and although the reproducibility is not always favorable, it can be seen that the tendencies unique to each taste substance are reflected in its data. The initial value of the potential is negative, so the upward response in the figure means that the response is in the positive direction.

The membrane potential changes at low concentrations for quinine (bitterness) and HC1 (sourness), whereas this changes at extremely high concentrations for sucrose (sweetness). A response begins to appear from a few mM for NaCl (saltiness), whereas the response to monosodium glutamate (MSG, umami) appears first at a negative potential, after which it switches to a positive value. The thresholds of these responses are similar to those of biological systems. The change from negative to positive values in umami substances is because the membrane first responds to glutamate (the main body of MSG), after which it responds to the Na^+ ion. A suppressive response seen in biological systems was also observed in the membrane, where adding a salty or sweet substance to a bitter substance reduced the bitterness response [15]. Furthermore, a synergistic effect that can be seen in biological systems was also observed, where adding a small amount of umami substances such as disodium inosinate and disodium guanylate to MSG dramatically increased umami [17]. From these facts, we succeeded in taking the first steps towards taste sensing using lipid membranes in the world.

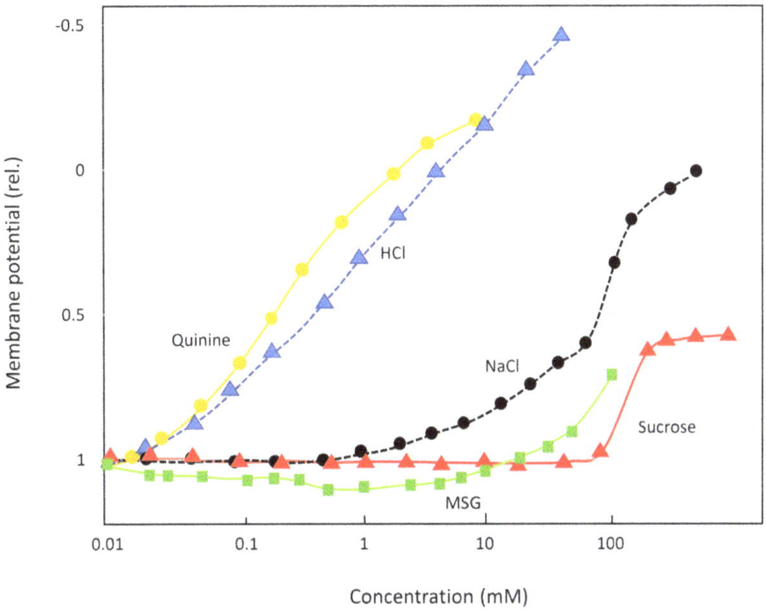

Figure 2.5. Changes in the electric potential of DOPH-adsorbed membrane with five types of taste substances.

Incidentally, the DOPH-adsorbed membrane is negatively charged because the lipid DOPH has a phosphate group. Therefore, it is generally predicted to respond to positively charged taste substances, but not negatively charged taste substances. Indeed, as shown in figure 2.5, positively charged bitter substances such as quinine hydrochloride have a large positive response, but negatively charged bitter substances like picric acid did not induce a response. Therefore, this time, we instead prepared a positively charged lipid molecule. This is dialkyldimethyl ammonium bromide ($(C_{18}H_{37})_2(CH_3)_2N^+Br^-$) (figure 2.6).

A small hole measuring approximately 50 μm in diameter was opened in a silicon film of 350 μm thickness, into which $(C_{18}H_{37})_2(CH_3)_2N^+Br^-$ and polymeric sodium polystyrene sulfonate (PSS^-) was cast. Figure 2.7 shows the results of adding a taste substance similar to that in the DOPH-absorbed membrane in this artificial membrane [19]. The membrane potential in the vertical axis here uses relative scale, and 1 is equivalent to approximately +100 mV. In addition, the reference point for electric potential was taken at 1 mM KCl. The response in the downward direction signifies the negative response. It can be seen from the figure that there was a large response to charged picric acid (bitterness). There was also a response to NaCl, but this only in response to the Cl^- ions. There was a response to the high-concentration quinine, but this was due to the Cl^- ions in quinine hydrochloride within the $(C_{18}H_{37})_2(CH_3)_2N^+Br^-$ cast membrane.

The aforementioned results were obtained from experiments conducted from 1985 to 1989 and can be summarized as follows.

(1) The electric potential of a DOPH-adsorbed membrane responds uniquely to each taste substance, and its threshold is similar to that in biological systems.

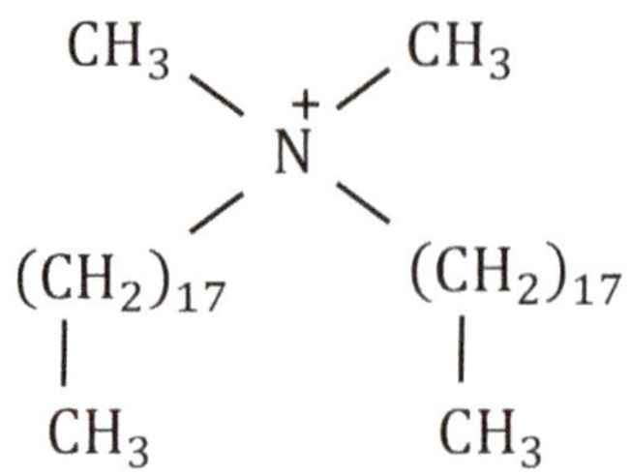

Figure 2.6. Chemical structure of dialkyldimethyl ammonium bromide.

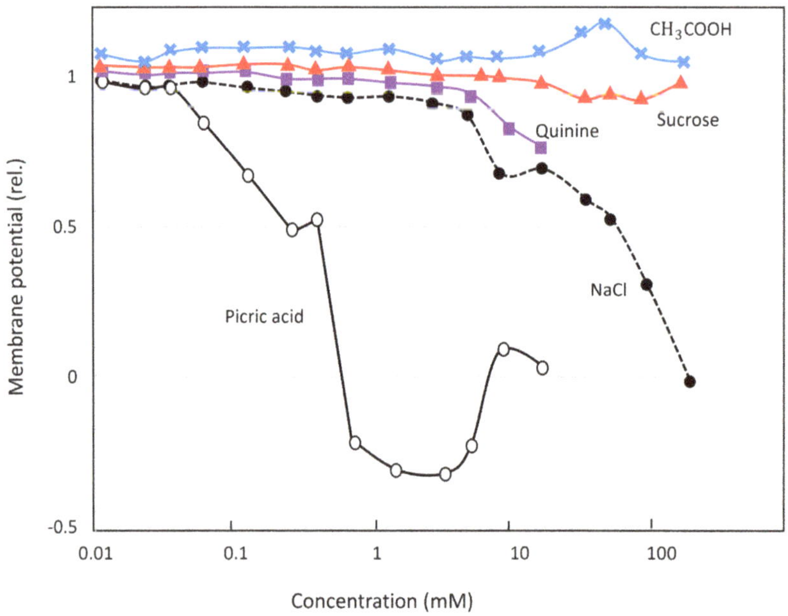

Figure 2.7. Changes in electric potential of $(C_{18}H_{37})_2(CH_3)2N^+Br^-$-cast membrane with five types of taste substances. Quinine hydrochloride and picric acid were used as typical positively charged and negatively charged bitter substances, respectively.

(2) This artificial membrane can reproduce the synergistic and suppressive effects seen between different taste substances and taste qualities.
(3) A different taste substance response can be seen when using a different lipid, $(C_{18}H_{37})_2(CH_3)_2N^+Br^-$, for the membrane.

Based on these results, we were able to use the lipid membrane to classify tastes into taste qualities such as bitterness, sourness, sweetness, saltiness, and umami, as well as demonstrate the potential for quantifying taste. In 1990, the development of the world's first multichannel taste sensor using lipid membranes was published [1].

2.3 Taste sensor: e-tongue with global selectivity

Humans receive taste substances in small organs called taste buds which are approximately 80 μm. Each taste bud is composed of approximately 100 taste cells. Taste substances are received by receptors and ion channels in the taste cells. Today, tastes are divided into five basic tastes: saltiness, sourness, umami, bitterness and sweetness. The existence of fatty acids as a sixth taste has become increasingly clear in recent years [20, 21]. Sweetness, umami, and bitterness are detected by G protein-coupled receptors such as the TAS1R2/TAS1R3 heterodimer [22], TAS1R1 /TAS1R3 heterodimer [23] and T2R family [24] (humans have approximately 25 types), respectively [25]. Receptors and candidates for sourness and saltiness include Otopetrin1 [26], polycystic kidney disease 2-like 1 protein (PKD2L) [27], acid-sensing ion channels (ASICs), hyperpolarization-activated cyclic nucleotide-gated channels (HCN) [28], epithelial sodium channels (ENaCs) [29]. Taste substances are ultimately transmitted to the brain from these receptors. The taste sensing mechanisms of the body, which involve receptors, intracellular signal transduction-related molecules, and neurotransmission, have been continuously identified through advances in molecular biological techniques. Please refer to the review papers on this subject for further details [30–32].

The initial phases of taste sensor development were described in section 2.2. In this section, we introduce a taste sensor that uses a lipid membrane developed by Toko *et al* of Kyushu University, which achieved engineering-based identification and quantification of taste based on the five basic tastes in humans. Like human receptors, this taste sensor has the taste quality-specific selectivity of the bitterness sensor responding to the bitter substance group and the astringency sensor responding to the astringency substance group. This is referred to as global selectivity, which makes this e-tongue unique.

2.3.1 Principle

The five basic tastes can be grouped according to their physicochemical characteristics, as shown in figure 2.8 [4]. From this perspective, we designed a membrane for determining the taste receptivity of the taste sensor. According to the Weber–Fechner law, the intensity of a sensory response in humans, E, is proportional to the logarithm of the intensity of the stimulus, R, for moderate stimuli. This can be expressed as follows:

$$E = C \log R \tag{2.1}$$

where C is a constant. The taste sensor response is also proportional to the logarithmic value of the concentration of the taste sample, so the amount of human taste sensation can be approximated from the response value of the taste sensor. The taste sensor is one of the chemical sensors that attempts to approximate the taste intensity perceived by humans by realizing the following concepts [11]:

(1) The threshold of the taste sensor electrode must be the same as the human taste threshold;
(2) the taste sensor electrode must respond consistently to the same taste, similar to the human tongue (global selectivity);

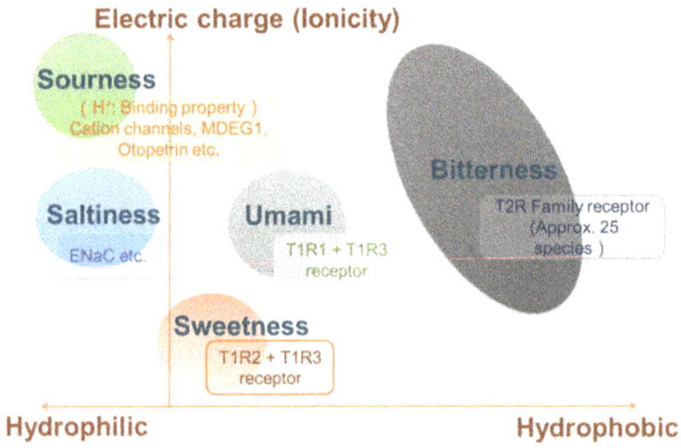

Figure 2.8. Physical/chemical characteristics of taste substances (reprinted with permission from Wu X *et al* [4]. Copyright (2020) Japan Society for Analytical Chemistry).

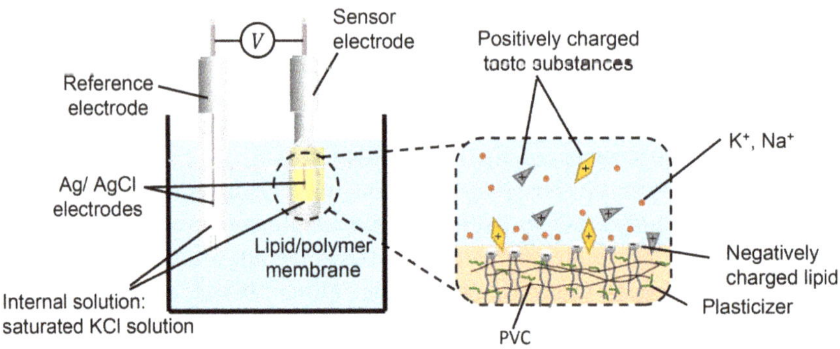

Figure 2.9. Taste sensor, sensor electrodes and membrane structure in sample.

(3) its electrode must have a high correlation with human sensory scores;
(4) there must be a clearly defined unit of information from the taste sensor electrode;
(5) the taste sensor electrode must detect the interaction between taste substances.

The taste sensor is composed of a lipid membrane attached to a hollow polyvinyl chloride (PVC) tube, a lipid membrane electrode and a reference electrode (Ag/AgCl saturated KCl) whose interior is filled with saturated KCl used in the sensing parts (figure 2.9), a handler where the sensing part is moved by a robot arm, and a data processing part. The lipid membrane is a mixture of lipid, a plasticizer, and PVC mixed with tetrahydrofuran cast into a petri dish to form a membrane of approximately 200 μm thickness, which can be manufactured very easily. Table 2.1 lists the details of each sensor part [10].

The lipid membrane interacts with the taste substance, which then changes the membrane potential as described below. When the concentrations of the internal and external solutions in contact with both sides of the lipid membrane are different, the difference between the potentials of the internal and external solutions is called the membrane potential. The membrane potential is the sum of the surface potential and diffusion potential generated at the interface between the membrane surface and the

Table 2.1. Chemical components of taste sensor membranes (reprinted from [10], copyright (2010) with permission from MDPI (CC BY)).

Taste sensor electrode	Artificial lipid	Plasticizer
Umami sensor AAE	Phosphoric acid di(2-ethylhexyl) ester Trioctylmethylammonium chloride	Dioctyl phenylphosphonate
Saltiness sensor CT0	Tetradodecylammonium bromide 1-Hexadecanol	Dioctyl phenylphosphonate
Sourness sensor CA0	Phosphoric acid di(2-ethylhexyl) ester Oleic acid Trioctylmethylammomium chloride	Dioctyl phenylphosphonate
Bitterness sensor C00 (for acidic bitter materials)	Tetradodecylammonium bromide	2-Nitrophenyl octyl ether
Bitterness sensor AC0 (for basic bitter materials)	Hexadecanoic acid	Dioctyl phenylphosphonate
Bitterness sensor AN0 (for basic bitter materials)	Phosphoric acid di-n-decyl ester	Dioctyl phenylphosphonate
Bitterness sensor BT0 (for bitter hydrochloride salts)	Phosphoric acid di-n-decyl ester	Bis(1-butylpentyl) adipate Tributyl O-acetylcitrate
Astringency sensor AE1	Tetradodecylammonium bromide	Dioctyl phenylphosphonate
Sweetness sensor GL1	Trimellitic acid Tetradodecylammonium bromide	Dioctyl phenylphosphonate

aqueous solution. The surface potential is caused by a diffusive electric double-layer that is formed in the solution layer near the membrane surface. The lipid membrane coming into contact with the electrolyte solution results in the membrane being charged due to ionization of the dissociating group of the lipid on the surface of the membrane and the adsorption of ions (figure 2.9). For positively charged lipid membranes, the anions are attracted to the membrane surface by electrostatic interactions, and the cations move away from it. Therefore, the positive and negative charges on the membrane surface form an electric double-layer. The diffusion potential is the potential difference caused by the difference in mobility of the cations and anions in the membrane.

2.3.2 Measurement process

The changes in membrane potential relative to the membrane potential in a tasteless standard solution (versus Ag/AgCl saturated KCl) (reference) are the taste sensor outputs. A standard solution consisting of 30 mM KCl and 0.3 mM tartaric acid was

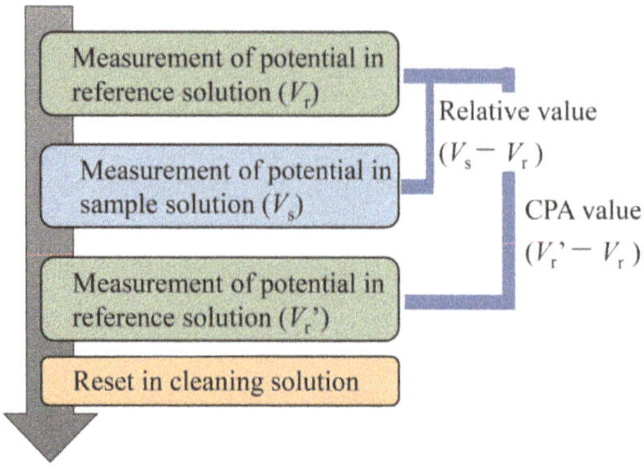

Figure 2.10. Measurement procedure by the taste sensor.

used as an electrolyte solution to make electrochemically stable measurements. The measurement procedure is illustrated in figure 2.10. It involves the following steps:

(1) Measure the membrane potential V_r in the standard solution.
(2) Measure the membrane potential V_s in the sample solution.
(3) Rinse the sensor electrode with the standard solution.
(4) Measure the membrane potential V'_r in the standard solution a second time.
(5) Clean the sensor electrodes in an ethanol-containing cleaning solution.

Considering the pH values of typical food products, 30 mM KCl and 0.3 mM tartaric acid were used as the standard solutions. Furthermore, 30 vol% EtOH, 100 mM HCl, 100 mM KCl or 30 vol% EtOH, 10 mM KOH, 100 mM KCl were used as cleaning solutions to remove contaminants and adsorbed taste substances in the membrane. For the sensor outputs, $V_s - V_r$ was defined as the relative value and $V'_r - V_r$ was defined as the change of membrane potential caused by adsorption (CPA) values for the first taste and after taste, respectively [10]. The CPA value reflects the changes in membrane potential caused by the taste substances adsorbed on the lipid membrane. Taste intensity was quantified using the relative values for saltiness sourness, umami, sweetness and CPA values for bitterness and astringency.

2.3.3 Response to six kinds of basic taste qualities

We used the fact that the charge and hydrophobicity of taste substances can be roughly grouped for each taste quality (figure 2.8) to quantify the first taste and after taste sensed by a human using the relative values ($V_s - V_r$) and CPA values ($V'_r - V_r$), respectively, of the changes in membrane potential of the taste sensor. Saltiness, sourness, umami and sweetness sensors detect changes in membrane potential using electrostatic interactions of the taste substances; therefore, relative values were primarily used. Meanwhile, for the bitterness and astringency sensors, the high hydrophobicity of their taste substances was used, and the CPA values reflecting the adsorption on lipid membranes through hydrophobic interactions were primarily used (see section 2.3.4). Taste intensity conversion was performed using simple regression analysis using the sensory or sensor output values for the five basic tastes

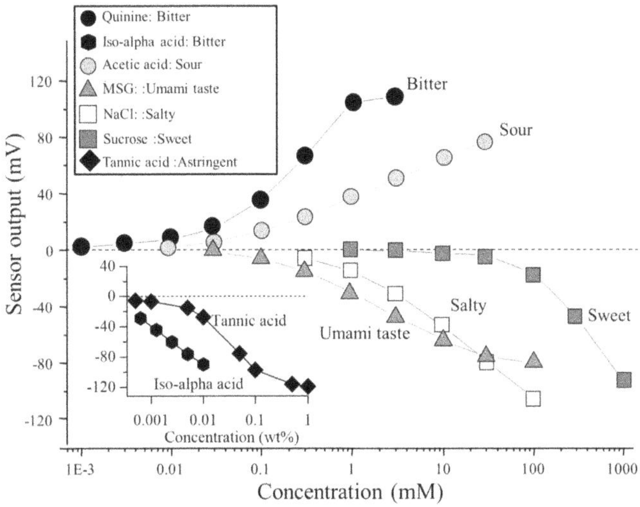

Figure 2.11. Responses of each taste sensor to basic taste substances.

or an estimation formula based on a multiple regression analysis obtained from the sensor outputs gained from two types of taste sensor electrodes (see section 2.4). Figure 2.11 shows the response of each taste sensor for the five basic tastes and astringency [8]. The threshold of the sensor is designed to match human and sensor thresholds; the response covers the human dynamic range (ranging from 1 to 10 times the threshold), which was presented as concept (1) in section 2.3.1.

2.3.4 Adsorption of taste substances onto the receptive membrane

The relative values and CPA values are used as indices of the first taste and aftertaste, respectively, as sensed by humans. The CPA value is the change in membrane potential due to the taste substances adsorbed on the lipid membrane and was described in section 2.3.1. Bitterness sensors and astringency sensor take advantage of adsorption and high hydrophobicity, respectively, of their respective taste substances. Figure 2.12 shows the CPA values of the bitterness and astringency sensors for each taste substance [33]. It can be seen that a high selectivity of the bitterness and astringency sensors is assured using the CPA values. This selectivity is not a one-to-one bond similar to the enzymatic reactions of antigen/antibody reactions. Instead, similar to the bonding abilities of human receptors, bitterness sensors respond to multiple groups that impart a bitter taste, signifying a rigid selectivity, and this is referred to as 'global selectivity'. The taste sensor introduced in this chapter is the only chemical sensor with global selectivity.

The composition ratios of lipids and plasticisers in membranes are important for achieving global selectivity. Bitterness and astringency sensors obtain global selectivity by utilising the adsorption characteristics of their respective taste substances. Figure 2.13(a) shows the CPA values and adsorption amounts of the bitter substance iso-α acid relative to the lipid concentrations in the lipid

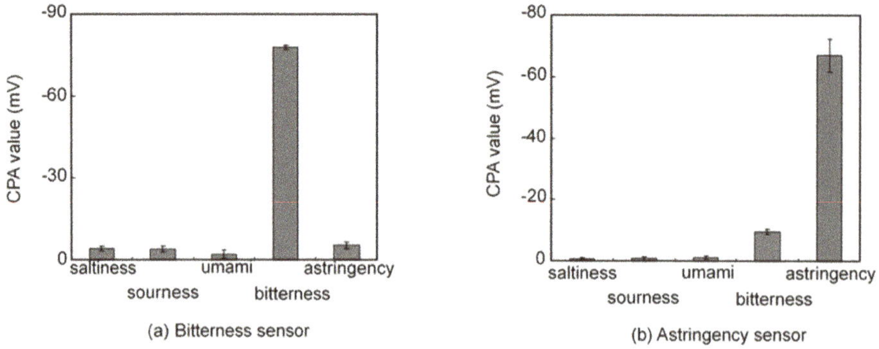

Figure 2.12. Sensor responses, (a) bitterness sensor, (b) astringency sensor. saltiness; 300 mM KCl, 0.3 mM tartaric acid, sourness; 30 mM KCl, 3 mM tartaric acid, umami; 10 mM MSG, bitterness; 0.01 vol% iso-α acid, astringency; 0.05 wt% tannic acid. Umami, bitterness, acidic bitterness and astringency samples include 30 mM KCl and 0.3 mM tartaric acid (reprinted with permission from Tahara [33]. Copyright (2020) Jenny Stanford Publishing).

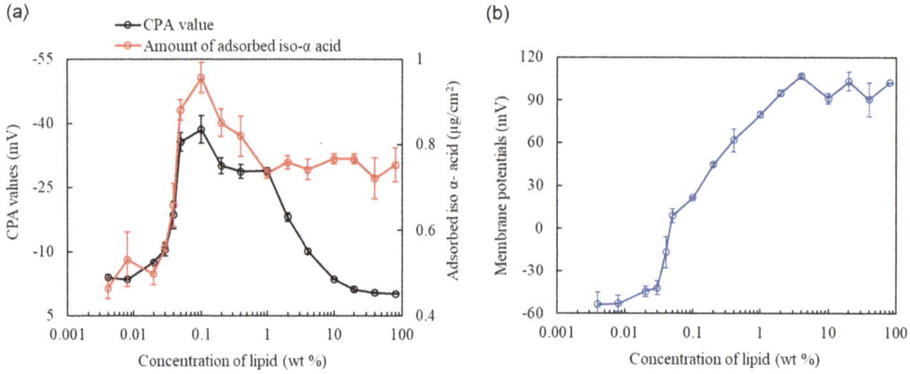

Figure 2.13. (a) Adsorption amount, CPA value, and (b) membrane potential of iso-α acid according to lipid concentration, reprinted from [36], copyright (2014) with permission from MDPI (CC BY).

membrane, and figure 2.13(b) shows the membrane potential (V_r) in the reference solution relative to the lipid concentration. The hydrophobicity and charge density of the membrane change depending on the lipid concentration, and the adsorption amount of the taste substance changes as well [34–36]. The increase in the lipid content of the membrane increases the surface charge density of the membrane and reduces its hydrophobicity. Increments in the surface charge density of the membrane result in the increased attraction of bitter substances; increase in the adsorption content also increases the CPA value. However, decrements in the hydrophobicity of the membrane surface decrease the adsorption content of the membrane and decrease the CPA value. Therefore, a CPA peak was observed. In other words, the sensitivity and dynamic range response to the taste substance can be changed by varying the lipid concentration. The composition of membranes is adjusted to match the threshold values and sensitivity of humans. Plasticizer types and their composition ratios also have an effect and are also important parameters in membrane design [37, 38].

2.3.5 Taste sensor for high-potency sweeteners

Sugars such as sucrose are uncharged sweet substances; advances have been made in developing sweetness sensors for such substances [39–42]. High-potency sweeteners such as sodium saccharin, acesulfame potassium, and aspartame are used in low-calorie foods and drinks because of their lower calorific values compared to those of sugars. The sweetness intensity of sweeteners was expressed with a sweetness scale using sucrose as the reference. Many high-potency sweeteners are charged and have a different chemical structure from that of sugars such as sucrose. It is challenging for taste sensors that work on the principle of membrane potential measurement to simultaneously measure taste substances with different charges using a single sensor. Therefore, high-potency sweeteners have been classified into two types (positively or negatively charged high-potency sweeteners) according to their charge state. Two types of sensors corresponding to these two types have been developed. Positively charged high-potency sweeteners include aspartame, and negatively charged ones include saccharin sodium [37, 42, 43]. To obtain selectivity, two types of sensors were developed considering the chemical structure of the high-potency sweeteners: one with negatively charged membranes having moderate hydrophobicity to measure positively charged sweeteners with moderate hydrophobicity, and the other with positively charged membranes and moderate hydrophobicity to measure negatively charged sweeteners with moderate hydrophobicity (table 2.2) [44]. Moderate hydrophobicity was used to avoid the adsorption of bitter and astringent taste substances onto the membrane. Figure 2.14(a) shows the response of high-potency sweetener sensors to negatively charged high-potency sweeteners, and figure 2.14(b) shows the response of high-potency sweetener sensors to positively charged high-potency sweeteners. Furthermore, figure 2.15 shows the response of aspartame and quinine hydrochloride when the negatively charged phosphoric acid di-n-decyl ester was used as a lipid in the membrane and eight different types of plasticizers were used for a high-potency sweetener sensor for positively charged high-potency sweeteners. The response characteristics of the bitter substance quinine hydrochloride changed according to the type of plasticizer, and a sensor response

Table 2.2. Designs of surface charge of membrane and hydrophobicity for lipid polymer membranes. Reprinted with permission from Onodera *et al* [44]. (Copyright (2020) American Scientific Publishers).

	Surface charge of membrane		
Hydrophobicity	*Positive*	*Near Neutral*	*Negative*
High	Acidic bitterness	–	Basic bitterness
Middle	**Negatively charged high-potency sweetener**	Umami	**Positively charged high-potency sweetener**
Low(hydrophilicity)	Saltiness (Cl^-)	Sourness	Saltiness (Na^+, K^+)

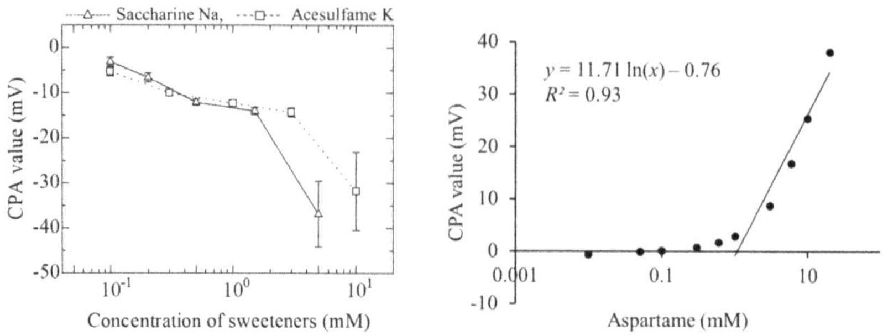

Figure 2.14. Sensor response to saccharine Na, acesulfame K and aspartame. Sweetness sensor for negatively charged high-potency sweeteners (a) (reprinted from Yasuura *et al* [42], copyright (2020) with permission from Elsevier). Positively charged high-potency sweeteners (b) (reprinted with permission from Wu X *et al* [4]. Copyright (2020) Japan Society for Analytical Chemistry).

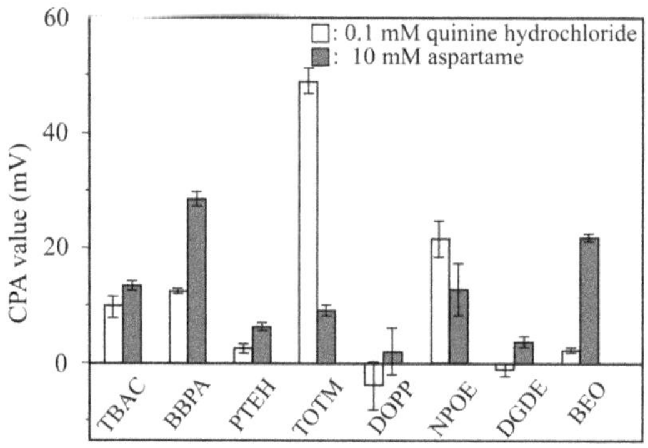

Figure 2.15. Differences in selectivity of sweetness sensor for positively charged high-potency sweeteners according to plasticizer.

with high selectivity was achieved using 2-butoxyethyl oleate. The responses from these sensors were proportional to the logarithm of the concentration of each sweetener and showed sufficient selectivity for high-potency sweeteners. Further advancements in sensitivity are needed, and some examples of potential solutions to this problem are introduced in section 2.4.2.

2.4 Expression of taste using two sensor electrodes

The taste sensor is configured to show a similar response for similar tastes (taste qualities) rather than quantifying the individual chemical substances. The amount of each taste quality is measured using sensors corresponding to that particular taste quality [10]. This mimics the human taste reception mechanism, but taste is processed in the brain in addition to the receptor. An example of this difference in processing is bitterness suppression due to high-potency sweeteners, which is caused by the activity of both sweetness and bitterness receptors and their processing in the brain. Bitterness suppression using commercially available substances that contain phospholipids occurs on the tongue. The bitterness suppression by phospholipids is achieved because phospholipids mask the bitter substances so that they do not contact the receptors. Therefore, the quantification of bitterness suppression is possible using a single sensor with a negatively charged lipid membrane [38]. In this chapter, the latest research on the suppression of bitterness using two taste sensors and improvements in its measurement sensitivity are discussed.

2.4.1 Bitterness evaluation of coffee containing high-potency sweeteners

Bitterness-masking materials use masking actions on bitter receptors or taste substances, which can be evaluated from decrements in the CPA value due to physical adsorption of bitterness-masking materials on the lipid membrane [38, 45]. Sweetening coffee not only increases its sweetness, but also decreases its bitterness. This bitterness suppression with sweetness is a phenomenon sensed in the human brain; therefore, it cannot be evaluated using conventional taste sensors that mimic the action of receptors and ion channels. Hence, an attempt was made to develop an evaluation system for bitterness suppression effects by focusing on the bitterness suppression/masking behavior of high-potency sweeteners in coffee using two high-potency sweetness sensors [37, 42], and a bitterness sensor that mimicked the bitterness suppression mechanisms in the human brain [46]. The bitterness scale for coffee was created by controlled roasting of coffee. It was a five-point scale (nine points, L*=18; seven points, L*=20; five points, L*=22; three points, L*=24; one point, L*=26). The chromatic parameter, L* value indicated the degree of coffee roasting. The smaller the L* value, the lower the degree of roasting. The sweetness scale was constructed with sucrose solutions adjusted to a seven-point scale (zero point: 0 mM, one point: 10 mM, two points: 100 mM, three points: 178 mM, four points: 346 mM, five points: 562 mM, six points: 1000 mM); the sweetness intensity was calculated from the sensory evaluation of acesulfame K solution. 0.5 g of instant coffee powder was dissolved in 50 ml of 1 mM KCl solution at 90 ± 5 °C. Instant coffee with different degrees of roasting (L*values of 30, 25, 20 and 15) were measured with bitterness and sweetness sensors in order to quantify the bitterness of each coffee and the sweetness due to high-potency sweeteners. A total of 16 samples were prepared by four types of L* value coffee solutions added to four types of acesulfame K concentrations (0, 0.5, 1, or 2 mM). The bitterness intensity of the coffee was evaluated by a sensory test using a total of 17 volunteers. Finally, multiple regression analysis was conducted on the response values of each sensor and the sensory test results to derive an equation that predicts the bitterness of coffee when a high-potency sweetener was added. Measurements showed that the bitterness sensor response value increased with the increase in the degree of roasting (i.e. the L* value became smaller), but the sweetness sensor response value did not change. Furthermore, in the coffee sample with acesulfame K added, an increase in the amounts of added acesulfame K increased the sweetness sensor response value, but the bitterness sensor response value was almost constant. Therefore, the bitterness of coffee and sweetness of

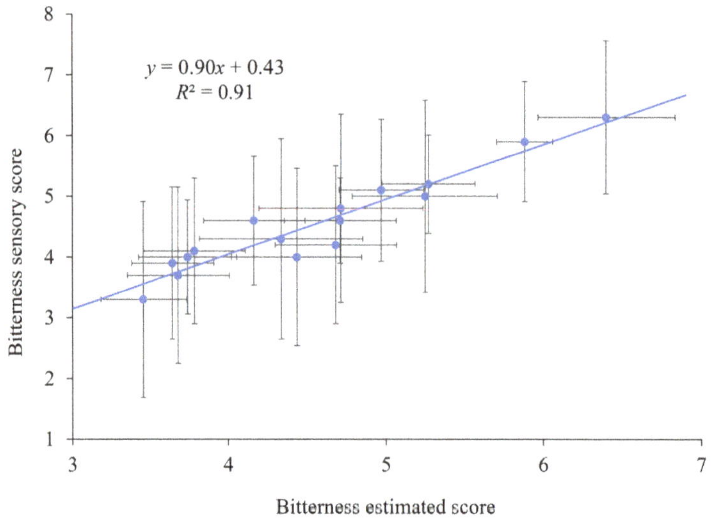

Figure 2.16. Relationship between bitterness sensory score and bitterness estimated score (reprinted with permission from Wu X *et al* [46]. Copyright (2020) Elsevier).

acesulfame K could be quantified in a mixed system of acesulfame K and coffee. Meanwhile, the sensory test results showed that adding acesulfame K to instant coffee suppressed its bitterness. Therefore, it was possible to predict the bitterness sensed by humans in a mixture of acesulfame K and coffee using the sweetness and bitterness sensors together. A multiple regression analysis yielded the following model formula [46]:

$$Y = Y_{\text{bitter}} - 0.64 \times Y_{\text{sweet}} - 0.08 \tag{2.2}$$

where Y is the coffee bitterness intensity, Y_{bitter} is the predicted coffee bitterness intensity calculated from the output of the bitterness sensor and Y_{sweet} is the predicted coffee sweetness intensity calculated from the output of the sweetness sensor. Figure 2.16 shows the sensor values and predicted values using the model formula. Therefore, the combined use of bitterness and sweetness sensors in the taste sensor is expected contribute to the production of the ideal and most pleasant taste of coffee according to consumer's taste.

2.4.2 Increased sensitivity for high-potency sweeteners

Section 2.3.5 introduced the development of high-potency sweetener sensors, and in section 2.4.1, we discussed the evaluation of bitterness suppression effects due to high-potency sweeteners in coffee using this sensor. This sensor uses CPA values to evaluate sweetness intensity; however, considering that the range of acesulfame K concentration used is 0.2–0.7 mM, there is a need to increase sensor sensitivity. As described in section 2.3, taste sensors use relative and CPA values. The relative values of high-potency sweetener sensors reflect the effects of the electrostatic and hydrophobic interactions of high-potency sweeteners and the electrostatic interactions due to salty substances such as NaCl in solution. These relative values have high selectivity and hardly respond to other taste qualities. CPA values were generated only from the hydrophobic interactions of high-potency sweeteners. Naturally, the relative values were higher than the CPA values. Therefore, a statistical model was created that used two sensors to evaluate high-potency sweeteners by subtracting the relative values of saltiness sensors from those of high-potency sweetener sensors [47].

Relative values are proportional to the logarithm of the concentration, so they were approximated using equations (2.3) and (2.4) (referring to the Weber–Fechner law in equation (2.1)).

$$F(x, y) = a \cdot \log(p \cdot x + h \cdot y) \tag{2.3}$$

$$G(x, y) = b \cdot \log(q \cdot x + i \cdot y) \tag{2.4}$$

where

F: relative value response of high-potency sweetener sensor;

G: relative value response of saltiness sensor;

x: acesulfame K concentration of high-potency sweeteners;

y: concentration of saltiness substance NaCl;

a, *b*, *p*, *q*, *h*, and *i* are coefficients, and *x* and *y* are as expressed in equations (2.5) and (2.6), respectively:

$$x = j \cdot \exp\left(\frac{F}{a}\right) k \cdot \exp\left(\frac{G}{b}\right) \tag{2.5}$$

$$y = m \cdot \exp\left(\frac{F}{a}\right) - n \cdot \exp\left(\frac{G}{b}\right) \tag{2.6}$$

where *j*, *k*, *m* and *n* are numerical constants.

The relative values of the high-potency sweetener sensor and saltiness sensor respond to both salty substances and acesulfame K; thus, equations (2.7) and (2.8) were used as correction terms to obtain the predictive model of equations (2.9) and (2.10), respectively:

$$c \cdot \left(\exp\left(\frac{F - G}{e}\right) - 1\right) \tag{2.7}$$

$$d \cdot \left(\exp\left(\frac{F - G}{f}\right) - 1\right) \tag{2.8}$$

The concentration of acesulfame K is expressed as

$$x = j \cdot \exp\left(\frac{F}{a}\right) + k \cdot \exp\left(\frac{G}{b}\right) + c \cdot \left(\exp\left(\frac{F - G}{e}\right) - 1\right) \tag{2.9}$$

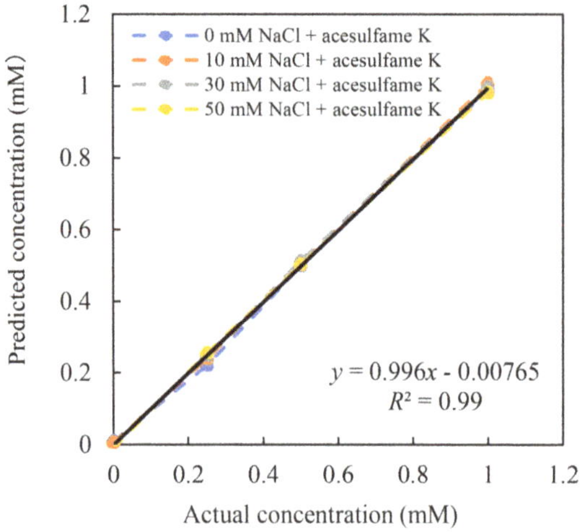

Figure 2.17. Relationship between the estimated concentration and the actual concentration of acesulfame K (reprinted from [47], copyright (2020), with permission from MDPI (CC BY).

whereas the concentration of NaCl is expressed as

$$y = m \cdot \exp\left(\frac{F}{a}\right) + n \cdot \exp\left(\frac{G}{b}\right) + d \cdot \left(\exp\left(\frac{F-G}{f}\right) - 1\right), \tag{2.10}$$

where *c*, *d*, *e* and *f* are numerical constants.

A total of 16 (4 × 4) samples, in which 0, 0.25, 0.5, and 1 mM of acesulfame K were mixed with 0, 10, 30 and 50 mM of NaCl, respectively, were measured with each sensor, and a regression analysis model was obtained. The relationship between the estimated and actual concentrations of acesulfame K is shown in figure 2.17. Thus, an estimation model that could quantify acesulfame K concentration without being affected by saltiness was obtained. The actual acesulfame K concentration used was 0.2–0.7 mM. Using the model, acesulfame K concentrations in the range of 0–1 mM could be quantified, including a maximum NaCl concentration of 50 mM (R^2 = 0.99). These results showed that it was possible to improve the sensor sensitivity using relative values instead of CPA through the use of two sensors. High-potency sweeteners were used at low concentrations; the model has been shown to be useful for evaluating food products with high-potency sweeteners. This experiment used solutions that included acesulfame K and NaCl, but future evaluations will be conducted on samples such as beverages with multiple components.

2.5 Application for coffee

This section describes recent examples of coffee taste quantification using a taste sensing system. Commercial coffee production primarily relies on two varieties of coffee, *Coffea arabica* L. and *Coffea canephora* P. For the *arabica* variety, in particular, two genetic bases (Typica and Bourbon) spread from Yemen after the beginning of the 18th century onwards; most of the commercially available *arabica* varieties cultivated worldwide

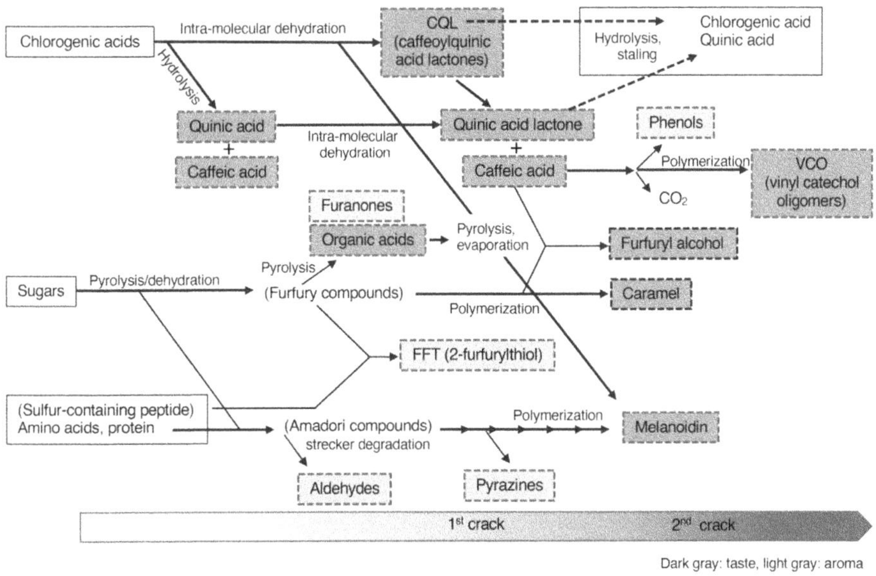

Figure 2.18. Major chemical changes due to the roasting process. The horizontal axis shows the degree of roasting, the dark and light grey boxes are substances relating to taste and aroma, respectively. Bitter substances such as caffeoylquinic acid lactones (CQL) and vinyl catechol oligomers (VCO) are produced through dehydration, hydrolysis and polymerisation (partially revised figure from Tanbe [50]. Reprinted with permission from Kodansha).

belong to these two bases [48]. The main factors affecting the flavor of coffee are the quality of the beans, including processing after harvesting and roasting. The precursors of flavor substances develop during the growth of the coffee cherry, and their quality can vary with climate, altitude, temperature and processing after harvesting [49]. The unique flavor of coffee is produced from the roasting process through various chemical reactions, as shown in figure 2.18 [50].

Figure 2.19 shows the extract obtained from coffee beans (*Coffea arabica*, Brazil) with different degrees of roasting that were fractionated by liquid–liquid extraction using an organic solvent. The results obtained by the taste sensor, liquid chromatography with tandem mass spectrometry (LC-MS/MS) and liquid chromatography-photodiode array (LC-PDA), were analyzed using partial least squares regression (PLS-R) [51–53]. A correlation was found between the beans-derived/roasted compounds and the sensor response. As shown in figure 2.20, this result corresponds to that of Yamaguchi, which showed that sensory tests using coffee with different degrees of roasting and taste sensor-based evaluations could reveal differences between bitterness and acidity at the same degree of roasting [54]. Ishiwaki reported a high correlation between sensory evaluation and taste sensor results when changing the blend ratios of Vietnamese *canephora* with Colombian *arabica* from 0% to 100% at 20% intervals [55].

The results of the evaluation of the coffee profile of each country of origin using taste sensors for acidity and bitterness are shown in figure 2.21. The average value of

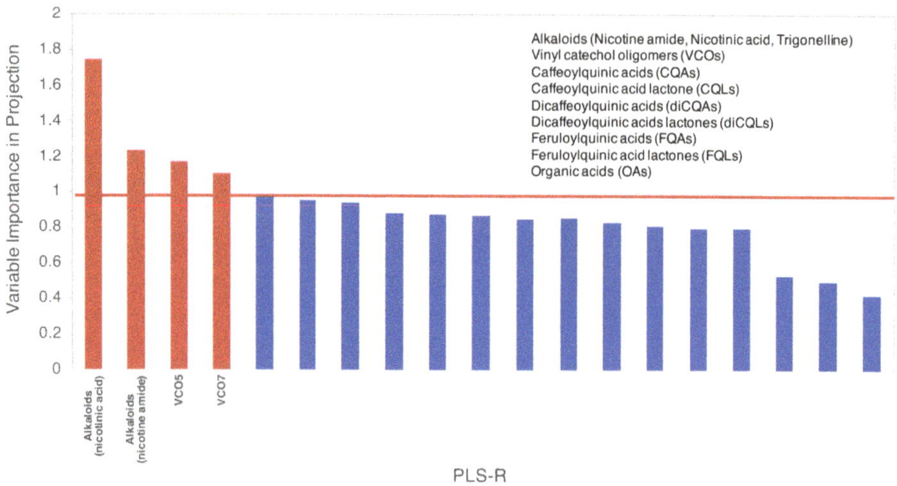

Figure 2.19. Correlation between the bitter substances included in coffee and taste sensor. The extracted solutions obtained from coffee beans (*Coffea arabica*, Brazil) with different degrees of roasting were fractionated from liquid–liquid extraction using an organic solvent, and the results obtained from the taste sensor, LC-MS/MS and LC-PDA were analyzed with PLS-R [51–53].

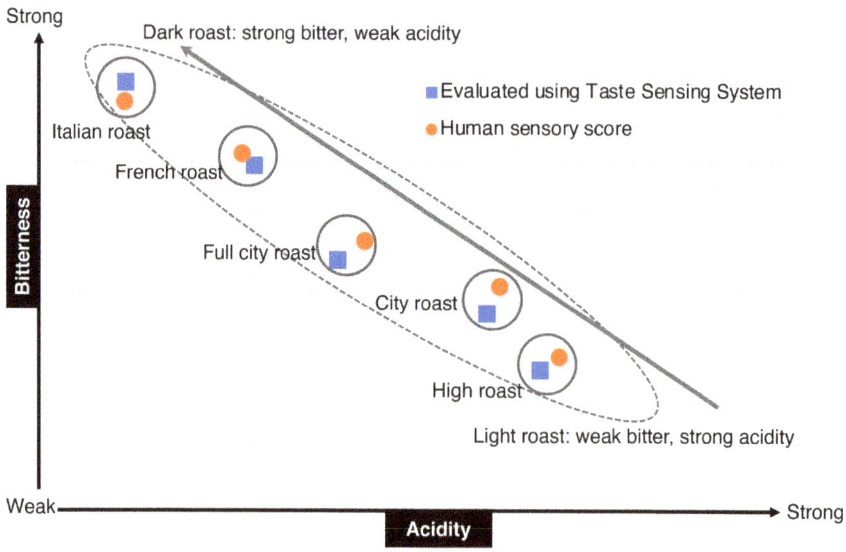

Figure 2.20. Correlation between the taste sensor and sensory evaluations for coffees with different degrees of roasting (partially revised figure from Yamaguchi [51]. Reprinted with permission from CMC Publishing).

all measured coffees was set to zero, and the bitterness and acidity intensities when changing the degree of roasting from light to dark are shown relative to the zero value. It can be seen that changing from light to dark roasting decreases the acidity and increases the bitterness, and that there are different taste characteristics according to production area and variety.

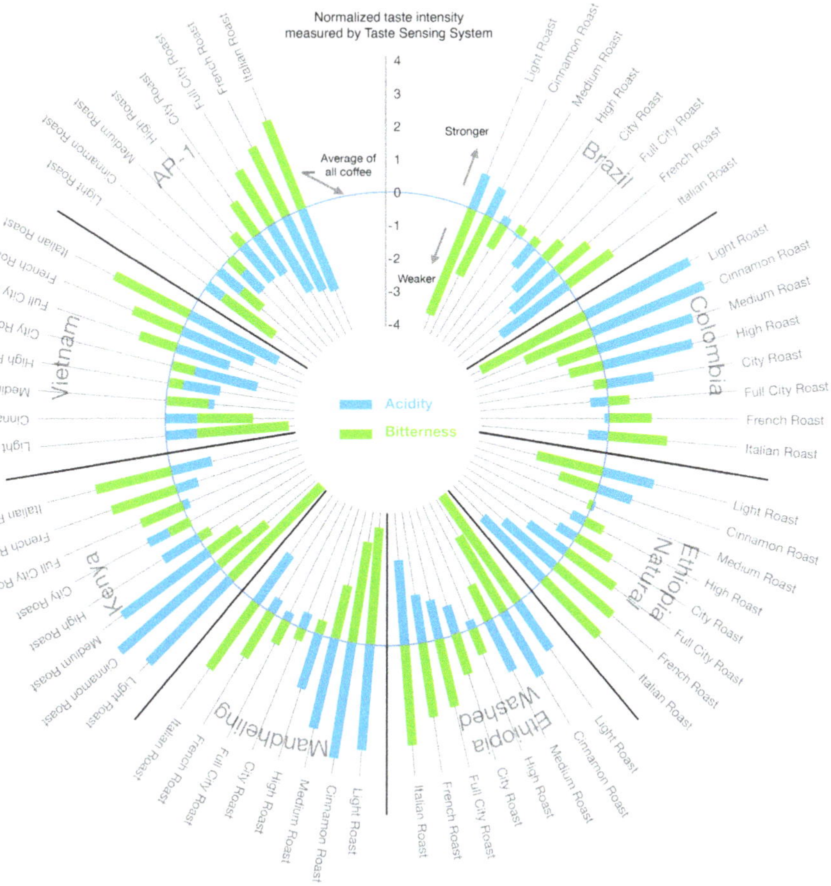

Figure 2.21. Taste profile of coffees of the world. Relative intensities of bitterness and acidity were shown for each coffee bean with different degrees of roasting, and the average of all coffees shown in the figure is set to zero.

The aforementioned results are based on the differences in sensitivity to basic taste qualities such as bitterness and acidity. There are also several reports relating to the evaluation of the 'taste quality' of coffee based on taste sensor results. 'Body' and 'cleanliness' have been defined by the Specialty Coffee Association (SCA) as coffee evaluation items [56]. On the Nestlé Nespresso website, for example, consumer information includes a range of characteristics for different types of coffee, such as body values, acidity and bitterness [57]. Regarding body and cleanliness evaluations based on taste sensors, Horiguchi *et al* found a correlation between sensory evaluations based on the Q-grader's results and taste sensor results; they proposed a metric obtained by multiplying the taste sensor value with a coefficient [58]. As shown in figure 2.22, Yamaguchi denoted cases where acidity was weak or strong with 'sharp bitterness' and 'rounded bitterness', respectively, in sensory evaluations even when the bitterness intensity was the same, and suggested that this can be evaluated by indexing this bitterness quality with a 'bitterness–acidity index' [54].

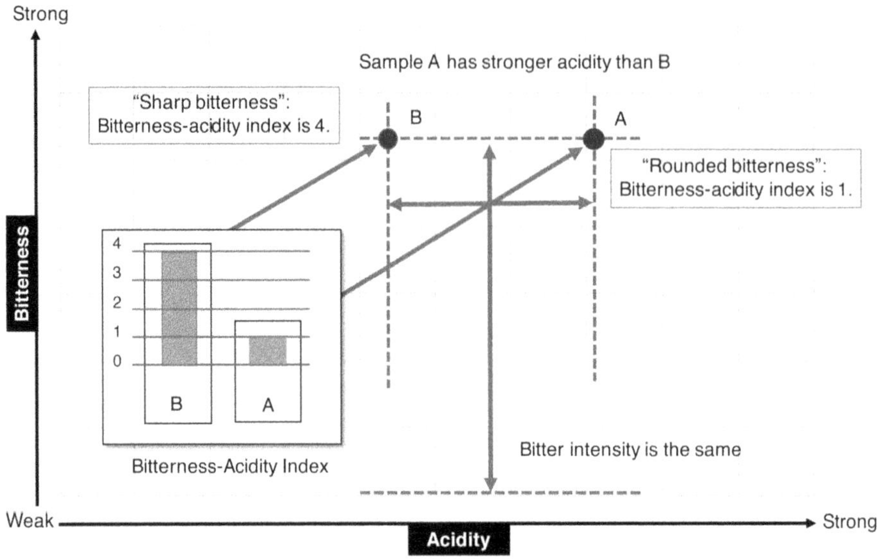

Figure 2.22. Scientific notation of bitterness quality. In cases where bitterness is constant and the acidity changes, the difference in their balance is indexed to quantify the bitterness quality (partially revised figure from Yamaguchi [51]. Reprinted with permission from CMC Publishing).

Ishiwaki highlighted the following benefits of being able to quantify and visualize the taste of coffee [55]: (1) the taste sensor map not only gives the customer an accurate idea of the taste of coffee but the results can also be used in marketing; (2) an accurate idea of taste can provide effective training for inexperienced employees; (3) advanced quality control is possible even without a skilled appraiser, enabling measures for customer complaints, setting expiration dates and managing manufacturing processes; (4) client and customer understanding can be deepened with data accumulation, leading to rapid and accurate product development; and (5) although only experienced blenders can create the desired taste by blending, the taste sensor can be used as a tool to not only optimize taste but also the price. Particularly with (5), which leads to cost reduction, an optimization calculation is used with the blending ratio as a parameter to find a blend that has the lowest price within the target numerical range; this was found to result in cost reductions of around 10% for instant coffee, which does not have much of an aroma [59]. Later, Sato *et al* proposed an algorithm for calculating the blend ratio and optimal price based on data of the taste of raw materials measured by a taste sensor [60].

Although commercial arabica coffee beans primarily come in two varieties (Typica and Bourbon), the size, thickness, density, hardness and water content of green beans differ according to the cultivated variety and location. As shown in figure 2.23, related to figure 2.18, the flavor substances produced during the roasting process are greatly affected by the physical properties of the green beans. Baggenstoss *et al* stated that 'attaining the same coffee bean color using different time–temperature conditions during roasting does not necessarily mean that coffees are equivalent in terms of aroma and physical properties'. This expresses a part of

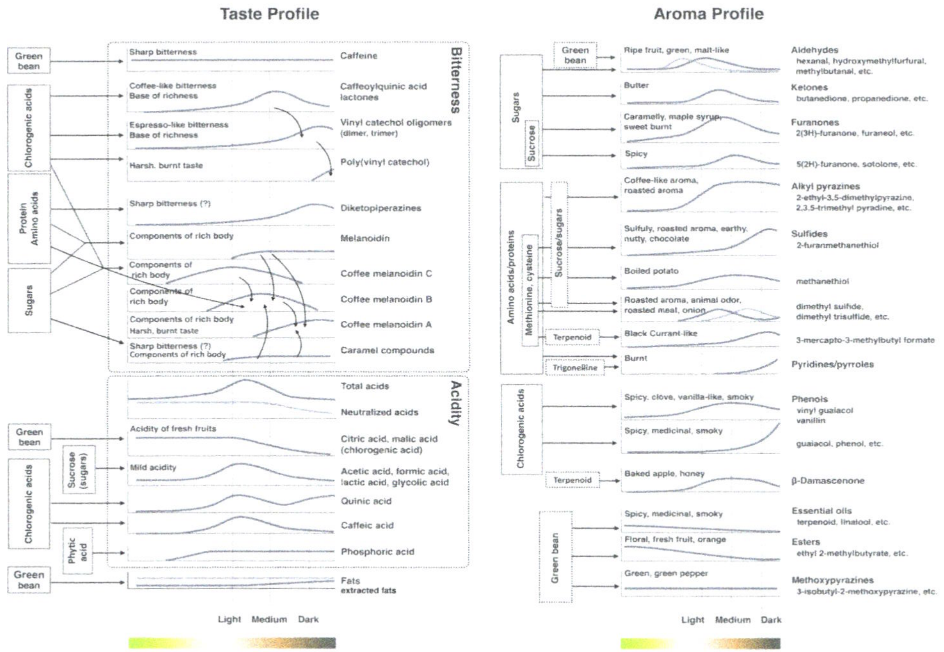

Figure 2.23. Formation of substances relating to aroma and taste. The horizontal axis shows the degree of roasting. The formation of substances that primarily contribute to bitterness, acidity and aroma are shown in the top left, bottom left and right, respectively (partially revised figure from Taguchi and Tanbe [61], reprinted with permission from NHK Publishing).

the complexity of the roasting process. Taguchi, who was formerly the president of the Specialty Coffee Association of Japan (SCAJ), stated that green beans can be broadly classified into four groups according to (1) green bean color, (2) changes in the wrinkles during roasting, (3) swelling of beans during roasting and (4) changes in the bean color after roasting, indicating that there is an appropriate degree of roasting for each group [61]. This assessment was based on many years of experience, but it also suggests that the flavor characteristics, which have to date been seen as features of each production area, could be reconsidered in terms of physical features, and that there may be an optimal roasting profile for each group. The Coffee Taster's Flavor Wheel from the Specialty Coffee Association was revised for the first time in 21 years in 2016. Three universities and one research institution were involved in this revision; the revision approach was based on scientific research results [62]. This revision suggested the potential for evaluating complex tastes worldwide using objective indices by reconsidering previously used rules of thumb from a scientific point of view. There is great potential for the use of taste sensors as a tool for such purposes.

2.6 Perspectives

This chapter provides a review of research and development conducted worldwide on potentiometric measurement-based taste sensors from the 1980s to the latest techniques. As discussed, there are various methods and materials such as voltammetry, impedance spectroscopy, conducting polymers, optical methods, and biological organs such as gustatory cells. Ultimately, they all have their advantages and disadvantages, and they should be put to practical use by utilising their advantages.

A very interesting phenomenon has recently been discovered in taste sensors that use lipid membranes. Potentiometric measurements are used in these sensors; therefore, they are not particularly proficient in the measurement of uncharged taste substances. Sweet substances such as sucrose and glucose could be successfully measured by modifying the membrane with chemical substance such as trimellitic acid and performing special cleaning (pre-processing). However, its mechanism has not yet been completely elucidated [39–42]. Meanwhile, no response was observed for uncharged bitter substances, such as caffeine or theobromine. Very recently, the selection of aromatic carboxylic acids (i.e. hydroxy-, dihydroxy-, and trihydroxybenzoic acids) as modifiers has resulted in the successful measurement of these substances [63]. The intramolecular hydrogen bond between the carboxy and hydroxy groups of the aromatic carboxylic acid plays an important role in this measurement. The bonding of taste substances such as caffeine to hydroxy groups of modifiers results in the protons in solution bonding to adjacent carboxy groups and changing the surface charge density of the membrane inducing a change in the membrane potential. This phenomenon signifies that caffeine bonding to hydroxy groups promotes the bonding of protons to adjacent carboxy groups and is simply the allosteric mechanism that is well known in biological systems.

Allostery (the adjective form is 'allosteric'), which works in many types of proteins, such as enzymes and receptors, in biological systems, is a concept that has been known for a long time. Here, the binding of a chemical substance (effector) on one site suppresses work at a different site (e.g., substrate affinity and enzyme activity). The same molecule may be an effector and a substrate; this can be interpreted more generally as the complex formation of a chemical substance and receptor promotes or suppresses the complex formation of the next stage.

Although various materials, methods and concepts (e.g. artificial enzymes, supramolecular allosteric signal amplification sensing, multi-step switches, and ion detection allosteric sensors) have been tested for various development targets in research on artificial production or utilising of the allosteric mechanism found in this taste sensor, many of these are still in the initial stages of development.

Research on taste sensors based on allosteric mechanisms that originated in Japan further enhanced the taste sensor. This research will create an allosteric sensing device that also uses self-organization to artificially reproduce the intelligence of classifying, identifying and quantifying chemical substances based on their attributes present in living organisms. Furthermore, these results promote the development of general biochemical sensors.

Sensors mimicking the physical senses of sight, hearing and touch have been developed. In this book, sensors related to taste are described in detail. The development of sensors related to the sense of smell, which is the remaining one among the five senses, has seen remarkable progress in recent years alongside the development of artificial intelligence. We are now entering an unprecedented new era in which we may have integrated sensors related to all five senses.

References

[1] Hayashi K, Yamanaka M, Toko K and Yamafuji K 1990 *Sens. Actuators* B **2** 205–13

[2] Riul A Jr, Dantas C A R, Miyazaki C M and Oliveira O N Jr 2010 *Analyst* **135** 2481–95

[3] Anand V, Kataria M, Kukkar V, Saharan V and Choudhury P K 2007 *Drug Discovery Today* **12** 257–65

[4] Wu X, Tahara Y, Yatabe R and Toko K 2020 *Anal. Sci.* **36** 147–59

[5] Toko K and Tahara Y 2016 Beer analysis using an electronic tongue *Electronic Noses and Tongues in Food Science* ed M L R Mendez (Oxford: Academic) pp 161–70

[6] Toko K, Tahara Y, Habara M, Kobayashi Y and Ikezaki H 2016 Taste sensor: electronic tongue with global selectivity *Essentials of Machine Olfaction and Taste* ed T Nakamoto (Singapore: John Wiley & Sons) pp 87–174

[7] Toko K, Onodera T and Tahara Y 2013 Nano-biosensors for mimicking gustatory and olfactory senses *Bio-Nanotechnology: A Revolution in Food, Biomedical and Health Sciences* ed B Debasis, B Manashi, M Hiroyoshi and S Fereidoon (Hoboken, NJ: Wiley-Blackwell) pp 270–91

[8] Tahara Y and Toko K 2013 *IEEE Sens. J.* **13** 3001–11

[9] Toko K 2013 *Biochemical Sensors: Mimicking Gustatory and Olfactory Senses* (Singapore: Jenny Stanford Publishing)

[10] Kobayashi Y, Habara M, Ikezazki H, Chen R, Naito Y and Toko K 2010 *Sensors* **10** 3411–43

[11] Toko K 2000 *Biomimetic Sensor Technology* (Cambridge: Cambridge University Press)

[12] Kobatake Y, Irimajiri A and Matsumoto N 1970 *Biophys. J.* **10** 728–44

[13] Yoshida M, Kamo N and Kobatake Y 1972 *J. Membr. Biol.* **8** 389–402

[14] Toko K, Nitta J and Yamafuji K 1981 *J. Phys. Soc. Japan* **50** 1343–50

[15] Iiyama S, Toko K and Yamaguji K 1986 *Agric. Biol. Chem.* **50** 2709–14
[16] Iiyama S, Toko K and Yamafuji K 1987 *Membrane* **12** 231–7
[17] Hayashi K, Matsuki Y, Toko K, Murata T, Yamafuji Ke and Yamafuji K 1989 *Sens. Mater* 321–34
[18] Iiyama S, Toko K, Hayashi K and Yamafuji K 1989 *Agric. Biol. Chem.* **53** 675–81
[19] Hayashi K, Yamafuji K, Toko K, Ozaki N, Yoshida T, Iiyama S and Nakashima N 1989 *Sens. Actuators* **216** 25–42
[20] Gaillard D and Kinnamon S C 2019 *Acta Physiol.* **226** e13246
[21] Yasumatsu K, Iwata S, Inoue M and Ninomiya Y 2019 *Acta Physiol.* **226** e13215
[22] Nelson G, Hoon M A, Chandrashekar J, Zhang Y, Ryba N J P and Zuker C S 2001 *Cell* **106** 381–90
[23] Nelson G, Chandrashekar J, Hoon M A, Feng L, Zhao G, Ryba N J P and Zuker C S 2002 *Nature* **416** 199–202
[24] Meyerhof W, Batram C, Kuhn C, Brockhoff A, Chudoba E, Bufe B, Appendino G and Behrens M 2010 *Chem. Senses* **35** 157–70
[25] Behrens M and Meyerhof W 2013 *Semin. Cell Dev. Biol.* **24** 215–21
[26] Zhang J, Jin H, Zhang W, Ding C, O'Keeffe S, Ye M and Zuker C S 2019 *Cell* **179** 392–402.e15
[27] Huang A L, Chen X, Hoon M A, Chandrashekar J, Guo W, Trankner D, Ryba N J P and Zuker C S 2006 *Nature* **442** 934–8
[28] Montell C 2018 *Science* **359** 991
[29] Chandrashekar J, Kuhn C, Oka Y, Yarmolinsky D A, Hummler E, Ryba N J P and Zuker C S 2010 *Nature* **464** 297–301
[30] Ahmad R and Dalziel J E 2020 *Front. Pharmacol.* **11** 587664
[31] Sanematsu K, Yoshida R, Shigemura N and Ninomiya Y 2014 *Curr. Pharm. Biotechnol.* **15** 951–61
[32] Roper S D and Chaudhari N 2017 *Nat. Rev. Neurosci.* **18** 485–97
[33] Tahara Y 2013 Portable taste sensor *Biochemical Sensors: Mimicking Gustatory and Olfactory Senses* ed K Toko (Singapore: Jenny Stanford Publishing) pp 423–33
[34] Hara D, Fukagawa T, Tahara Y, Yasuura M and Toko K 2014 *Sens. Lett.* **12** 1172–6
[35] Harada Y, Tahara Y and Toko K 2015 *Sensors* **15** 6241–9
[36] Toko K, Hara D, Tahara Y, Yasuura M and Ikezaki H 2014 *Sensors* **14** 16274–86
[37] Yasuura M, Tahara Y, Ikezaki H and Toko K 2014 *Sensors* **14** 7359–73
[38] Kobayashi Y, Hamada H, Yamaguchi Y, Ikezaki H and Toko K 2009 *IEEJ Trans. Electr. Electron. Eng.* **4** 710–9
[39] Toyota K, Cui H, Abe K, Habara M, Toko K and Ikezaki H 2011 *Sens. Mater* **23** 475–82
[40] Toyota K, Cui H, Abe K, Habara M, Toko K and Ikezaki H 2011 *Sens. Mater* **23** 465–74
[41] Yasuura M, Shen Q, Tahara Y, Yatabe R and Toko K 2015 *Sens. Mater* **27** 351–8
[42] Yasuura M, Okazaki H, Tahara Y, Ikezaki H and Toko K 2014 *Sens. Actuators* B **201** 329–35
[43] Tahara Y, Hattori T, Wu X, Yatabe R, Ikezaki H, Habara M and Toko K 2017 Development of sweetness sensor for high-potency sweeteners using lipid polymer membrane *Proc. ISOCS/IEEE Int. Symp. on Olfaction and Electronic Nose (ISOEN) (Montreal, Canada, May 2017)* 256–66
[44] Onodera T, Tahara Y, Yatabe R and Toko K 2018 Nanobiosensors for gustatory and olfactory senses *Encyclopedia of Nanoscience and Nanotechnology* ed H S Nalwa (Stevenson Ranch, CA: American Scientific Publishers) pp 363–85

[45] Takagi S, Toko K, Wada K and Ohki T 2001 *J. Pharm. Sci.* **90** 2042–8
[46] Wu X *et al* 2020 *Sens. Actuators,* B **309** 127784
[47] Liu Y, Wu X, Tahara Y, Ikezaki H and Toko K 2020 *Sensors* **20** 400
[48] Anthony F, Combes M, Astorga C, Bertrand B, Graziosi G and Lashermes P 2002 *Theor. Appl. Genet.* **104** 894–900
[49] Sunarharum W B, Williams D J and Smyth H E 2014 *Food Res. Int.* **62** 315–25
[50] Tambe Y 2016 *Science of Coffee* (Tokyo: Kodansha) p 193 ISBN-13: 978-4062579568 [in Japanese]
[51] Fujimoto H *et al* 2018 *Proc. The 27th Int. Conf. on Coffee Science (Portland, OR, Sep. 2018)* 46
[52] UCC Research Activities https://ucc.co.jp/company/research/taste/bittertaste.html Accessed 18 September 2020
[53] Fujimoto H *et al* 2021 *Food Chem.* **342** 128228
[54] Yamaguchi A 2017 Embodiment of coffee of stable quality and the expected taste *Forefront of the Science of Palatability and their Business* ed K Toko and M Kashiwayanagi (Tokyo: CMC Publishing) p 397–403 ISBN-13: 978-4781312491 [in Japanese]
[55] Ishiwaki T 2013 Application of taste sensor to blending of coffee *Biochemical Sensors: Mimicking Gustatory and Olfactory Senses* ed K Toko (Singapore: Jenny Stanford Publishing) p 83–90
[56] Specialty Coffee Association 2020 Cupping protocols, individual component score https://sca.coffee/research/protocols-best-practices Accessed 1 July 2020
[57] Nespresso https://nespresso.com/uk/en/order/capsules/original/fortissio-lungo-coffee-capsule Accessed 1 July 2020
[58] Horiguchi T 2019 Study on correlativity of physicochemical evaluation and sensory evaluation to construct criterion of specialty coffee *PhD Thesis Tokyo University of Agriculture* (Tokyo: Tokyo University of Agriculture)
[59] Ishiwaki T 2012 Creation of coffee using the taste sensor *Science and Technology to Produce Deliciousness of Foods/Medicines and Ensure the Safety* ed K Toko (Tokyo: CMC Publishing) p 208–13 ISBN-13: 978-4781306087 [in Japanese]
[60] Satoh M, Satoh H and Ikezaki H 2015 *IEEJ Trans. Sens. Micromach.* **135** 57–64
[61] Taguchi M and Tambe Y 2014 *Equation for the Taste of Coffee* (Tokyo: NHK Publishing) p 162–63 ISBN-13: 978-4140332825 [in Japanese]
[62] Specialty Coffee Association News https://scanews.coffee/2016/04/27/the-science-behind-the-coffee-tasters-flavor-wheel/ Accessed 18 September 2020
[63] Yoshimatsu J, Toko K, Tahara Y, Ishida M, Habara M, Ikezaki H, Kojima H, Ikegami S, Yoshida M and Uchida T 2020 *Sensors* **20** 3455

Chapter 3

Guidelines on the development of sensors and application of data analysis tools for potentiometric electronic tongues

Karthika Kappalakandy Valapil, Elżbieta Jarosińska, Marcin Szymon Filipiak, Wojciech Mazurkiewicz and Emilia Witkowska Nery

Potentiometric sensors are the most popular transducers in electronic tongue systems and the only ones used commercially. As well as some examples of interesting experimental setups and a short review of different applications of potentiometric e-tongues, this chapter presents the development of such sensor arrays. An electronic tongue consists of a set of sensors and chemometrics used for the data analysis. Together with the fabrication of the sensor array, we describe the machine learning algorithms used to refine the choice of sensors and for the classification of samples in the final application of the system.

3.1 Introduction

The first electronic tongues proposed in the early 1990s of the last century were based on potentiometry. They gained wide recognition for their similarity to human taste recognition. A complex interaction pattern of all the different molecules present in the sample results in a single change of potential of an electrode, or a single firing neuron in the case of our taste cells. Contributions of individual species cannot be resolved from just one cell or a single sensor. The whole matrix is needed to obtain a sufficient amount of information.

Electrochemical sensing of ions can be achieved through potentiometric ion-selective electrodes (ISE) and ion-selective field-effect transistors (ISFETs), which will be briefly presented at the beginning of this chapter. There is unfortunately still a knowledge gap between electrochemists working on ISEs and electrical/material engineers working on ISFETs—that is why we also include a practical comparison of both types of systems.

Other electrochemical techniques that can be used for ion sensing, such as voltammetry at the interface of two immiscible solutions or measurements based on impedance, are used less frequently and will not be included in this discussion. Next, after gaining an understanding of the functioning of basic sensors, we describe the concepts of the mammalian taste and compare the natural and artificial senses, taking into account the differences between the number of recognition sites, redundancy, type of output signal, etc. Those similarities and differences are of utmost importance for researchers designing new electronic tongue systems.

We also take a closer look at the different setups presented until now in the literature, which can guide the reader to the development of assemblies tailored to a specific application. Given so many advantages of the general idea of e-tongue, we describe and compare two commercially available systems and discuss why after 40 years of research they are still a not well-recognized niche. The chapter finishes with a practical description of methods of signal preparation, analysis and validation.

3.2 Potentiometric sensors

3.2.1 History of potentiometric sensors

Potentiometry is one of the oldest instrumental techniques used in chemical analysis. It uses a relationship between the electrode potential and the activity of the determined ion. Popularity comes partially from the sensing principle that does not require current flow, which significantly reduces the complexity of the equipment needed. Historically, the first glass ion-selective electrode designed for pH measurements was created in 1908 by Fritz Haber and Zygmunt Klemensiewicz [1], however, almost 15 years earlier the first use of a mercury ion-sensitive electrode in mercurometric titration of halogen was described [2].

A typical glass electrode consists of a glass tube ending in a thin-walled glass bulb made of doped glass. Inside the glass tube, there is an internal electrolyte solution (0.1 M HCl) in which the silver chloride electrode is immersed [3]. Many years of research have allowed the development of glass electrodes sensitive to other ions. The specificity of the glass membrane was obtained by modifying its composition, mainly through the introduction of oxides whose oxidation state is lower than the coordination number, e.g., aluminum or boron [4].

In the following years, crystalline materials such as LaF_3, AgCl or Ag_2S were introduced as sensing materials, increasing the number of measurable ions. However, they were mostly characterized by poor ionic conductivity which significantly limited the interest of the scientific community. The most notable example was fluoride electrode based on a LaF_3 crystal membrane doped with europium designed by Orion Research Inc. Ross and Frant, founded by James Ross and John Riseman [5] in the early 1960s.

An important breakthrough in the development of ion-selective electrodes was the discovery of ion migration between intracellular membranes, which completely transformed the field of potentiometric sensors and started the era of host-guest chemistry. Moore and Pressman have thoroughly investigated the first ionophores isolated from bacteria species [6]. They indicated that the antibiotic effect of valinomycin is related to the presence of potassium cations that induce a selective potentiometric response. The principle of operation is based on changing the conformation of the molecule and the system of hydrogen bonds. Carrier ionophores, as with most ionophores applied in ion-selective electrodes, complex the metal cation in a hydrophilic cavity, directing the hydrophobic part (alkyl groups) to the outside of the molecule. This unique structure allows the cation to be isolated from the external environment. The new complex is electrically neutral and soluble in polar and non-polar solvents [7].

About the same time, Pedersen was researching on catalysts for the polymerization reaction and accidentally received synthetic ionophore—dibenzo-18-crown-6, capable of complexing alkali metal cations [8]. Afterwards, Simon and Stefanac introduced the first potassium-selective electrode with valinomycin as an ionophore and polymeric membrane based on polyvinyl chloride [9], which Bloch and coworkers earlier designed [10]. This innovative solution gained the recognition of many scientists due to the easy preparation and durability of membranes, and it is still used today. In the case of electronic tongue systems, the similarity to the natural mechanism of molecular recognition of such electrodes is also notable [11, 12].

One of the most important achievements in the field of potentiometry was the development of glucose biosensors. In the second half of the 20th century, Leland C Clark introduced the concept of the first biosensor—the oxygen electrode with immobilized glucose oxidase [13]. This idea was also later adapted to potentiometric electrodes, which could be used to detect changes in ammonia concentration as in the case of urea biosensors coated with urease enzyme.

The end of the 20th century brought significant miniaturization of ion-selective electrodes. The internal electrolyte used in the standard sensor configuration (figure 3.1(A)) had to be replaced to achieve a considerable scale reduction. In the new 'solid-contact' ISEs (figure 3.1(B)), an inner filling solution was replaced by a transducer element providing ionic and electron conductivity at the same time [14]. It eliminated problems with inner solution, such as drying, increasing storage times, and was more favorable with transport, sterilization and miniaturization. Different layers, such as electrolyte-filled hydrogel, carbon nanotubes and conducting polymers were tested, the latter being the most popular approach. Remarkably, electrodes without the intermediate layer, so-called 'coated wire' sensors (figure 3.1(C)), in which the ion-sensitive membrane is deposited directly on the electrical contact still perform very well, although till today it is not fully understood why. Subsequently, the research was conducted all over the world, introducing ion-selective electrodes for more than 50 ions, microelectrodes for *in vivo* measurements, improving parameters of ion-selective membranes, lowering the detection limit of sensors and simplifying the design aimed at obtaining cheap disposable electrodes [5, 15, 16].

Nowadays, potentiometric sensors are widely used globally, especially in healthcare, where ion-selective electrodes are crucial in the medical analysis of major electrolytes, blood gases, and some metabolites [17]. From the early 1900s, the pH glass electrode became a standard in determining blood acid-base balance, replacing the earlier method of titration and color indicator.

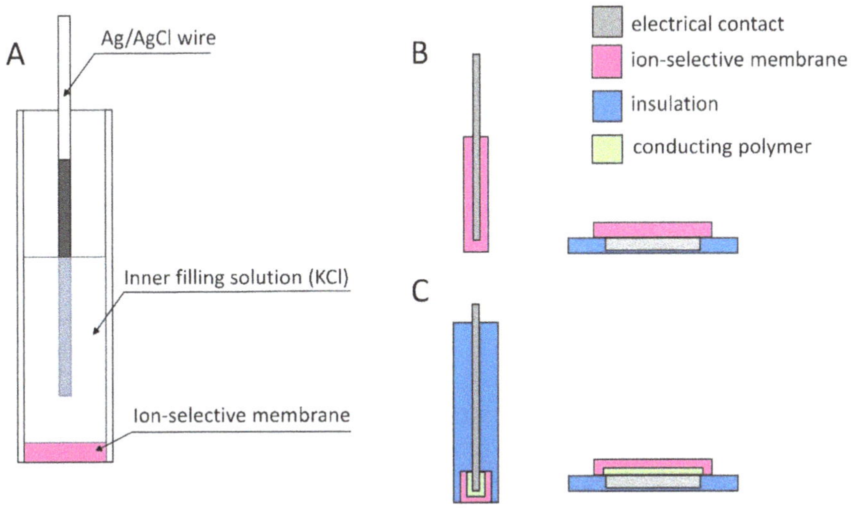

Figure 3.1. Main types of construction of ion-selective potentiometric sensors. (A). glass body electrode with internal electrolyte and the membrane deposited at the bottom. (B). coated wire ISE, in which the membrane is deposited directly on the electrical contact. (C). solid contact ISE, with a layer of ion to electron transducer deposited between the membrane and the electrical contact. Both miniaturized approaches (B) and (C) are shown in standard vertical (left) and planar versions (right).

The next landmark event was introducing a gasometric apparatus for measuring CO_2, followed by the development of an electrode for direct measurement of blood pCO_2 [18]. The detection was achieved through combining pH glass electrode and silver chloride reference separated from the blood sample by a carbon dioxide gas-permeable membrane. This approach led to the commercial success of the first pH and pCO_2 analyzers—the AME1 and the IL 100 introduced by the Radiometer and Instrumentation Laboratory [17]. New solutions appeared, such as a continuous-flow chemistry analyzer for measuring Na^+ and K^+ in diluted serum by Technicon company. NOVA1 analyzer by Nova Biomedical company followed Na^+/K^+ quantification in the whole blood with a shortened time of analysis.

Another milestone was achieved with the integrated blood gas/electrolyte analyzer introduced by Nova Biomedical. NOVA Stat Profile 1 simultaneously measures pH, pCO_2, pO_2, Hct (hematocrit), Na, K, and Ca, the system was expanded with glucose, lactate, and BUN (blood urea nitrogen) analysis in the following years [17]. Although based on amperometric measurements, Clark electrode biosensors can also be included in this list of important breakthroughs as they are also based on a selective membrane, in this case, similar to that used in pCO_2 analyzers. The YSI, as well as AKRKAY companies, introduced the first handheld biosensors for glucose monitoring, replacing color indicators [19, 20]. Ease of use and reliability made them widely recognized in the world. However, the membrane design was soon replaced with a conducting polymer directly wiring the enzyme to the electrode (e.g. FreeStyle Libre by Abbott). Although glucose biosensors are indisputably the electrochemical probes that achieved the highest commercial success, their development did not result in similar popularity of potentiometric biosensors.

3.2.2 Working principle

The fundamental part of a potentiometric sensor is its membrane, which allows selective extraction of ions from the solution. The response is related to the linear dependence of the potential change on the activity of the determined ion, described by the Nernst equation [21]. This technique does not provide a possibility of direct measurement; therefore, a setup consists of a two-electrode cell measuring the potential difference of the reference electrode against an indicator electrode (ISEs), performed in 'zero' current condition. The Nernst model was the first one used, however, it is applied to ideal conditions, which include only the activity of one type of ion in the solution, while the real samples are composed of many. A more precise model describing the potential of the ion-selective electrode is the Nikolski–Eisenmann equation (3.1) [22], which combines the activity of the interfering ions.

$$E = E^{\circ} + \frac{RT}{z_i F} \ln \left[a_i + \sum_{\mathrm{j}} K_{ij} (a_j)^{\frac{z_i}{z_j}} \right] \tag{3.1}$$

The included selectivity coefficient K_{ij} for the primary ion i in the presence of an interfering ion j defines the selectivity of the potentiometric sensors. There are several different ways to determine selectivity [23]—separate solution method (SSM), fixed interference method (FIM) or matched potential method (MPM). Each of them has some advantages and drawbacks; therefore, one should compare the selectivity coefficient obtained using at least two of the above-mentioned means, although there are no guidelines on which methods provide the best results.

The activities of primary and interfering ions (a_i, a_j) in the formula are related to their molar concentration through the activity coefficient, which depends primarily on the concentration of solutes but can be calculated from the Debye and Hückel approximation that uses the ionic strength of the solution. Another method to estimate the activity coefficient uses the Pitzer equation, which is implemented in computer programs such as PHREEQC or Visual MINTEQ, thanks to which it is possible to estimate the coefficients for multi-component solutions.

Parameters of ion-selective electrodes change over time; therefore, it is necessary before each measurement to perform the electrode characterization and determine the current sensitivity of the sensor. It comes down to determining the calibration curve—a graphical representation of the dependence of the electromotive force change (EMF) of the electrode in terms of the logarithm of the ion activity $E = f(\log(a))$. In potentiometric sensors, two- or three-point calibration can be carried out when the slope is known; however, multiple-point calibration provides better accuracy and precision of the measurements [24]. A correctly assigned calibration curve allows obtaining the ISE detection limit (LOD) [25], which is defined by the intersection of the Nernstian (linear) range with the approximation of points lying outside the linear range.

A linear calibration is a standard approach for the electrochemistry community, although increasingly ineffective, especially for environmental samples when the crucial ions are at low concentrations. Additionally, interfering ions modify sensor characteristics, decreasing the linear range of the sensor and thus increasing the LOD, making detection impossible. Therefore, non-linear regression methods have been used to improve ISE measurements and limit bias near LODs. However, these methods do not provide measurement uncertainty, which is essential in environmental and health monitoring. It is necessary to apply approximation methods such as bootstrapping, asymptotic or Bayesian algorithm [26], which allow obtaining the prediction interval.

3.3 Field-effect transistors

The first article about a field-effect transistor used for ion activity determination was described as early as 1970 by Bergveld [27]. Since then, numerous field-effect transistor-based devices have been developed. Here, for the sake of simplicity, we will talk about ISFETs—ion-sensitive field-effect transistors in comparison to ISE—ion-selective electrodes. It seems that there is no comprehensive review and comparison of both technologies.

3.3.1 Working principle of ISFETs

The working principle of an ion-sensitive field-effect transistor would become much clearer if we started describing a simple metal oxide semiconductor FET, which is present in most of today's electronic devices. In fact, Bergveld himself described an ISFET as 'a special type of MOSFET with a gate at a certain distance' [28].

The substrate of a MOSFET can be a p-type semiconductor (e.g., silicon doped with boron, aluminum, gallium or indium) with two n-doped regions (if we stick to silicon—phosphorus, arsenic, antimony, bismuth, lithium) that act as source and drain terminals. The gap between two n-doped regions is called channel, and a metal gate (third terminal) is placed on top of it isolated by a metal oxide (figure 3.2(A)). When a positive voltage is applied to the gate terminal, negative charge carriers (electrons) are attracted to the semiconductor/oxide interface, thus creating an inversion layer. If no current flows at $V_{SD} = 0$, the transistor is 'normally-off'. When an additional positive source–drain voltage V_{SD} is applied, the current I_{SD} starts to

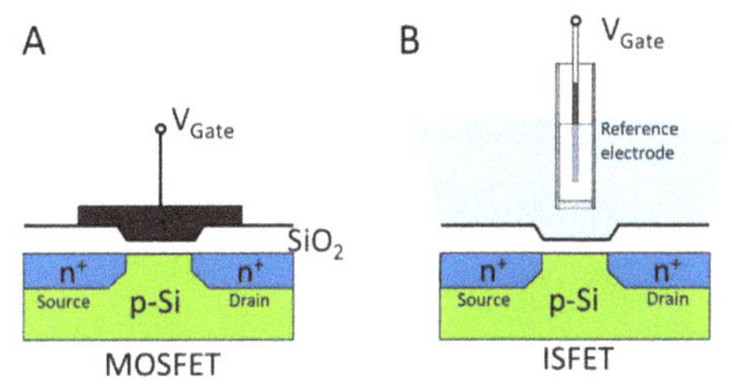

Figure 3.2. Schematic representation of (A) MOSFET and (B) ISFET.

flow between those two terminals. The source–drain current I_{SD} through the channel is described by the equation:

$$I_{SD} = \mu C_{ox}\frac{W}{L}QV_{SD} = \mu C_{ox}\frac{W}{L}(V_G - V_{th})V_{SD}$$

where μ—charge mobility, C_{ox}—capacitance of the gate oxide per area, W—channel width, L—channel length, Q—charge density in the channel, V_{SD}—source-drain voltage, V_G—gate voltage and V_{th}—threshold voltage. The consequence of the above equation is that if the gate voltage is much less than the threshold voltage $V_G \gg V_{th}$, there is no current flowing between the source and drain I_{SD}.

If the gate terminal is separated from the chip and placed in the form of a reference electrode (e.g., well-known Ag/AgCl) in an aqueous solution, we are starting to talk about an ion-sensitive field-effect transistor (figure 3.2(B)). The insulating oxide that is placed on top of the semiconducting channel has properties similar to those of a glass membrane in a glass electrode. An electrical double-layer is then formed at the electrolyte|oxide interface in the same way as in the case of electrolyte|glass membrane interface. Provided that the oxide thickness is low enough, the electrolyte|oxide double-layer will interact with the oxide|semiconductor junction. Consequently, the induced conducting channel in the silicon will depend on the ion activity in the electrolyte in contact with the oxide. A change of ion activity will be measured as a change of channel conductance [27]. This basic architecture of an ISFET is often called electrolyte-insulator-semiconductor ISFET.

However, there are other architectures worth mentioning. One of them, a floating gate ISFET can be imagined in principle as MOSFET with two input gates—sensing gate (SG) and a control gate (CG) that are capacitively coupled to a common floating gate (figure 3.3). The semiconducting channel is covered with an insulator layer. On top of it, a conducting layer is placed that serves as the floating gate, which is insulated from the SG and CG with another insulator. The floating gate has no ohmic contact with the surroundings and thus transfers the signal from the electrolyte|insulator interface by electrostatic interactions. Both the SG and CG can be used to modulate the current flow through the channel. In chemical sensing, the control gate can be used for biasing, whereas the sensing gate gives a response from the desired target.

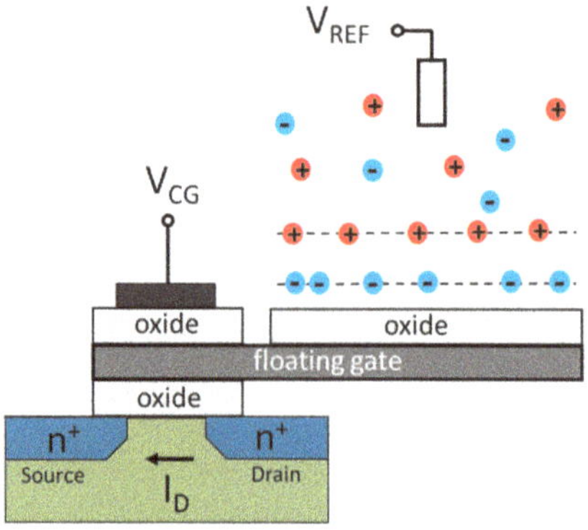

Figure 3.3. A scheme of floating gate FET, adapted with changes from [29].

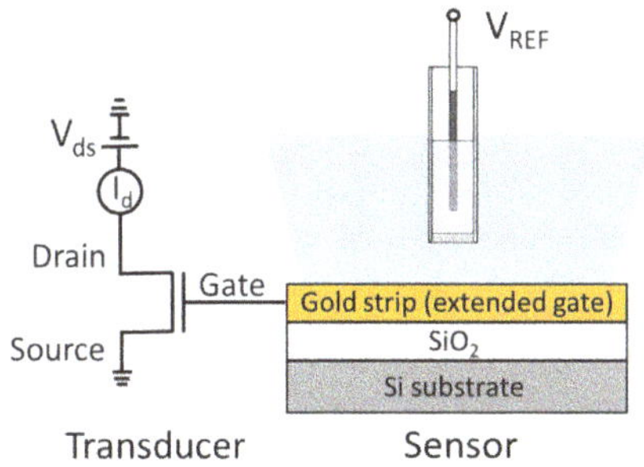

Figure 3.4. A scheme of extended gate FET adapted with changes from [30].

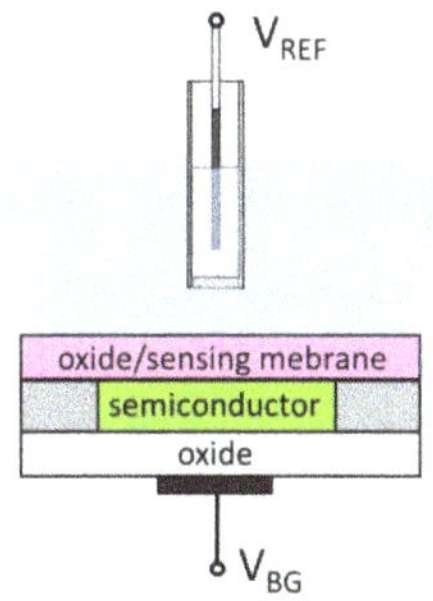

Figure 3.5. A scheme of a dual-gate FET adapted with changes from [31].

Another interesting ISFET architecture is extended gate FET. In this structure, the sensing surface (e.g., gold) is electrically connected to the gate terminal of a MOSFET but extends off the chip (figure 3.4). Only the sensing surface is exposed to the electrolyte solution. This solution makes the device fabrication significantly simpler (as commercially available MOSFETs can be used) and allows convenient post-processing steps as its surface can be independently engineered from the transducer.

In a dual-gate FET (figure 3.5), a modification of a conventional electrolyte-insulator-semiconductor ISFET, the semiconducting channel is insulated from the bottom with a thick layer of oxide and from the top with oxide as well or some other insulator serving as a sensing surface. A conductive layer serving as a bottom gate is placed on the bottom oxide and the top sensing surface is exposed to the liquid with an immersed reference electrode serving as a top gate. The additional bottom gate yields an increased slope (above the Nernstian limit) in the sensor response, caused by the capacitive coupling between the top gate and bottom gate.

In a simpler junctionless modification of ISFET, the p-n junctions are eliminated causing low leakage current (current flowing between S and G). The channel and the source and drain terminals are of the same type of doping with the channel being much more heavily doped.

3.3.2 ISE versus ISFET

We can consider a 'simple' example of pH detection. In a pH glass electrode (ISE), there is an ion exchange inside the membrane. This can be described by the

Nikolsky–Eisenman equation (3.1), an extension of the Nernst equation. In the Si-based ISFETs, the ions bind to the surface of the oxide. The latter mechanism is described by the site-binding model (SBM) [32]. In brief, this model is based on thermodynamical fundamentals of the equilibrium reactions of the surface groups and the Gouy–Chapman–Stern theory of the ion distribution at an electrolyte|solid interface. The SBM also yields a simple Nernst equation if there are many binding sites.

Contrary to the ion-sensitive electrodes, where the potential difference between the ISE itself and a reference electrode is measured, for an ISFET several different techniques of measurement can be used. If the applied gate potential (V_G) is constant, then changes in the ion activity and thus charge changes at the electrolyte|oxide interface are reflected in changes in source–drain current I_{SD}. However, if the source–drain current is maintained at a constant value by means of an operational amplifier, which directly controls the applied gate potential using a negative feedback loop, then the output voltage, V_{out}, is a potential difference that varies with change in ion activity governed by the Nernst equation. The output (analytical signal) is, therefore, effectively the same as that from an ISE [33].

Classical ISE architecture requires an inner electrolyte. Although there have been quite substantial efforts in the development of solid-state ISE, current sensors still rely on a direct deposition of membrane on the electric contact or inclusion of a layer of conducting polymer. In both cases, the exact mechanism of functioning of such sensors is not known [34]. In contrast, the process of FET devices preparation is compatible with silicon technologies that have been developed for over 50 years. Those processes allow ISFETs to be mass-manufactured, miniaturized and integrated into a CMOS (complementary metal-oxide-semiconductor) circuit.

Both ISE and ISFETs can be easily multiplexed—the same RE (reference electrode) can operate the gate voltages of several ISFETs, or serve as a pair for an ion-selective electrode. Interestingly, multiplexing paved the way for the development of REFET—the reference field-effect transistor. The idea is pretty simple—one of the two FET devices (called active) is specific to the analyte (with additional background noise), whereas the other—REFET with an ion-blocking layer (called passive)—responds only to the mentioned background noise. By subtracting the signals obtained from the two devices (active and passive), a more specific analytical signal can be obtained. REFETs can also partially solve the problem with the stability of the reference electrode since both active and passive devices are influenced by it. On the other hand, it is known that noise and drift are limiting issues for the application of the device as a platform for chemical and biomedical sensors. This is yet to be overcome with the ISE devices.

When it comes to the measurement equipment, ISEs can be essentially measured with a simple electrometer. Since the ISFETs require applying two different voltages (SD and G) as well as measuring the source–drain current, they require more cumbersome laboratory equipment (parameter analyzer, source meters, etc). In commercial ISFET solutions, this obstacle can be easily overcome by a skilled electrical engineer. On the other hand, the performance of ISFETs is limited by the signal-to-noise ratio with the most crucial interfering noise in the low-frequency regime, since it is in the same order of magnitude as biomedical signals. The source of noise outside the ISFET itself (external noise) is due to environmental interference and instrument noise. Another source of noise is the 'electronic' nature of ISFET itself (intrinsic noise), which has its source mainly in trapping and de-trapping of charge carriers at the insulator|semiconductor interface. The last source of noise is extrinsic—generally originating from the electrolyte|insulator interface caused by the ion exchange at this interface [35]. In these terms, ISEs are much less prone to interference.

3.4 The tongue and the taste mechanism

Natural human recognition systems are superior to the artificial ones, developed so far in their potential to perceive, analyze, and respond towards stimuli [12, 36]. Considering many attempts to mimic chemical senses (taste and smell) since the 1980s, the task of building analogues of the human nose or tongue is clearly complicated and challenging; nevertheless, the possibility of rapid sensing, unbiased, and inexpensive would be a significant advancement [37]. Being certain that the sensors employed in the electronic tongue are inspired by biological systems, understanding the main features and working principles of our tongue is inescapable.

The tongue is the primary organ for gustation that helps in chewing food. In addition to tasting, it supports phonetic articulation and aids as natural means of cleaning teeth [38]. It is one of the softest, most flexible, and sensitive muscular portions of our body with a pink surface tissue called mucosa, where thousands of mechanical receptors, taste receptors, and ion channels reside. Five principal tastes constitute the human sense of taste: sweet, sour, bitter, salty (the most common), and umami (a typical response to glutamate) [36, 37, 39].

Receptors are located in the membranes of taste cells, which in turn are clustered in taste buds located on tiny bumps called papillae on the tongues' epithelium (figure 3.6). Present in tens of thousands, the taste buds are connected to an array of nerve cells running into the brain. They serve as chemoreceptors and transduce chemical signals into electrical ones called action potentials. Each of these buds is built from 50–100 tightly packed taste cells. The saliva, composed of 99% of water and a mixture of electrolytes, helps in binding the tastants to the specific taste receptors present in the plasma membrane of the taste cells and hence perception of different kinds of tastes.

The taste cells in each taste bud can be organized into three types. As a single taste bud is built from a number of different taste cells, each expressing different receptors, its response is a superposition of more than one taste stimuli.

Type I cells are most abundant and form at least half of the taste buds. Their cell membrane includes transporters of neurotransmitters and K^+ channels. Similar to astrocytes in the nervous system, the type I cells are generally believed to have a support function in the taste buds, as their main function is the clean-up of the ATP released by the type II cells. They do not generate action potentials [40, 41].

One-third of the taste bud cells are of type II. Those cells express taste G protein-coupled receptors and can recognize bitter substances, sugars and amino acids.

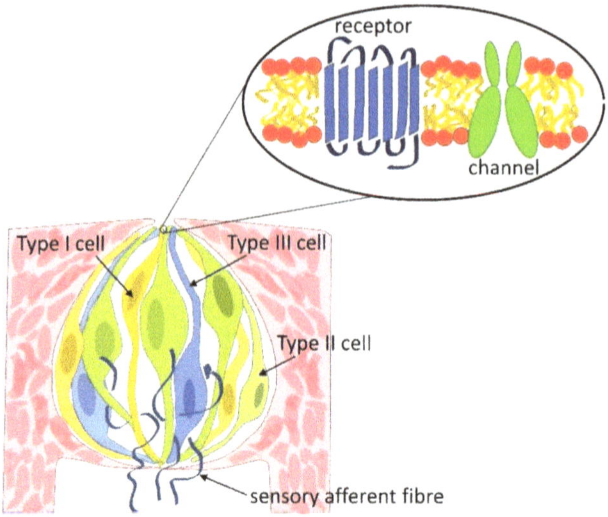

Figure 3.6. Taste bud located in the papillae formed from few (number of cells reduced for clarity purposes) taste cells of different types. Signal from the cells is transmitted to the brain through the nerve fibres. The inset shows a close-up of the taste cell membrane, which incorporates receptors, channels and transporters responsible for the recognition of different taste stimuli (adapted with changes by permission from Macmillan Publishers Ltd [40], copyright (2017)).

Although umami and sweet receptors can be co-expressed at a single cell, bitter receptors' presence is always unique, as the sweet/umami and bitter receptors are mutually exclusive. Those cells communicate through ATP release, activating receptors on afferent nerve fibers and adjacent taste cells [42].

Type III cells are the least numerous (2%–20%), and they are able to recognize sour taste, but do not express G protein-coupled receptors (GPCRs). The presence of the non-selective cation channel PKD2L1 is another exclusive characteristic of these cells; it has been hypothesized as the sour receptor. They can also modulate the signal transduction from type II cells via P2Y adenosine receptors by release of serotonin. Type III cells are the only ones forming conventional synapses with the nerve fibers. It is not yet known which cell type is responsible for the recognition of salty taste [40].

The tastes are ultimately registered in the brain by the cranial nerves, but each of the taste cell types has a different mechanism of communication with the surrounding nerve cells. For example, type III cells send information through synaptic vesicles (i.a. serotonin or acetylcholine secretion) and can be inhibited by GABA, whereas type II cells communicate through non-vesicular ATP release [40, 42].

Not only do we have three cell types, which make up almost endless variations ($> 3^{100}$) of possible taste buds, with each cell type having a different mechanism of signal transduction to the central nervous system, but there are also numerous ways of recognition of the stimuli. The taste recognition is based on ion channels, transporters, and the family of transmembrane G-protein coupled receptors. Although dozens of different structures are responsible for the recognition, it takes place mainly through two mechanisms; one based on the extracellular recognition and resulting intracellular activation of the G-protein, and the other on the direct inflow of ions to the cell.

G-protein receptors are responsible for the recognition of taste and form the vastest group of membrane receptors in eukaryotes (more than 1200 different GPCRs found in humans) able to recognize peptides, lipids, sugars, proteins, or even light. Although the class is immense, all receptors share common structural features, being composed of a single polypeptide chain, with seven segments spanning the entire cell membrane. The transmembrane portions of the protein are connected by loops (inset of figure 3.6), which on the extracellular side bind molecules of interest. Two classes of GPCR connected to taste are T1Rs and T2Rs. T1Rs mediate sweet (sugars, D-amino acids, synthetic sweeteners, and some sweet proteins) and umami (glutamate) tastes, while T2Rs mediates bitter taste (alkaloids). The T1R function as heterodimers and numerous combinations of the functional subunits exist. On the other hand, T2R responsible for the bitter taste are monomers with around 25–30 different receptors found in mammals [38, 42, 43]. Except for the two classes, other GPCR were found in taste buds and are considered to contribute to taste sensation through recognition of medium and long-chain fatty acids or calcium and are responsible for potentiation of other tastes [42]. As the name suggests, upon binding of the signalling molecule they undergo a conformation change, which results in the interaction with a nearby G-protein, situated on the inside part of the cell membrane. Activation of a single G-protein can result in the production of hundreds to thousands of secondary messenger molecules, including diacylglycerol, cyclic AMP, and inositol 1,4,5-triphosphate, all of which initiate and coordinate their own signalling pathways [44].

Another mechanism is based on the direct flow of sodium or hydrogen ions through the ion channels in response to the salty (sodium) and sour (organic and inorganic acids) tastes [39, 43, 45]. A greater membrane permeability of the undissociated organic acid molecules and the associated proton release in the cytoplasm makes the weak acids, like acetic acid or citric acid, elicit a stronger sour taste sensation than strong mineral acids such as HCl. In the sour case taste, intracellular acidification is the direct trigger for action. The organic acids are noted to permeate type III cells, acidify the cytoplasm and block leak K^+ channels to depolarize the cell membranes. On reaching the threshold for action potential initiation, the cells stimulate the release of neurotransmitters; cholecystokinin and neuropeptide Y are the neurotransmitters involved. Epithelial sodium channels (ENaCs), hyperpolarization-activated cyclic nucleotide-gated channels, and acid-sensing ion channels are among several membrane ion channel transducers proposed for sour sensations [38, 40, 43].

Understanding the mechanism of the salty taste has been the most difficult among the five basic tastes and is being deeply investigated. Cracking of the sweet, bitter and umami tastes was possible with the cloning of their specific G-protein-coupled receptors, and sour by discovering the proton entry channel that serves as sour receptors. While other tastes are largely restricted to a single cell type or single taste cells, the identification and analysis of salty taste can be considered as a cooperative effect. The studies conducted on the basis of diuretic amiloride sensitivity indicated that Na^+ ions enter the taste cells via the ENaCα channels, which results in the depolarization and triggers the action potentials. In response to the strong membrane depolarization, the voltage-activated ATP release channel CALHM1/3 discharges ATP to afferent nerve fibers [46].

The inclusion of taste qualities like fat, metallic, kokumi, as well as stimuli responsible for cooling sensations (menthol or eucalyptol) in the list of tastes is a topic of discussion in recent years. Dietary fats are surely associated with texture and viscosity. Kokumi flavour elicited by glutathione and related di- and tripeptides is related to mouthfulness, thickness, long-lasting savory sensation, and astringency. Capsaicin, which is the main molecule responsible for spicy flavour, is detected by a thermoreceptor, which in principle should be activated by elevated temperature. The spicy flavour is elicited by a large number of non-related compounds that, except for thermoreceptors, can also be detected by activating transient receptor potential cation channels associated with irritation and pain [40].

The natural sense of taste is highly evolved, but it varies between individuals and it can be subjective, adaptive, get tired or infected, and can consequently influence human panelists [12, 47]. The idea of reproducing the human response by artificial means has been introduced and extended so far that it finds application beyond the realms of food, beverages, and pharmaceuticals.

3.5 Potentiometric electronic tongues as mimics of the mammalian taste

In an electronic tongue, an array of sensors imitates the taste buds. The signal is transformed from chemical to electrical information, like in the nerve cells, and analyzed using machine learning algorithms which, similarly to our brain, compare the obtained signal with a database of known patterns.

The key objective of the electronic tongue is quite different from the chemical sensors. E-tongues aim to obtain general information about the sample, while single sensors usually aim to be highly selective. Receptors in the taste cells in our tongue can be specific, as in the case of umami GPCR, which is triggered solely by glutamate, or some of the bitter receptors with very limited recognition abilities. Taste cells also possess recognition ways sensitive to whole classes of compounds as the sour ion channels or different sweet taste GPCRs. Apart from the recognition ability of the single receptor, we need to remember that the final information sent to the brain from a single taste bud is a combination of the response of numerous cells that form it. The question should be posed whether the chemical sensors composing the electronic tongue matrix should resemble the recognition element of a single cell (receptor, ion-channel, transporter) or the whole taste bud.

We already know that potentiometric electronic tongues need to include both selective and cross-sensitive sensors [48]. Most electrode modifications are non-specific and usually termed cross-selective and cross-sensitive, as their response is not only governed by the analyte in question but also by the temperature, pH, interferents, etc. Just as in our tongue measurement and examination of complex liquid matrices, providing fingerprint details for each sample regarding the concentration of its chemical constituents is still possible despite the non-specific singularity of the sensors [49]. If the array is composed only of sensors that in the experiment conditions are specific to single analytes, the amount of data obtained from such a matrix can be too low for the electronic tongue to work properly. Similarly, the array cannot be composed of only extremely cross-sensitive electrodes as in this case we will not be able to filter out the noise. Ideally, the array should include sensors specific to the most relevant species for the discrimination of our samples, as API (active pharmaceutical ingredient) and the masking species in the case of determining taste of drugs and include some cross-sensitive sensors, for example for cations. This would enable mimicking the diverse range of recognition elements of the taste cells.

The sensors constituting an e-tongue interact with analytes and generate reversible changes in the electrical signals, which are then used for pattern recognition [37]. The total number of sensors varies from 4 to 40, and depends on the analytical task at hand, availability of the detection interface, and general concept of the device. On the other hand, the human tongue possesses 25–30 bitter receptors, a few sweet and umami ones, different ion-channels and transporters, thermoreceptors, pain receptors, and others associated with the sensations, which just now are being recognized as tastes. In total, we can assume there are 60–100 different recognition structures. Although in the case of bigger electronic tongue arrays, those numbers seem comparable, the number of copies should also be taken into account.

It has been found that the set of redundant sensors in the array provides many additional advantages. They can be used for diagnosing malfunctions, obtaining concentration profiles or 2D imaging for designed analytes [50]. Further, it has been found that the set of equivalent ion-selective electrodes in the array improves the precision and accuracy of the measurement [26, 51]. Redundancy is also essential in environmental and medical applications when results oscillate in a non-linear manner. Usually, 2–4 repetitions of the most critical sensors in the matrix are used. In the tongue, we have up to 10 000 taste buds, each composed of 50–100 taste cells.

Before deciding what should be the number of different sensors and their copies in the electronic tongue, let us analyze the types of signals in both natural and artificial systems. Again, we should distinguish two parts of the process: the recognition of the stimuli and the signal that is finally being analyzed. As said in the previous section, taste recognition is based on two main mechanisms: recognition of specific molecules through the extracellular parts of the GPCRs and the direct influx of certain ions, regulated through channels and transporters. Apart from those, there are also pain receptors responsible for the spicy taste, which include different thermo- and mechanoreceptors.

Potentiometric sensors are based on selective binding of analytes that takes place at the sensor–solution interface or the transport of ions inside the membrane based on their lipophilicity. The membrane is usually made of plasticized poly(vinyl chloride), which is an organic phase and can be treated as the simplest approximation of the cell membrane. The selectivity and the detection limits vary according to the composition and properties of the materials used for the sensor coating. In the case of ion-selective electrodes with ionophore-doped polymeric membranes, the sensitivity is determined by the formation of an ion-exchange complex [49]. The recognition molecules used in membranes vary from very simple organic compounds to natural transporters such as monensin and valinomycin extracted from bacteria. To take the similarity to natural recognition a step further, professor Toko and his group included lipids in the polymeric membranes (i.e. dioctyl phosphate, trioctyl phosphate, methyl ammonium chloride and a molar mixture of dioctylphosphate and trioctyl methyl ammonium chloride) [12, 38, 52].

Potentiometric probes can also be coupled with other sensors. Hybrid potentiometric-voltammetric systems will be commented on in more detail in the subsequent chapter but, in principle, potentiometric sensors can be coupled with any other sensor types. The only important thing to remember is that care should be taken during the pre-treatment of such data to not drown a single data point coming from potentiometric systems in the hundreds composing a single UV–vis spectrum or a cyclic voltammogram.

A wide range of recognition means provided by potentiometric sensors, and the possibility of coupling them with other detection methods show the comparability of such a constructed electronic tongue with natural recognition at this level. Although the machine learning algorithms are similar to what happens in our brain, as they rely on extensive training and constant updates, the signal processing that happens in-between is quite different.

Despite having non-neuronal precursors, taste cells are the only cells generating regular action potentials spontaneously or responding to an electrical or chemical stimulus. These responses can drive changes in the membrane potential such as depolarization, depolarization followed by hyperpolarization, or hyperpolarization only. A number of laboratories have reported features of voltage-gated transduction in rodents, the whole-cell recordings reported that Na^+ and Ca^{2+} currents contribute to the depolarizing phase, while the repolarization phase involves K^+ and probably Cl^- species. Potential bursts generated by Na^+, K^+ and Ca^{2+} channels are propagated towards the basal region of the taste cells and are capable of opening the voltage-gated Ca^{2+} channels located there. The series of actions leads to the release of neurotransmitters by type III taste cells that subsequently diffuse across the synaptic cleft and lead to the initiation of action potentials in afferent nerve fibers. The frequency of action potentials in taste cells is proportional to the amplitude of the stimulus applied and their firing pattern may be quality-dependent [41, 53].

An example of the response generated by a taste cell and the pattern later transmitted by the corresponding nerve fiber is shown in figure 3.7. Similarities between the breadth of tuning in the taste cell to that in the nerve fiber suggest that the coding of information may start already at a taste cell level [41]. It should be remembered that the observed pattern is not connected to a single receptor but numerous receptors triggered, inhibited and potentiated at that instant in the taste cell as well as interactions between type I, II and III taste cells and ATP communication of the type II cell with the nerve fiber.

As the membranes of ion-selective electrodes and transistors are composed of numerous ionophore molecules, just as taste cells express numerous copies of a

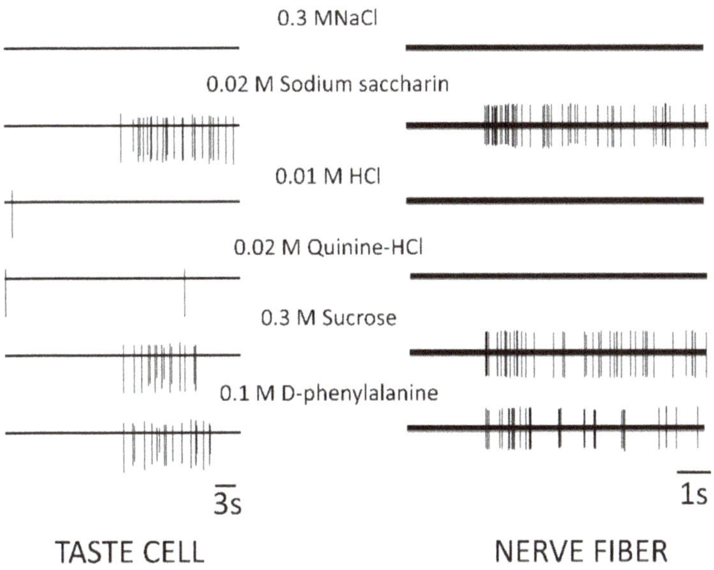

Figure 3.7. Example of a firing pattern of a taste cell (on the left) and a corresponding nerve fiber (on the right) upon stimulation of the taste bud with different tastants. Figure adapted with changes from [41], CC BY 4.0.

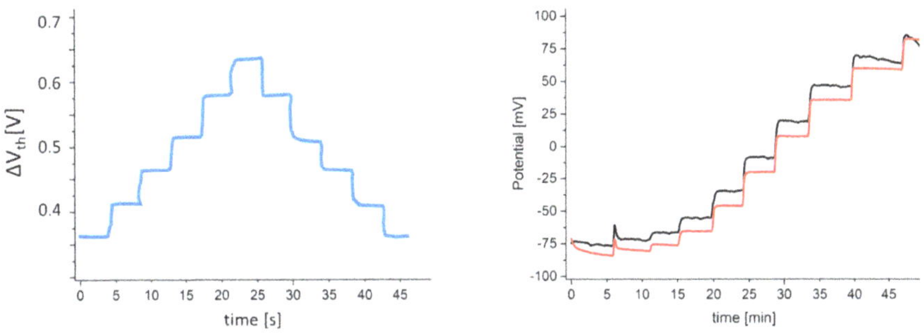

Figure 3.8. Example of a raw signal measured with ISFET (left) and ISE sensor (right) in time upon the change of the activity of an analyte.

single receptor, we can treat each sensor as a single cell. Although, it should be held in mind that chemical sensors are usually based on a single type of recognition molecule and cells can express even more than 100 different receptors [54].

In chemical sensors, the increasing activity of primary ions will result in a steady increase in the potential response. In general, potentiometric sensors are characterized by relatively fast response time (ISE even under miliseconds in a wall-jet cell [34]) and steady response. However, some drift of the signal linked to changes of the membrane–solution interface is often observed. The typical response time observed is on the order of seconds (figure 3.8) and not under miliseconds values as in taste cells. The signal registered is usually transformed to a single point per measured sample/concentration. In this way, instead of a train of spikes that codes a considerable amount of information about the sample, our algorithm is left with a single point per sample.

Briefly, it is known that electronic tongues should be composed of both selective and cross-sensitive sensors mimicking natural receptors and channels. Although, we should make an effort to check what is the adequate number of sensors (possible increase to 60–100 different sensors) in the matrix and test different signal processing strategies before the machine learning step. Enhanced mimicking of the natural taste in those two aspects might result in the growth of the recognition capabilities of the electronic tongue beyond what is known today.

3.6 Construction of potentiometric e-tongues

There are several approaches to taste technology even if we only consider potentiometric systems. We can divide systems by the type of sensors used (dynamic or stationary), sample introduction, number of sensors or the footprint of the device.

3.6.1 Sensors

Sensors comprising the matrix can be based on selective polymer membranes, chalcogenide glass porphyrin derivatives or conducting polymers. Except for hybrid systems combining different detection techniques (e.g. voltammetry–potentiometry), we can also think of all potentiometric matrices, which are composed of sensors using distinct sensing principles (e.g. chemosensors and biosensors). Through the use of sensors, which differ in the recognition mechanism, it is possible to address a wider range of molecules.

An example of the above is a bioelectronic tongue for urea determination [55]. The biosensor developed by del Valle group consists of all-solid-state ISE with membranes made from PVC and PVC-COOH polymers, plasticized with DOS, BPA or NPOE. The array was composed of two electrodes sensitive to ammonium and two for hydrogen, a single electrode for each potassium and sodium sensing and one generic. Additionally, urease was immobilized on the membrane surface of both ammonium and hydrogen electrodes, thus obtaining 12 different electrodes in total. The proposed system provides a simple procedure without the need for sample preparation; however, it needs further development to be more suitable for 'real' sample analysis.

Electronic tongues were also applied in food and beverage analysis, especially wines, beers, juices, teas, honey, fruits, vegetables, and cooking oils. However, taste, flavor, quality and intensity evaluation should not be based only on specific substances, but the general characteristics of samples. Therefore, the sensor array should consist of non-specific sensors of low-selectivity and cross-sensitivity [22], which is, according to IUPAC, the ability to obtain a stable response with high reproducibility to various analytes in solution. Sensors in the electronic tongues should provide different responses to the same set of substances, allowing identification of characteristics of the analyzed product [21].

Dias *et al* presented an electronic tongue for semi-quantitative and quantitative analysis of commercial fruit juices [56]. They applied two different sensor arrays with 18

Table 3.1. List of plasticizers and additives used in sensor arrays to obtain cross-selectivity in order to analyze fruit juices [56].

Number identification	Additive substance	Number identification	Plasticizer substance
A1	Octadecylamine	P1	Bis(2-ethylhexyl) phthalate
A2	Bis(2-ethylhexyl)phosphate	P2	Bis(1-butylpentyl) adipate
A3	Oleyl alcohol	P3	Tris(2-ethylhexyl) phosphate
A4	Methyltrioctylammonium chloride	P4	Dibutyl sebacate
A5	Ridodecylmethylammonium chloride	P5	2-Nitrophenyl-octylether
A6	Oleic acid	P6	Dioctyl phenylphosphonate

sensors composed of non-specific lipid/polymer (PVC) membranes. Arrays consisted of all combinations of additives A1–A6 with plasticizers P1–P3 in the case of the first system and plasticizers P4–P6 for the second set (table 3.1). Authors were able to differentiate four beverage groups based on fruit juice content using linear discrimination analysis. Multiple linear regression and PLS calibration models provided estimates of the concentration of two major compounds, fructose and glucose, present in the tested beverages, although neither of the sensors is truly sensitive to either of the sugars.

Wróblewski's group [57] demonstrated another approach, combining selective and cross-sensitive types of sensors. Their system consisted of standard pH electrode and eight types of ion-selective electrodes, two probes of each type (table 3.2), resulting in a 17-sensor array. Standard membrane composition and ionophores were used for selective sensors (E1–E4), however, in the case of partially-sensitive probes, membranes included mixtures of ionophores (E5–E6). The authors successfully determined various non-alcoholic beverages such as juices, milk and tonic with a high degree of accuracy.

E-tongue has also been applied for the analysis of alcoholic beverages. The 23 potentiometric cross-sensitive chemical sensors were employed to discriminate Italian red wines [58]. The matrix should be able to differentiate inorganic cations (K^+, Ca^{2+}) and anions (Cl^-), and organic substances such as phenols, organic acids and alcohols. Sensor arrays consisted of 14 sensors with chalcogenide glass and crystalline membranes, two metallic sensors (Pt and Sb), six sensors with plasticised polymer membranes and a standard pH glass electrode. The e-tongue was able to distinguish the wines from different geographical areas, which is crucial in quality and authenticity control of the origin of wines. Moreover, it was capable of identifying wines of the same vintage and denomination. In another study, 26 potentiometric sensors were used for the analysis of Madeira wines [59]. The core sensors were based on plasticized

Table 3.2. Different types of sensors used in a sensor array employed in discrimination of non-alcoholic beverages.

Electrode number	Electrode type	Ionophore
E1	Ca^{2+}	ETH 1001
E2	NH_4^+	Nonactine
E3	Cl-	TPPClMn
E4	HCO_3^-	ETH 6010
E5	Na^+/K^+	Ionophore X/Valinomycin
E6	$F^-/H_2PO_4^-$	Ionophore $H_2PO_4^-$/ Ionophore F^-
E7	'cation selective'	
E8	'anion selective'	

PVC membranes sensitive to organic cations and anions (phenols in particular), chalcogenide glass sensors displaying redox response and a conventional pH glass electrode. The authors made an attempt to determine vintage, grape variety and their interaction but only the age of wine could be predicted from the e-tongue data, while all factors could be determined from the HPLC data. The concentration of most of the organic acids present in the samples was determined with a 5%–18% error.

Interesting architecture of e-tongue has been proposed by Peres's group [60]. Instead of standard sensors with internal electrolyte or planar solid-state electrodes, the solid-state sensors were positioned on two cylinders. The two cylindrical sensor arrays were applied as a tool assessing perfume olfactory family and storage time-period, allowing them to discriminate between women's and men's perfumes. The array consists of 20 cross-selective sensor membranes containing 3% of lipid additive, 32% plasticizer, and 65% of polyvinyl chloride. The hydrostatic and hydrophobic interaction of lipid membranes with polar compounds present in perfumes, combined with chemometric techniques and variable selection algorithm, allowed discriminating women's and men's perfumes. Previously the same architecture of sensors was used to determine the quality of olive oils based on single-run assays [61]. It is worth noting that it was the first e-tongue introduction as a practical tool used in the perfume industry.

Moreno *et al* developed ISFET sensors that were bagged in a single six-channel device. Transistors were coated with photocurable polyurethane membranes sensitive to K^+, Na^+, Ca^{2+} and Cl^-. After the incorporation of pH and redox potential sensing elements, this array has been used for the discrimination of mineral waters, grape juice and wine samples [62].

3.6.2 Hydrodynamic systems

The systems presented above were intended for stationary measurements. However, there are applications where the flow-through analysis would be more convenient, for example to introduce the system for in-line process monitoring or to achieve high sample throughput.

Having in mind future process monitoring, modular flow-through setups were applied to the analysis of methane fermentation, including miniaturized sensors of traditional construction [63] and solid-state probes [64]. Easy assembly of the individual modules allowed for cleaning of the cell. For prolonged measurements, single sensors could be also readily exchanged in the case of malfunction.

In another setup, a flow cell permitted achieving lower sample volume, which, given the type of the sample—cell culture medium available in very small amounts (up to 1 mL), was the main advantage of the system. This electronic tongue allowed one to analyze the number of dead cells in the culture and estimate the cytotoxic effect of 1,4-dioxane and 5-fluorouracil [65].

Very interesting results can be achieved by injection analysis. Although the recognition process is very fast (microseconds [34]), potentiometric sensors provide a steady-state response, which can be usually resumed with one data point per electrode. When a potentiometric matrix is coupled with flow injection systems, additional dynamic data can be obtained, which to some extent describes the kinetics of the interaction of ions and the membrane components.

Through the years, two systems for flow injection analysis were developed—flow injection analysis (FIA) and sequential injection analysis (SIA), both introduced by Růžička and his colleagues [66, 67].

A typical injection system [68] consists of a pump, an injection valve, a reaction cell and a detector [69]. In the case of flow injection analysis, the sample is inserted into the sample loop of the two-position injection valve, in the meantime the carrier is continuously transported through the channels, the sample is repeatedly introduced into a carrier stream and finally, when the mixture reaches the detector, the dynamic response is measured. Sequential injection analysis [68] allows carrying out measurements with higher precision and preparing samples in a more sophisticated way enabling a wider range of assays. SIA includes a single-channel high precision bi-directional pump, a holding cell, a multiposition valve, which is the fundamental part of SIA and a flow-through detector (figure 3.9) [70]. Contrary to FIA, the setup allows inserting different volumes of the reagents and the sample in the holding cell. However, both injection methods allow a significant redution in the volume of reagents needed, and shorten the response time.

The simplicity of flow analysis systems makes them a great alternative to other more expensive analytical techniques, such as mass spectroscopy or HPLC. Electronic tongue detection was also coupled with such setups and used for ions [69, 72–75] and noble metal detections, as well as evaluation of different beverages [76, 77].

3.6.3 Miniaturized and low-cost systems

Except for the developments made in the membranes and methods of molecule recognition, research has been carried out in providing more miniaturized systems. There are many advantages related to the reduction of reagents and materials used, but at the same time difficulties with the preparation of such sensors must be taken into account. Lowering the scale generates critical limitations, noticeable

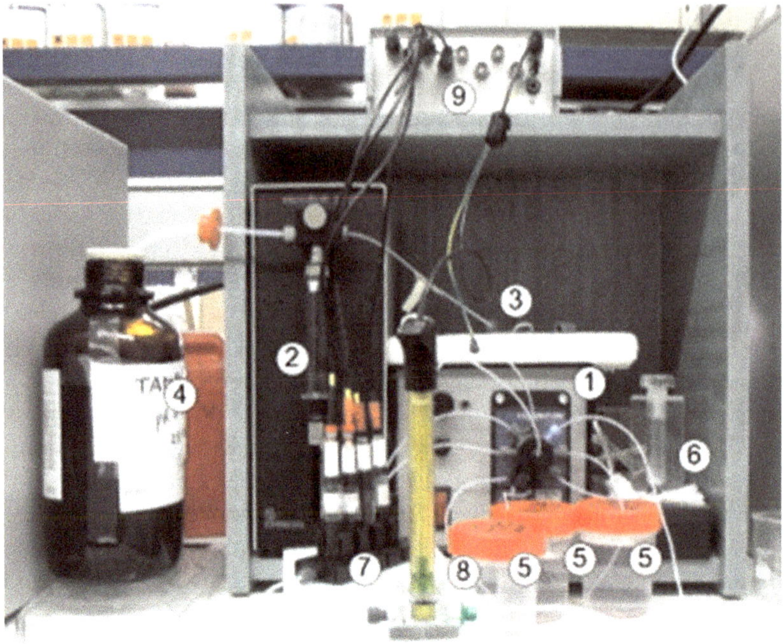

Figure 3.9. Picture of an SIA flow system. (1) 8-way valve; (2) automatic microburette; (3) holding coil; (4) buffer reservoir; (5) stock solutions of metals; (6) mixing cell; (7) ISEs array; (8) reference electrode; (9) multipotentiometer. Reprinted from [71], copyright (2010) with permission from Elsevier.

evaporation of sample, and problems with signal transduction [12]. There are two main strategies: the miniaturization of single sensors associated with the elimination of internal electrolyte and the development of integrated sensor arrays [78].

Ciosek and Wróblewski [77] introduced an electronic tongue with single miniaturized sensors, which was used for the recognition of various beverages: milk, orange juice, and beer. They used a simple technique of casting membrane solution on Au transducers combining classical ion-selective, partially-selective and pH microelectrodes. 69.7% of corrected classifications were obtained using PLS regression; however, the combination of PLS and ANN analysis enabled enhancing the recognition of beer (83%). The adopted technique of miniaturization brings sensor arrays a step closer to the real application in quality control in the foodstuff industry.

In recent years, various materials had been applied to fabricate the planar-type potentiometric e-tongue. All-solid-state screen-printed sensor arrays with PVC membranes over polyester support [79] and metallic ink screen printed electrodes on ceramic support [80] have been reported. Toko *et al* presented multichannel electrodes with a liquid inner solution and lipid/polymer membranes [81]. Subsequently, e-tongues with an integrated array of microelectrodes made either from conductive epoxy polymers [82] and epoxy-glass laminate [83] or with silicon technology [84] have been presented.

An unusual approach to e-tongue fabrication was the paper-based device introduced by Kubota's group. Both ion-selective and cross-sensitive sensors were drawn by standard 6B graphite pencil and used for discrimination of bottled water [85]. Furthermore, in order to improve the sensitivity of the e-tongue, previous electrodes were replaced by conductive paper strips coated with silver paint [86], creating a four-electrode array designed for the analysis of wines and beers.

3.7 Commercial systems

Although electronic tongue systems have truly amazing capabilites: they are able to discriminate samples, diminish the impact of interfering species, or provide reliable quantification even in samples with changing background signals, their market penetration is not high. At the moment only two companies provide such equipment, which in both cases is based on ion-selective sensors. The commercial electronic tongues are ASTREE provided by Alpha MOS (Toulouse, France), and TS-5000Z by Intelligent Sensor Technology Inc. (Insent, Atsugi-shi, Kanagawa, Japan).

ASTREE electronic tongue was introduced in 2000. Its working principle is based on chemical field-effect transistor (ChemFET) liquid sensors. Each silicon transistor is covered with a unique proprietary organic membrane responsible for different interactions with compounds dissolved in a solution, and thus regulating the different behavior of individual sensors. The molecules of diverse concentrations interact with the membrane altering its potential, and the created voltage difference between sensors and an Ag/AgCl reference electrode is being measured. These sensors are said to be sensitive and cross-selective—the analysis of a specific compound is based on a signal from all sensors rather than a particular one [87].

Alpha MOS offers a fully automated device composed of an array of seven sensors consecutively immersed in investigated solutions, controlled by a 16- or 48-position autosampler ensuring uniform distribution of analytes throughout the sample and reproducible time of measurement (figure 3.10). The data are monitored, acquired, and processed on a computer via enclosed software [88]. The individual sensors are combined in dedicated sets, e.g., the pharmaceutical [89, 90] or the food and beverage set [91, 92] but if needed, respective sensors can be replaced with different ones [93]. Research groups from Corvinius University, Tokyo University of Agriculture, University College Dublin and Optimum Nutrition, Inc. claim the usefulness of ASTREE e-tongue in characterizing food products and improving their quality [88]. There is no public data on all available sensors and sets, but up to 25 different sensors are available [94].

TS-5000Z was manufactured for the first time in 2006, a decade after the confirmation of the global selectivity concept by Toko's group [95]. Each sensor consists of a lipid polymer membrane reflecting the lipid bilayer from which the tongue of living organisms is built. It allows the substances in the investigated solution to change the electric potential of the membrane through electrostatic and hydrophobic interactions. This change of membrane potential being a carrier of

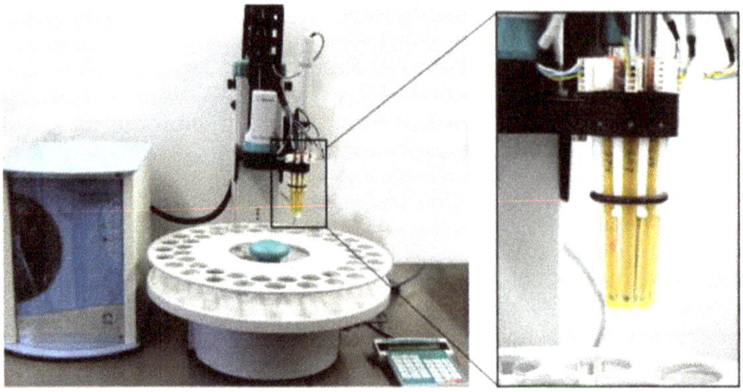

Figure 3.10. Astree2 electronic tongue (AlphaMOS, Toulouse, France), reprinted from [93], copyright (2011) with permission from Elsevier.

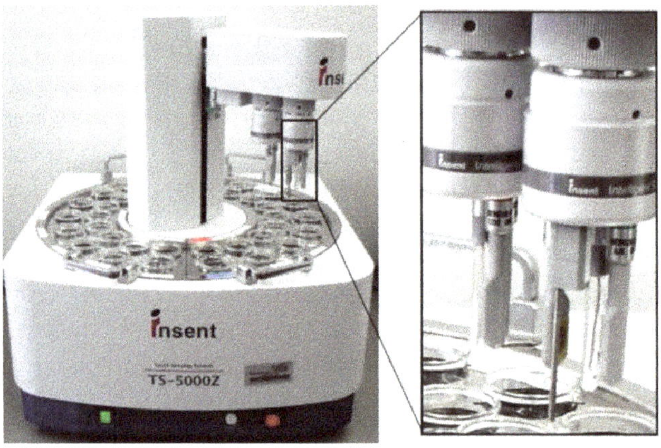

Figure 3.11. TS-5000Z electronic tongue (Insent, Atsugi-shi, Kanagawa, Japan), reprinted from [93], copyright (2011) with permission from Elsevier.

taste perception is transmitted through an internal conducting solution to an electrode that mediates it to the sensing system equipped with a sensing unit and a 14-position autosampler (figure 3.11). Each sensor is assigned to a specific taste recognized by human tongue receptors. To this moment, TS-5000Z is said to recognize and assess seven tastes: sweetness, saltiness, sourness, bitterness and umami as the five basic tastes as well as astringency and pungency being complementary ones that affect the way people interpret taste [96]. In contrast to ASTREE by Alpha MOS, the data concerning sensors' construction and materials used for their production as well as their indication are public. The TS-5000Z electronic tongue is applied for quality control of foods and beverages both in liquid and solid form (after proper liquefaction of solids) and pharmaceuticals, including drugs bitterness and the ability of its masking [97]. Among the users, scientists from

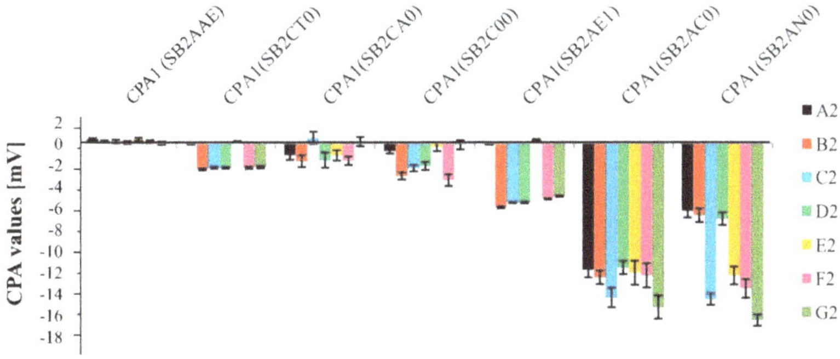

Figure 3.12. Mean CPA values for seven different sensors tested by six independent research groups, reprinted from [98], copyright (2014) with permission from Elsevier.

Taste & Aroma Strategic Research Institute Co., Ltd, Mukogawa Women's University and Sojo University can be distinguished. In 2020 the TS-5000Z taste sensor with seven sensing units and three reference electrodes along with installation and training was quoted as 99,000.00 €.

Since electronic tongues are being used for pharmaceuticals assessment, the idea of data comparison obtained by different laboratories has emerged. The first experiment was conducted by five independent centers. The tested equipment was SA402B and TS-5000Z taste sensing systems provided by Insent, both with different sensors sets.

Tested systems were declared comparable, although a modified response from aged sensors was reported [98], as shown in figure 3.12. The obtained data seem to be comparable rather than repeatable, and it is unknown whether the signal could be credibly distinguished if there were more samples of a similar nature.

Woertz *et al* compared Insent TS-5000Z with Alpha MOS ASTREE in terms of the provided software, applied sensors and the measurement process, pointing to a longer lifetime of TS-5000Z sensors at the cost of bigger sample volume and analysis duration [93]. However, both devices require dissolution and filtration of a solid sample and large volumes in comparison to a saliva sample (< 2 mL). Moreover, the analysis is still relatively long, taking ~10 min per measurement. Furthermore, researchers describe a high baseline drift of some of the sensing units included (especially JB sensor from Alpha MOS [91]).

From our own experience in a startup, we can conclude that the biggest hurdle in monetizing e-tongue technology is the database. Although clients and investors are interested in the technology itself, there is a complete lack of understanding of the need for algorithm training for each application.

3.8 Data analysis and validation

Data analysis is an indispensable part of the electronic tongue system. We can divide the process into two parts: preparation and training/validation. In this section, we will describe the most used and simplest algorithms that can be used for e-tongue data analysis.

3.8.1 Preparation

During the preparation of the system, we should ascertain that all our sensors are working correctly. Sensors used in the array should be repeatable in terms of providing constant response during the experiments but also that each sensor of the same type gives the same response. This is especially important in the case of potentiometric ion-selective electrodes as they cannot be baseline corrected based on a single measurement. To ensure the proper functioning of the system, first they should be checked according to some established protocol, for example by plotting a calibration curve for the main species for which each sensor is sensitive, and by a short repeatability study in which the sensor is several times transferred between samples of low and high concentration of a certain ion. This procedure allows us to remove unreliable sensors from the matrix and provides data for check-ups that should be performed regularly during the operation of the system.

Next, the matrix of sensors should be used to measure model samples, and depending on the application we can choose model solutions of varying concentration of principal constitutents of the future samples (nitrates, magnesium, calcium, iron etc for water analysis, active pharmaceutical ingredients and masking agents for pharmaceuticals), or we can simply measure different samples that well represent the population (clean and contaminated water from a few different sources). After joining the data from all samples and sensors, we can perform exploratory data analysis, usually done through PCA (principal component analysis). Ideally, we should measure the samples using at least two or three sets of sensors, to make sure that the data is not biased and that we will obtain the same results in the future when the matrix will have to be exchanged for a new one due to use.

PCA allows visualization of the variance, so the information in our data in a lower number of variables. If our matrix is composed of 16 sensing units it would be quite impossible to visualize the data in a straightforward manner, but after PCA we can visualize usually more than 90% of the information in just three dimensions, which can be easily plotted. The Scores plot usually shown in scientific publications shows the position of all measured samples in the new data space. The first principal component describes the highest amount of variance, the second lower, and so on. As PCA is an unsupervised method, which means that the samples are not ascribed to any groups before the analysis, we can see if our data has a tendency to form clusters. Although it does not mean that the future samples will be classified correctly, using supervised methods, well-defined clusters in the Score plots are the first indication of the proper choice of sensing units and data points. The next equally important is the Loadings plot, which is too often omitted. It describes the contribution of each original variable to the information of a given principal component. Therefore, by the analysis of the Loadings plot we can choose which sensing units should be excluded from the matrix. Figure 3.13. presents a hypothetical Scores and Loadings plot. PCA is an unsupervised method, but most software allows colouring each sample, here different sample types are marked with different colours and shapes. Let us assume that the analysis was

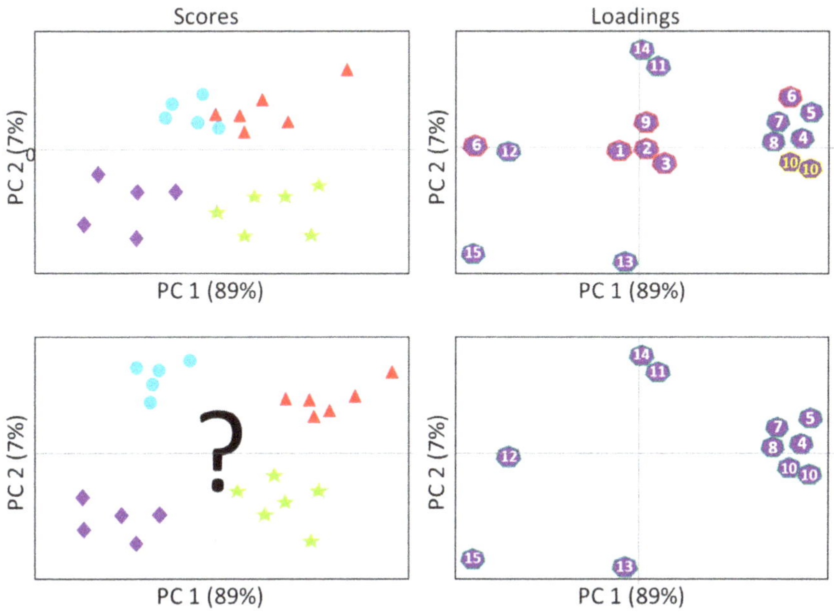

Figure 3.13. Example of a PCA Scores and Loadings plot before (top) and after (bottom) electrode pruning procedure.

made with a matrix of 15 different sensing units, with two of them, number 10 and 6 in two copies each. The Loadings plot on the right presents the amount of information provided by each sensor (zero is marked with the thin horizontal and vertical lines in the middle of each graph). Sensing units 1, 2, 3 and 9 are situated close to the 0,0 point of the Loadings plot, therefore, they provide very little information and can be probably excluded without influencing the separation of the samples. Both copies of sensing unit number 10 are close to each other, which is the first indication that they are working correctly. However, one copy of sensing unit number 6 is on the far right of the graph and the other on the far left, showing that they provide completely opposite information. It might be an indication of some malfunction or wrong connection of the electrodes. Apart from that, there is a wide range of sensors on the right side of the plot, additional analysis should be made to see if they are all necessary because in many cases exclusion of a few of them will not only not worsen the analysis but also might make the separation of samples better. Of course, no analysis can be done blindly, and if only some preliminary data is being analyzed the situation when a full dataset is acquired might be completely different.

By careful analysis of the Loading plot, we can decrease the complexity of the sensor matrix or the measurement (if only a specific range of potentials/wavelength etc is contributing to the measurement) and most often better define the clusters of our samples. Although this step is one of the most important in the whole analysis, it is often omitted.

3.8.2 Training and validation

After the preparation, we can pass to the training and validation of our model. Numerous methods allow predicting information about unknown samples. In the case of regression, the simplest are partial least squares (PLS)—if we are interested in the prediction of continuous values, and for classification K-nearest neighbours (KNN)—if we want to assign samples to certain groups (bitter/sweet; original/falsified; clean/contaminated). Neural networks are gaining popularity but readers should take into account that the final result cannot be connected to a specific sensor, and their performance is often better only for very large datasets, which are rather rare in research laboratories.

Regardless of the algorithm used, we will have to measure a considerable number of samples that are representative of the whole population. Measurements should include any changes in the measurement protocol indicated by the preparation step (exclusion of sensors etc). In the case of water, we should measure bottled mineral water from different brands and lots, tap water from different places, water from lakes, rivers, puddles, etc. This is probably the biggest impediment in successful commercialization, as clients need to tailor the database for their specific needs, and it requires not only the equipment but also experience and a considerable amount of work. The data should be divided into three parts: training, validation and test, where training is the most numerous. Regardless of the algorithm chosen, we will need to adjust certain parameters, number of neighbours in the case of KNN, number of components in the case of PLS or principal component regression, etc. The choice of parameters will be made based on the ability of the algorithm to correctly predict values of the samples from the validation set, but as we were using the validation set to obtain the best values, we cannot be certain that they are not biased by the choice of validation data. In this case, the overall performance of the algorithm is given based on the prediction of the test set.

It is possible to perform cross-validation in which instead of dividing the whole dataset once, data is divided multiple times and the training and testing with different combinations of the data is repeated. Very often leave-one-out cross-validation is applied, but this method is very limited and should not be used to judge the final performance of the algorithm.

There is no easy method to choose the best algorithm for a given application. Many scientific publications compare algorithms, although those results cannot be easily transferred to other datasets. The most popular methods used are: PLS, principal components regression, support vector machines, KNN and random forest.

3.8.3 External validation

Apart from internal validation, using the data from the electronic tongue, our model should be validated by some other method, for example through quantification of different species by means of mass spectrometry, high-performance liquid-chromatography or through comparison with expert panels [99, 100], if the analyzed samples can be tested by people. The number of people chosen for tests differs greatly, ranging from few to more than a hundred [100]. In the case of drug testing, except for taste assessment, other parameters are also analyzed including solubility and dispersibility. *In vitro* drug release studies and pharmacokinetic studies are also performed [101, 102]. Very interesting study was performed by Rudnitskaya *et al* [103] in which taste of model substances was assessed with an electronic tongue and compared to data obtained from human panellists and rats. Although animal taste panels are gaining market due to lower price and possibility of testing samples that could be dangerous to humans, it was shown that electronic tongue predicts taste perceived by humans better than the animal model.

3.9 Conclusions

As artificial intelligence, deep and machine learning are currently hot topics and electronic tongues are still considered a niche, we should expect that the concept will be reinvented. Data analysis tools are more available for example through open source, high-level programming languages such as Python, which includes libraries for operations on tables, multivariate analysis, neural networks and 2D and 3D plots. As a few hours of an online course are usually enough to perform basic analysis, more researchers can apply those methods to their data, not being aware that what they are actually creating are electronic tongues.

By definition, an e-tongue is composed of a set of cross-selective sensors and multivariate data analysis, and such systems, without the e-tongue/e-nose keywords are currently being described for example for optical analysis of amino-acid mixtures [104] or to diagnose COVID-19 patients based on breath analysis using an array of chemiresistors [105].

After 40 years of development, only two companies are providing potentiometric electronic tongues. As concepts such as the internet of things are getting more popular, we can expect that soon companies will be joining data from all the sensing systems and treating them as a whole, thus applying the most fundamental concept behind the electronic tongue and nose systems. It is up to us to still be part of this new, growing community.

We hope that the information provided in this chapter, especially the comparison of the mechanisms governing the artificial and natural senses presented herein, will help to develop new, more robust intelligent sensor arrays able to respond to commercial needs.

References

[1] Haber F and Klemesiewicz Z 1909 Electrical forces at phase boundaries *Z. Phys. Chem.* **67** 385–431

[2] Behrend R 1893 A F R 1893 Elektrochemische analyse *Z. Phys. Chem.* **11** 466–91

[3] Cygański A *Metody Elektrochemiczne* (Warsaw)

[4] Bakker E and Meyerhoff M E 2004 Ion-selective electrodes for measurements in biological fluids *ChemInform* **35**

[5] Lubert K H and Kalcher K 2010 History of electroanalytical methods *Electroanalysis* **22** 1937–46

[6] Pressman B C 1976 Biological applications of ionophores *Annu. Rev. Biochem.* **45** 501–30

[7] Antoszczak M, Huczyński A and Brzezinski B 2017 Synteza i aktywność biologiczna pochodnych salinomycyny *Wiadomości Chem.* **71** 629–61
[8] Pedersen C J 1988 The discovery of crown ethers (Nobel lecture) *Angew. Chemie Int. Ed. English* **27** 1021–7
[9] Štefanac Z and Simon W 1967 Ion specific electrochemical behavior of macrotetrolides in membranes *Microchem. J.* **12** 125–32 .
[10] Bloch S A and Saroff H A 1967 Fabrication and evaluation of membranes as specific electrodes for calcium ions *Biophys. J.* **7** 865–77
[11] Nery E W 2019 Use of a potentiometric and hybrid electronic tongue for the analysis of beer and wine. *Encyclopedia of Food Security and Sustainability* ed P Ferranti, E M Berry and J R Anderson (Amsterdam: Elsevier) pp 424–32
[12] Ciosek P and Wróblewski W 2007 Sensor arrays for liquid sensing—electronic tongue systems *Analyst* **132** 963–78
[13] Clark L C Jr, Noyes L K, Grooms T A and Moore P E 1984 *Ann. N* **4** 434–512
[14] Cadogan A, Gao Z, Lewenstam A, Ivaska A, Diamond D, Lewenstam A and Gao Z 1992 All-solid-state sodium-selective electrode based on a calixarene ionophore in a poly-(vinyl chloride) membrane with a polypyrrole solid contact *Anal. Chem.* **64** 2496–501
[15] Maksymiuk K and Michalska A 2015 Elektrody jonoselektywne—klasyka i nowe koncepcje *Chemik* **69** 373–82
[16] Bakker E and Pretsch E 2002 The new wave of ion-selective electrodes *Anal. Chem.* **74** 420A–6A
[17] Young c.c 1997 Evolution of blood chemistry analyzers based on ion selective electrodes *J. Chem. Educ.* **74** 177–82
[18] Severinghaus J W and Bradley A F 1958 *J. Appl. Physiol.* **13** 515–22
[19] https://www.ysi.com (accessed 1 January 2021)
[20] http://www.arkray.pl/ (accessed 2 January 2021)
[21] Legin A, Rudnitskaya A and Vlasov Y 2002 Electronic tongues: sensors, systems, applications *Sens. Updat.* **10** 143–88
[22] Vlasov Y, Legin A, Rudnitskaya A, Di Natale C and D'Amico A 2005 Nonspecific sensor arrays ('electronic tongue') for chemical analysis of liquids (IUPAC Technical Report) *Pure Appl. Chem.* **77** 1965–83
[23] Bakker E, Pretsch E and Bühlmann P 2000 Selectivity of potentiometric ion sensors *Anal. Chem.* **72** 1127–33
[24] Dybko A 2001 Errors in chemical sensor measurements *Sensors* **1** 29–37
[25] Buck R P and Lindner E 1994 Recomendations for nomenclature of ion-selective electrodes (IUPAC recommendations 1994) *Pure Appl. Chem.* **66** 2527–36
[26] Dillingham P W, Radu T, Diamond D, Radu A and McGraw C M 2012 Bayesian methods for ion selective electrodes *Electroanalysis* **24** 316–24
[27] Bergveld P 1970 Short communications: development of an ion-sensitive solid-state device for neurophysiological measurements *IEEE Trans. Biomed. Eng.* **BME-17** 70–1
[28] Bergveld P 1985 The impact of MOSFET-based sensors *Sens. Actuators* **8** 109–27
[29] Kaisti M, Zhang Q, Prabhu A, Lehmusvuori A, Rahman A and Levon K 2015 An ion-sensitive floating gate FET Model: operating principles and electrofluidic gating *IEEE Trans. Electron Devices* **62** 2628–35
[30] Tarasov A, Gray D W, Tsai M-Y, Shields N, Montrose A, Creedon N, Lovera P, O'Riordan A, Mooney M H and Vogel E M 2016 A potentiometric biosensor for rapid on-site disease diagnostics *Biosens. Bioelectron.* **79** 669–78

[31] Kaisti M 2017 Detection principles of biological and chemical FET sensors *Biosens. Bioelectron.* **98** 437–48
[32] Yates D E, Levine S and Healy T W 1974 Site-binding model of the electrical double layer at the oxide/water interface *J. Chem. Soc. Faraday Trans. 1 Phys. Chem. Condens. Phases* **70** 1807
[33] Covington A K 1994 Terminology and conventions for microelectronic ion-selective field effect transistor devices in electrochemistry (IUPAC Recommendations 1994) *Pure Appl. Chem.* **66** 565–9
[34] Bühlmann P and Chen L D 2012 Ion-selective electrodes with ionophore-doped sensing membranes *Supramolecular Chemistry* ed P A Gale and J W Steed (New York: Wiley)
[35] Bedner K *et al* 2014 Investigation of the dominant 1/f noise source in silicon nanowire sensors *Sens. Actuators* B *Chem.* **191** 270–5
[36] Wasilewski T, Kamysz W and Gębicki J 2019 Bioelectronic tongue: current status and perspectives *Biosens. Bioelectron.* **111923**
[37] Tan J and Xu J 2020 Applications of electronic nose (e-nose) and electronic tongue (e-tongue) in food quality-related properties determination: a review *Artif. Intell. Agric.* **4** 104–15
[38] Latha R S 2012 Electronic tongue: an analytical gustatory tool *J. Adv. Pharm. Technol. Res.* **3** 3–9
[39] Yeom J, Choe A, Lim S, Lee Y, Na S and Ko H 2020 Soft and ion-conducting hydrogel artificial tongue for astringency perception *Sci. Adv.* **6** eaba5785
[40] Roper S D and Chaudhari N 2017 Taste buds: cells, signals and synapses *Nat. Rev. Neurosci.* **18** 485–97
[41] Vandenbeuch A and Kinnamon S C 2009 Why do taste cells generate action potentials? *J. Biol.* **8** 42
[42] Kinnamon S C 2012 Taste receptor signalling—from tongues to lungs *Acta Physiol.* **204** 158–68
[43] Gravina S A, Yep G L and Khan M 2013 Human biology of taste *Ann. Saudi Med.* **33** 217–22
[44] Adams J U and O'Connor C 2014 G-protein-coupled receptors play many different roles in eukaryotic cell signaling *Essentials of Cell Biology* (London: NPG Education)
[45] Id Z W and Id K Z 2020 Taste of time: a porous-medium model for human tongue surface with implications for early taste perception *PLOS Comput. Biol.* 1–13
[46] Desimone J A and Lyall V 2020 Taste receptors in the gastrointestinal tract III: Salty and sour taste: sensing of sodium and protons by the tongue *Am. J. Physiol. Gastrointest. Liver Physiol.* **291** 1005–10
[47] Woertz K, Tissen C, Kleinebudde P and Breitkreutz J 2010 Performance qualification of an electronic tongue based on ICH guideline Q2 *J. Pharm. Biomed. Anal.* **51** 497–506
[48] Witkowska Nery E and Kubota L 2016 Integrated, paper-based potentiometric electronic tongue for the analysis of beer and wine *Anal. Chim. Acta* **918** 60–8
[49] Małgorzata Ł, Patrycja C-S and Wojciech W 2019 Critical evaluation of laboratory potentiometric electronic tongues for pharmaceutical analysis—an overview *Sensors* **19** 5376–91
[50] del Valle M 2010 Electronic tongues employing electrochemical sensors *Electroanalysis* **22** 1539–55
[51] Dillingham P W, Alsaedi B S O, Granados-Focil S, Radu A and McGraw C M 2020 Establishing meaningful limits of detection for ion-selective electrodes and other nonlinear sensors *ACS Sens.* **5** 250–7

[52] Wang W and Liu Y 2019 *Electronic tongue for food sensory evaluation* Evaluation Technologies for Food Quality *(Woodhead Publishing Series in Food Science, Technology and Nutrition) ed J Zhong and X Wang (Cambridge: Woodhead Publishing) pp 23–36*

[53] Hutchins M O 1997 Chemical senses: olfaction and gustation *Neuroscience Online* (Houston, TX: The University of Texas Health Science Center at Houston)

[54] Insel P, Snead A, Murray F, Zhang L, Yokouchi H, Katakia T, Kwon O, Dimucci D and Wilderman A 2012 GPCR expression in tissues and cells: are the optimal receptors being used as drug targets? *Br. J. Pharmacol.* **165** 1613–6

[55] Gutiérrez M, Alegret S and del Valle M 2007 Potentiometric bioelectronic tongue for the analysis of urea and alkaline ions in clinical samples *Biosens. Bioelectron.* **22** 2171–8

[56] Dias L G, Peres A M, Barcelos T P, Sá Morais J and MacHado A A S C 2011 Semi-quantitative and quantitative analysis of soft drinks using an electronic tongue *Sens. Actuators,* B *Chem.* **154** 111–8

[57] Ciosek P, Augustyniak E and Wroblewski W 2004 Polymeric membrane ion-selective and cross-sensitive electrode-based electronic tongue for qualitative analysis of beverages *Analyst* **129** 639–44

[58] Legin A, Rudnitskaya A, Lvova L, Vlasov Y, Di Natale C and D'Amico A 2003 Evaluation of Italian wine by the electronic tongue: recognition, quantitative analysis and correlation with human sensory perception *Anal. Chim. Acta* **484** 33–44

[59] Rudnitskaya A, Rocha S M, Legin A, Pereira V and Marques J C 2010 Evaluation of the feasibility of the electronic tongue as a rapid analytical tool for wine age prediction and quantification of the organic acids and phenolic compounds. The case-study of Madeira wine *Anal. Chim. Acta* **662** 82–9

[60] Jarboui A, Marx Í M G, Veloso A C A, Vilaça D, Correia D M, Dias L G, Mokkadem Y and Peres A M 2020 An electronic tongue as a classifier tool for assessing perfume olfactory family and storage time-period *Talanta* **208** 120364

[61] Rodrigues N, Marx Í M G, Casal S, Dias L G, Veloso A C A, Pereira J A and Peres A M 2019 Application of an electronic tongue as a single-run tool for olive oils' physicochemical and sensory simultaneous assessment *Talanta* **197** 363–73

[62] Ha D, Hu N, Sun Q, Su K, Wan H, Li H, Xu N, Sun F and Wang P 2014 Recent achievements in electronic tongue and bioelectronic tongue as taste sensors *Sens. Actuators* B *Chem.* **207** 1136–46

[63] Witkowska E, Buczkowska A, Zamojska A, Szewczyk K W and Ciosek P 2010 Monitoring of periodic anaerobic digestion with flow-through array of miniaturized ion-selective electrodes *Bioelectrochemistry* **80** 87–93

[64] Buczkowska A, Witkowska E, Górski Ł, Zamojska A, Szewczyk K W, Wróblewski W and Ciosek P 2010 The monitoring of methane fermentation in sequencing batch bioreactor with flow-through array of miniaturized solid state electrodes *Talanta* **81** 1387–92

[65] Witkowska Nery E, Jastrzębska E, Żukowski K, Wróblewski W, Chudy M and Ciosek P 2014 Flow-through sensor array applied to cytotoxicity assessment in cell cultures for drug-testing purposes *Biosens. Bioelectron.* **51** 55–61

[66] Růžička J and Hansen E H 1980 Flow injection analysis. principles, applications and trends *Anal. Chim. Acta* **114** 19–44

[67] Ruzicka J and Marshall G D 1990 Sequential injection: a new concept for chemical sensors, process analysis and laboratory assays *Anal. Chim. Acta* **237** 329–43

[68] Gutés A, Céspedes F and del Valle M 2007 Electronic tongues in flow analysis *Anal. Chim. Acta* **600** 90–6

[69] Gallardo J, Alegert S and Del Valle M 2004 A flow-injection electronic tongue based on potentiometric sensors for the determination of nitrate in the presence of chloride *Sensors Actuators,* B *Chem.* **101** 72–80

[70] Calvo D, Durán A and del Valle M 2007 Use of sequential injection analysis to construct an electronic-tongue. Application to multidetermination employing the transient response of a potentiometric sensor array *Anal. Chim. Acta* **600** 97–104

[71] Mimendia A, Legin A, Merkoçi A and del Valle M 2010 Use of Sequential Injection Analysis to construct a potentiometric electronic tongue: application to the multidetermination of heavy metals *Sens. Actuators* B *Chem.* **146** 420–6

[72] Cortina M, Duran A, Alegret S and Del Valle M 2006 A sequential injection electronic tongue employing the transient response from potentiometric sensors for anion multi-determination *Anal. Bioanal. Chem.* **385** 1186–94

[73] Cortina M, Gutés A, Alegret S and Del Valle M 2005 Sequential injection system with higher dimensional electrochemical sensor signals: part 2. Potentiometric e-tongue for the determination of alkaline ions *Talanta* **66** 1197–206

[74] Durán A, Cortina M, Velasco L, Rodríguez J A, Alegret S and Del Valle M 2006 Virtual instrument for an automated potentiometric e-tongue employing the SIA technique *Sensors* **6** 19–29

[75] Ipatov A, Ivanov M, Makarychev-Mikhailov S, Kolodnikov V, Legin A and Vlasov Y 2002 Determination of cyanide using flow-injection multisensor system *Talanta* **58** 1071–6

[76] Gallardo J, Alegret S and Del Valle M 2005 Application of a potentiometric electronic tongue as a classification tool in food analysis *Talanta* **66** 1303–9

[77] Ciosek P, Brzózka Z and Wróblewski W 2006 Electronic tongue for flow-through analysis of beverages *Sens. Actuators,* B *Chem.* **118** 454–60

[78] Jiang H, Zhang M, Bhandari B and Adhikari B 2018 Application of electronic tongue for fresh foods quality evaluation: a review *Food Rev. Int.* **34** 746–69

[79] Lvova L, Kim S S, Legin A, Vlasov Y, Yang J S, Cha G S and Nam H 2002 All-solid-state electronic tongue and its application for beverage analysis *Anal. Chim. Acta* **468** 303–14

[80] Martínez-Máñez R, Soto J, Garcia-Breijo E, Gil L, Ibáñez J and Llobet E 2005 An 'electronic tongue' design for the qualitative analysis of natural waters *Sens. Actuators,* B *Chem.* **104** 302–7

[81] Sakai H, Iiyama S and Toko K 2000 Evaluation of water quality and pollution using multichannel sensors *Sens. Actuators,* B *Chem.* **66** 251–5

[82] Saurina J, López-Aviles E, Le Moal A and Hernández-Cassou S 2002 Determination of calcium and total hardness in natural waters using a potentiometric sensor array *Anal. Chim. Acta* **464** 89–98

[83] Ciosek P and Wróblewski W 2008 Miniaturized electronic tongue with an integrated reference microelectrode for the recognition of milk samples *Talanta* **76** 548–56

[84] Verrelli G, Francioso L, Paolesse R, Siciliano P, Di Natale C, D'Amico A and Logrieco A 2007 Development of silicon-based potentiometric sensors: towards a miniaturized electronic tongue *Sens. Actuators,* B *Chem.* **123** 191–7

[85] Witkowska Nery E, Guimarães J A and Kubota L T L T 2015 Paper-based electronic tongue *Electroanalysis* **27** 2357–362

[86] Nery E W and Kubota L T 2016 Evaluation of enzyme immobilization methods for paper-based devices-a glucose oxidase study *J. Pharm. Biomed. Anal.* **117** 551–9
[87] Łabańska M, Ciosek-Skibińska P and Wróblewski W 2019 Critical evaluation of laboratory potentiometric electronic tongues for pharmaceutical analysis—an overview *Sensors (Switzerland)* **19** 5376
[88] Alpha MOS Website https://www.alpha-mos.com/ (accessed 9 February 2021)
[89] Raithore S, Bai J, Plotto A, Manthey J, Irey M and Baldwin E 2015 Electronic tongue response to chemicals in orange juice that change concentration in relation to harvest maturity and citrus greening or huanglongbing (HLB) disease *Sensors (Switzerland)* **15** 30062–75
[90] Pan M, Tong L, Chi X, Ai N, Cao Y and Sun B 2019 Comparison of sensory and electronic tongue analysis combined with HS-SPME-GC-MS in the evaluation of skim milk processed with different preheating treatments *Molecules* **24** 1650
[91] Beullens K, Mészáros P, Vermeir S, Kirsanov D, Legin A, Buysens S, Cap N, Nicolaï B M and Lammertyn J 2008 Analysis of tomato taste using two types of electronic tongues *Sens. Actuators,* B *Chem.* **131** 10–7
[92] Sipos L, Kovács Z, Sági-Kiss V, Csiki T, Kókai Z, Fekete A and Héberger K 2012 Discrimination of mineral waters by electronic tongue, sensory evaluation and chemical analysis *Food Chem.* **135** 2947–53
[93] Woertz K, Tissen C, Kleinebudde P and Breitkreutz J 2011 A comparative study on two electronic tongues for pharmaceutical formulation development *J. Pharm. Biomed. Anal.* **55** 272–81
[94] Tan T, Schmitt V and Isz S 2001 Electronic tongue: a new dimension in sensory analysis *Food Technol.* **55** 44
[95] Toko K 1996 Taste sensor with global selectivity *Mater. Sci. Eng.* C **4** 69–82
[96] Wu X, Tahara Y, Yatabe R and Toko K 2020 Taste sensor: electronic tongue with lipid membranes *Anal. Sci.* **36** 147–59
[97] Makita Y, Ishida T, Kobayashi N, Fujio M, Fujimoto K, Moritomo R, Fujita J and Fujiwara S 2016 Evaluation of the bitterness-masking effect of powdered roasted soybeans *Foods* **5** 44
[98] Pein M, Gondongwe X D, Habara M and Winzenburg G 2014 Interlaboratory testing of Insent e-tongues *Int. J. Pharm.* **036** 1–9
[99] Maniruzzaman M and Douroumis D 2015 An in-vitro-in-vivo taste assessment of bitter drug: comparative electronic tongues study *J. Pharm. Pharmacol.* **67** 43–55
[100] Lorenz J K, Reo J P, Hendl O, Worthington J H and Petrossian V D 2009 Evaluation of a taste sensor instrument (electronic tongue) for use in formulation development *Int. J. Pharm.* **367** 65–72
[101] Maniruzzaman M, Boateng J S, Bonnefille M, Aranyos A, Mitchell J C and Douroumis D 2012 Taste masking of paracetamol by hot-melt extrusion: an *in vitro* and *in vivo* evaluation *Eur. J. Pharm. Biopharm.* **80** 433–42
[102] Rao M R P, Bhutada K and Kaushal P 2019 Taste evaluation by electronic tongue and bioavailability enhancement of Efavirenz *AAPS PharmSciTech* **20** 1–12
[103] Rudnitskaya A, Kirsanov D, Blinova Y, Legin E, Seleznev B, Clapham D, Ives R S, Saunders K A and Legin A 2013 Assessment of bitter taste of pharmaceuticals with multisensor system employing 3 way PLS regression *Anal. Chim. Acta* **770** 45–52

[104] Duan Q, Lee J, Zheng S, Chen J, Luo R, Feng Y and Xu Z 2020 A color-spectral machine learning path for analysis of five mixed amino acids *Chem. Commun.* **56** 1058–61
[105] Shan B *et al* 2020 Multiplexed nanomaterial-based sensor array for detection of COVID-19 in exhaled breath *ACS Nano* **14** 12125–32

IOP Publishing

Electronic Tongues
Fundamentals and recent advances
Flavio M Shimizu, Maria L Braunger and Antonio Riul Jr

Chapter 4

Voltammetric electronic tongues

Marta Bonet-San-Emeterio, Xavier Cetó and Manel del Valle

This chapter aims to introduce voltammetric electronic tongues (e-tongues) by covering first their fundamentals, followed by a review of some of the most relevant applications over different scenarios reported so far. In this regard, special mention will be given to bioelectronic tongues, that is, when one or several biosensors are incorporated into the sensor array. Furthermore, given the larger dimensionality of the generated data when voltammetric sensors are employed, an overview of some of the more common strategies that might be used for data pre-processing will be presented. Finally, the detection of biogenic amines using a voltammetric e-tongue will be taken as a study case to illustrate the differences between the different methods.

4.1 Introduction

Analytical chemistry is the science that deals with methods for the determination of the chemical composition of samples [1]. Since its early years, most analyses consisted in the separation of analytes from other components of a sample by precipitation, extraction, or distillation (classical methods). Early, in the 20th century, with the appearance of instrumental methods, such procedures were replaced by more efficient techniques such as chromatography or electrophoresis, which combines separation with the application of a source of energy that upon stimulation of the sample generates a measurable response from the analyte.

However, such methods do not suffice to fulfil the new challenges that arise in environmental monitoring, food safety or public health; for example, when not requiring the quantification of a certain analyte, but an overall effect on a biological system (e.g., toxicity or quality indices). In order to tackle such scenarios in which classical approaches fail to succeed, biomimetic devices inspired in the senses of mammals have appeared.

Common principles, shared by different animal senses, rely on broad response receptors (rather than highly specific ones) and the use of a bunch of slightly

doi:10.1088/978-0-7503-3687-1ch4

different receptors, which by combinatorial principles are able to discriminate between thousands of targets. This trend is somehow opposite to the usual trend in analytical chemistry, where researchers aim for highly selective and sensitive devices (unfortunately, both challenging features that are not always possible to achieve).

Such an approach was first introduced in 1982 by Persaud and Dodd with the concept of electronic nose (e-nose), a device that aims to mimic the odour discrimination ability of animals' olfactory system [2]. This was the first biomimetic system based on sensors that sought to imitate animal senses, and that allowed the rapid development of interesting applications in different fields [3]. A few years later, around the 1990s [4, 5], there appeared also the concept of the electronic tongue (e-tongue), which was based on the same principles, but applied to the analysis of liquids (figure 4.1).

Initial studies with what nowadays is known as e-tongues were performed employing potentiometric sensors and aimed to overcome possible interferences of the ion-selective electrodes (ISEs) used [5]. A few years later Toko *et al* attempted to actually reproduce humans' taste sense, in what they referred to as 'taste sensor' [6]. Again, such a system was based on a potentiometric e-tongue using electrodes modified with different lipid membranes rather than an ionic selective membrane.

A common feature between early applications of e-noses and e-tongues is that the sensing principle in both cases was based on potentiometric measurements. The former were based on metal-oxide-semiconductor field-effect transistors (MOSFETs) and the latter on different types of ISEs (including ion-sensitive field-effect transistors, ISFETs). It was not until 1997 that Winquist *et al* reported the application of an e-tongue based on voltammetric measurements [7].

Despite the approach principles being the same, the information sought by the sensors is entirely different, and the nature of the measurements is also significantly distinct. Voltammetric sensors provide information about the compounds present in the sample that can be oxidized or reduced, while potentiometric sensors provide information about the ionic composition of the samples (either cations or anions). Furthermore, voltammetry offers further advantages over potentiometry such as its versatility (derived from the actual technique chosen among the different voltammetric techniques and the possibility to adjust the parameters used to carry out the measurements conveniently), high sensitivity or possibility to carry out the

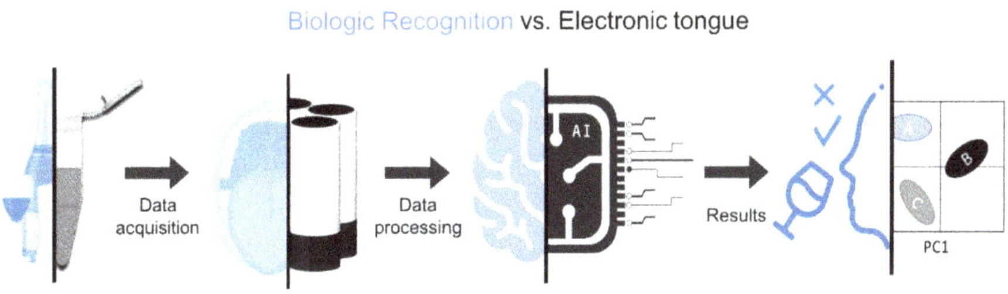

Figure 4.1. Comparison of the biological recognition process of a sample with the biomimetic e-nose/e-tongue approach.

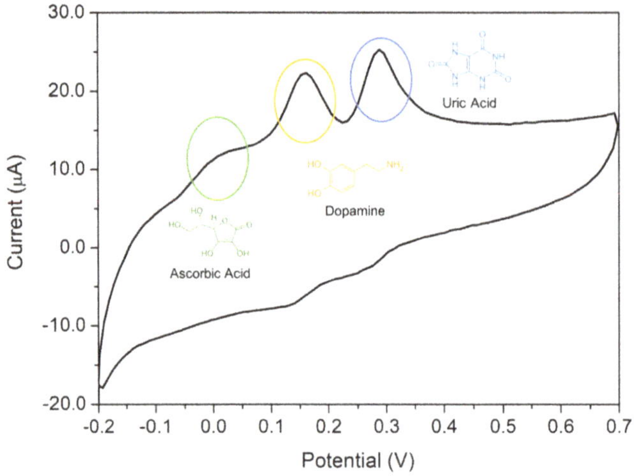

Figure 4.2. Example of a cyclic voltammogram for a sample containing ascorbic acid, dopamine, and uric acid (peaks from left to right).

simultaneous determination of several analytes (from the same voltammogram, e.g., figure 4.2, where a voltammogram for a ternary mixture is shown) [1]. However, despite the advantages that voltammetric methods and sensors offer, the larger dimensionality of the generated data (e.g., if compared to potentiometric sensors) is one of the drawbacks that may hinder its application. In this direction, the usage of a pre-processing stage for data compression can help to overcome such problems [8], while at the same time it offers many advantages (as will be discussed in section 4.4.2).

Nowadays, the principle of operation of such biomimetic systems has been further expanded by incorporating one or several biosensors into the sensor array, usually employing the same transduction principle to facilitate the measurements, in what is known as bioelectronic tongue (bioe-tongue) [9, 10]. Although biosensors may be expected to be very selective towards a specific analyte and do not suffer from interferences, the reality is that bioreceptors often cross-react with different species and might suffer undesired matrix effects. For example, enzymes might show group specificity (e.g., alcohol dehydrogenases or phenol degrading oxidases), and their kinetics might be affected by the presence of certain compounds that can act as activators or inhibitors [11]. Thus, although the usage of biosensors may seem to defer from e-tongue principles, they do not only fit perfectly, but the synergies between them improve the performance of both. That is, the concept exploits the higher selectivity of biosensors to address the lack of selectivity of e-tongues, and the capabilities of e-tongues to extract meaningful data from complex responses to compensate biosensors' interference problems and/or to allow the multi-determination of several species.

Lastly, despite being out of the scope of this chapter, it is worth mentioning that although such biomimetic systems are usually classified depending on the nature of the samples measured (e.g., e-noses dealing with gases, e-tongues with liquid samples, etc) and the sensing technique employed (e.g., potentiometric, voltammetric, impedimetric, etc), it is also possible to combine sensors of different nature in order to obtain more information about the samples [12, 13]. Such hybrid systems mimic again the perception of animals, where the response from the brain derives from the combination of the stimuli perceived by different senses. However, as it could be expected, such combination is not straightforward, and it requires some tools that permit one to make compatible the measurements originated from very different natures. Such strategies are known as data fusion, and deal not only with the compatibilization of the multisource data, but on taking advantage of their characteristics in order to improve the representation of sample information [14].

After this brief introduction on the background of voltammetric e-tongues, this chapter aims to cover the fundamentals of voltammetric (bio)e-tongues as well as present some relevant applications to better illustrate their potential. Finally, the application of a voltammetric e-tongue for the detection of biogenic amines (BAs) will be shown as an example to provide an overview of some of the most common strategies that might be used for signal compression.

4.2 Voltammetric (bio)e-tongues

As already introduced, these multisensory systems are based on (bio)sensors, with different selectivity towards the analytes, providing complementary data about the sample, which after appropriate data analysis, allow the classification of samples and/or the quantification of selected analytes. The idea behind this concept is to exploit the synergies between (bio)sensors and chemometrics to infer underlying relationships between the measured analytical signals and the properties of the samples. Thus, it is clear that the system has two differentiated stages: the voltammetric measurements and the data analysis.

4.2.1 Voltammetric (bio)sensors array

Voltammetric methods are based on the measurement of the current generated upon polarization of the working electrode as a function of time [1]. By changing the waveform of the excitation signal, characteristic responses dependent on the chosen technique will be obtained, which would lead to distinct responses for the same sample as different information is being extracted. Precisely, the huge versatility of voltammetric methods arises on the different techniques that may be employed (e.g., amperometry at fixed potential, cyclic voltammetry, differential pulse voltammetry, etc) and the parameters that define those (e.g., step potential, scan rate, amplitude, etc).

Although the choice of the voltammetric method has a direct impact on the voltammetric response, this is not a factor that is usually evaluated, but researchers seem to stick to their preferred technique and slightly tune only the measuring parameters. Conversely, much attention is paid to the choice of the sensor array, as depending on the nature of the chosen electrodes, different electroactive compounds may be oxidized or reduced at a specific potential and/or the associated current might differ significantly. Such changes arise in the different catalytic properties that may appear within the electrode surface.

In this sense, initial studies with voltammetric e-tongues employed arrays of different bare metals (e.g., gold, platinum, palladium, copper, etc), which were enough to generate different sensitivity and selectivity patterns required for e-tongue studies [15]. However, in order to further improve the performance of e-tongues, chemically modified electrodes arose as a better alternative to fit the detection of the different analytes and/or to achieve an improvement of its analytical performance. In this direction, many different modifiers have been reported in the literature, including the use of noble metals such as gold or platinum, inorganic materials such as phthalocyanines, and conducting polymers such as poly-ethylenedioxythiophene (PEDOT) or pyrrole [16–19]. Especially relevant is the usage of nanoparticles (NPs), which have emerged as an alternative to the respective bulk materials given their higher surface-to-mass ratio and improved electrochemical properties. Lastly, special mention has to be given to the use of biorecognition elements such as enzymes, antibodies or oligos for the preparation of biosensors, connecting these with the development of bioe-tongues.

Although different sensor assemblies can be found in biosensors arrays, the most important biological modifier is enzymes. Enzymes can be found in biosensor arrays both as a single receptor in one of the electrodes and as a general biorecognition element of the set of electrodes; in the last case, other modifiers must be added, e.g., different electrochemical mediators. One example of a biosensor array that uses the same enzyme in all the electrodes is the work of Medina-Plaza *et al* [20]. The use of different arrays of modified screen-printed electrodes with glucose oxidase (GOX) or tyrosinase (tyrs) in the presence of different electrochemical mediators is presented, providing a large variety of responses. Another option is the use of enzymes with group specificity that have cross-reactivity with the studied analytes, as shown in Lange's *et al* paper [21]. They worked with an array composed of three different amine oxidases (diamine oxidase, tyramine oxidase, and plasma amine oxidase) for the detection of biogenic amines in food samples. However, enzymes are not the only biological modifier used in bioe-tongues, as nowadays a wide range of new proposals can be found in the literature, for example, the use of peptides, aptamers, and specific tissues or cells, among others [22–25].

For the preparation of such (bio)chemically modified electrodes several procedures have been reported in the literature, but among those, the more common approaches include physical adsorption, covalent binding, polymer film coating or the incorporation of those into composite materials. Precisely, the latter include carbon paste electrodes (CPEs) and graphite-epoxy composites (GECs), which are two of the most widely employed electrodes when developing voltammetric (bio)e-tongues. More recently, screen-printed electrodes (SPEs, either based on the printing of different materials or upon their modification) have arisen as a popular choice for the development of (bio)e-tongues due to their inherent advantages [26, 27].

Despite the importance that the selection of the sensor array has in the final performance of the e-tongue, most papers dealing with e-tongues focus on the developed application itself or the data treatment stage, but very few reports on the choice of the sensors. In this direction, the development of a simple methodology that allows the *a priori* selection of the optimal sensor array to carry out a specific application is of utmost interest. To this aim, the autocorrelation between the signals of the different sensors or the use of principal component analysis (PCA) have been recently proposed [28–30], but further research is still required in order to better establish a robust methodology.

4.2.2 Data analysis

Unlike potentiometric methods in which a unique value is obtained for each sensor, with voltammetric devices normally a vector of readings is obtained for each sensor. On the one hand, this represents an advantage as richer data is generated, and higher selectivity might be attained if, e.g., two species show response with a specific electrode, but at a different potential. On the other hand, this may hinder the data treatment stage as the dimensionality of the data is increased.

The straightforward solution is the use of trilinear (multi-way) approaches such as parallel factor analysis (PARAFAC) or Tucker models for qualitative analysis [31], and multi-way partial least squares regression (nPLS) for quantitative models [32]; or even quatrilinear methods for much more complex data sets [33]. However, despite the advantages that those offer and the fact that they represent new trends in chemometric analysis, their usage is still very limited in voltammetric e-tongues.

Alternatively, one can employ the simpler multivariate data treatment procedure in which one vector of input information is modelled. In this instance, the first step for data analysis is the unfolding at the switching potential of the generated data from a 3D to a 2D matrix, by combining different current readings in a single vector independently of which polarization potential they are related to and to which sensor they do correspond (figure 4.3). This data arrangement can be the easy procedure to make voltammetric e-tongues compatible with the more common multivariate modelling tools.

The data processing stage can then be divided into three different steps: weighting, signal compression and modelling. Although only the last one is

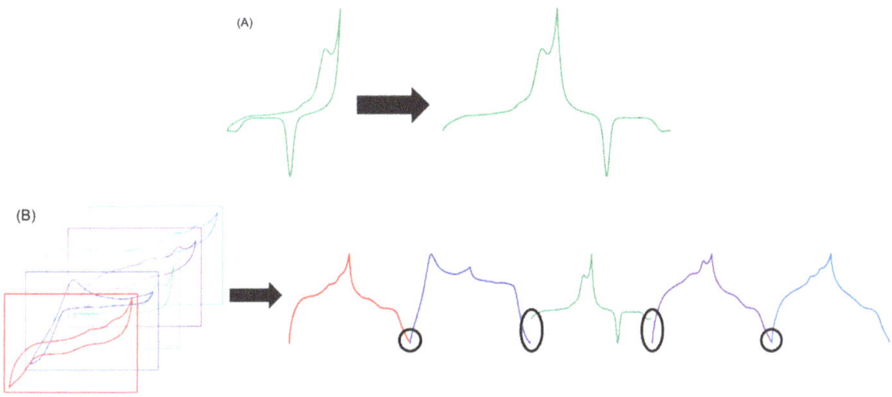

Figure 4.3. Schematic of the 'unfolding' of (a) a single voltammogram and (b) the voltammograms from a sensor array.

mandatory, the usage of the other two improves e-tongues performance significantly and is therefore recommended [8, 34, 35].

The first step is directed to ensure consistency among the raw data and to equate the *a priori* contribution of each sensor from the array to the final model [34, 35]. Some of the strategies that can be used are: centering, standardization, normalization, or auto-scaling of the data.

The second step is directed towards the reduction of the high dimensionality of the data, while preserving the relevant information [8]. Furthermore, this data reduction provides further advantages as a reduction in training time and computing requirements, avoids redundancy in input data and might even contribute to a reduction in noise, which ultimately allows an improvement of the performance of the system thanks to the obtaining of simpler models (i.e., less risk of over-fitting) with better generalization ability. Some of the strategies that can be used are PCA, *kernels*, fast Fourier transform (FFT), discrete wavelet transform (DWT), genetic algorithms (GAs), etc.

The last step is the building of the actual chemometric model itself. Among the different multivariate methods that can be implemented, PCA, partial least squares regression (PLS), and artificial neural networks (ANNs) are the most widely used for e-tongues applications [36]. Concretely, PCA is the most commonly used for qualitative applications as a better visualization tool to assess samples (dis) similarities; whereas most advanced quantitative applications may be achieved with the use of PLS or ANNs. More recently, the use of support vector machines (SVMs) is growing in popularity, both for qualitative and quantitative applications.

For a more in-depth revision and detailed explanation on the data analysis stage, readers may consult the following reviews and comparative studies [8, 37–40].

4.3 State of the art and examples in the literature

Usage of voltammetric e-tongues has been reported over a wide range of scenarios to develop very diverse applications. Some of those applications will be reviewed to provide an overview of the capabilities of such systems, both for the identification and classification of different samples (qualitative applications) or for the determination of a certain index or concentration of different compounds (quantitative applications).

A quick search on the bibliography easily demonstrates the complexity of the voltammetric e-tongues topic, which leads the research to a few groups. From table 4.1 it can be confirmed that despite the wide range of available chemometric tools, in most cases, the same modelling methods are always used; ANNs and PLS for the quantitative applications, and PCA for the qualitative ones. Regarding the compression step, given it is closely related to the data source, the methods in use are more heterogeneous.

One of the most studied topics employing (bio)e-tongues approach is the study of alcoholic drinks and their derivatives. Concretely, the quantification of the total phenolic content and the alcoholic degree in wine using an array containing laccase

Table 4.1. Relevant applications employing voltammetric e-tongues over different scenarios.

Electrode array	Analytes/Samples	Data treatment	Reference
GEC + lac, tyrs, and CuNPs	Catechol, *m*-cresol, and guaiacol	FFT + ANN	[56]
ITO with a Langmuir-Blodgett film + tyrs, lac, and Lutetium bisphthalocyanine	Vanillic acid, catechol, caffeic acid, hydroquinone, gallic acid, and pyrogallol	Windowed slicing method + PCA	[57]
GEC + lac, tyrs, and CuNPs	Gallic acid, ferulic acid, and sinapic acid	Windowed slicing method + ANN	[50]
GEC + lac, tyrs, and CuNPs	Gallic acid, (±)-catechin, *p*-coumaric acid, caffeic acid, catechol, phenol, *m*-cresol, ferulic acid, chlorogenic acid, and quercetin	FFT + ANN + PCA	[41]
GEC + lac, tyrs, and CuNPs	Wine samples[1]	Windowed slicing integral + PCA + ANN	[58]
CPE + tyrs	Hydroquinone and guaiacol	PLS	[59]
CPE + lac, Botryosphaeran, and PtNPs	Chlorogenic acid (5-O-caffeoylquinic acid)	PCA	[60]
Polycrystalline gold electrodes + Cellobiose dehydrogenases	Lactose, glucose, and Ca(II)	DWT + ANN	[61]
SPE + GOX, and tyrs	Glucose, catechol, and must[1]	Kernels + PCA + PLS	[20]
Pt and SS disks + GOX, tyrs, polypyrrole, and AuNPs	Must[1]	Kernels + PARAFAC + PCA + PLS + SVM	[42]
GEC + SH, γ-Glu– Cys, Cys–Gly	Cd(II), Pb(II), and Zn(II)	FFT + ANN	[62]
Carbon resin + NiONPs, LuPc2, and tyrs	Catechol and wine[1]	Genetic algorithms + PLS	[63]
Metallic Pt and Au	Orange drinks[1]	PCA	[7]
Metallic Au + TA, TA-Gly-Gly-His, TA-glutathione, and TA-angiotensin I	Cu(II), Cd(II), and Pb(II)	PLS	[25]
Metallic Cu and Au	Wine and whisky samples[1]	PCA	[64]
GEC + MIPs	4-ethylguaiacol and 4-ethylphenol	DWT + ANN	[65]
SPE + CNF, *ex situ*-Sb-CNF, GSH-CNF, and Cys-CNF	Pb(II), Cd(II), Tl(I), and Bi(III) in presence of In(III) and Zn(II)	PLS	[66]

Metallic Au, Pt, Ir, Rh, Ag, Cu, Ni, and Co	Soluble chemical oxygen demand, soluble biological oxygen demand, ammonia, orthophosphate, Sulphate, acetic acid, and alkalinity	PLS	[67]

[1]Analysed real samples; *ANN*, artificial neural networks; *CNF*, carbon nanofibers; *DWT*, discrete wavelet transform; *FFT*, fast Fourier transform; *GOX*, glucose oxidase; *lac*, laccase; *MIP*, molecularly imprinted; *NPs*, nanoparticles; *PCA*, principal components analysis; *PLS*, partial least squares; *SS*, stainless steel; *TA*, D,L-6,8-thioctic acid; *tyrs*, tyrosinase; *GEC*, graphite epoxy composite; *CPE*, carbon paste electrode; *SPE*, screen printed electrode.

(lac), tyrs and/or GOX are two of the most common cases. One example of the use of ANNs for the quantification of phenols in wine samples is the work of Cetó *et al* [41], where an array of four sensors (laccase, tyrosinase and CuNPs modified electrodes plus a bare one) was used. The data recorded with the array was used to feed an ANN model able to predict the Folin-Ciocalteu (FC) and I_{280} indexes. For this, the reduction of the data was first done using FFT. Data compression not only reduces the training time but also helps to achieve a better generalization ability of the model. However, a compromise must be taken when the Fourier coefficients are chosen as we will see in section 4.4.2. A similar idea was also developed in the work of Garcia-Hernández *et al* [42], where a bioe-tongue based on tyrs and GOX enzymes was applied for the determination of the total polyphenol index (TPI), aiming to correlate the characteristics of wines and musts according to the grape varieties. For that purpose, different chemometric tools (quantitative and qualitative) were compared. First, a compression step based on the application of the 'kernel' method was used to transform the voltammograms down to only 10 variables per sensor. The sample classification was based on the use of PCA and PARAFAC tools. From the PCA score plot, wines were grouped in separated clusters according to the grape variety and by the TPI value. In contrast, for must only the grape variety was clustered. In the PARAFAC models, the root mean square error (RMSE) was the parameter selected to optimize the final structure, which corroborated the PCA results. On the other hand, PLS-1 and SVMs were used for the quantification part, concretely, to correlate the outputs of the bioe-tongue with the chemical analytical parameters: the alcoholic degree and the TPI. The statistical results demonstrated that a better modelling was achieved with SVMs than with PLS-1, as could be expected due to the superior performance of SVMs. In this way, table 4.1 summarizes some illustrative works in the field, with special emphasis on bioe-tongues applications.

4.4 Study case: detection of biogenic amines

In order to better illustrate some of the common dimensional reduction methods and their importance to e-tongue analysis, the detection of biogenic amines by means of a voltammetric e-tongue will be taken as an exemplary study case over the next sections.

BAs are endogenous nitrogen-containing organic compounds of low molecular weight. Although being naturally present in the human body, high concentrations can produce Scombrotoxin fishing poisoning (SFP), also named histamine poisoning. Despite SFP is normally related to the ingestion of fish and fish derivatives, also other food products such as dairy products, fermented foods from vegetables to beverages, soy products, and meat contain BAs [43, 44]. Food products have microorganisms that catalyze the production of BAs, with proliferation closely related to poor conservation conditions. Basically, long periods of storage at temperatures above the recommended will cause microbial growth, and so as the concentration of BAs. For this reason, BAs concentration is also important as an indicator of inadequate product handling and preservation [45].

Histamine (His) is the most commonly studied BA, but putrescine (Put), cadaverine (Cad), and tyramine (Tyr) are also other BAs important to understand the relevance of this family of compounds. Recently, the effect of these together with histamine has been studied, concluding that they can potentiate its toxicity [46]. Since there is little information about the concrete contribution of each BA in the human body, the existing legislation focuses only on His concentrations. Food and Agriculture Organization of the United Nations (FAO) and the World Health Organization (WHO), both reported in 2012 a homogenization of the legal limits of His in fish and fishery products. The official announcement declares a general maximum level of 200 $mg{\cdot}kg^{-1}$ [46].

4.4.1 Voltammetric measurements and sensor array

For the multichannel voltammetric measurement, a six-channel Autolab PGSTAT20 (Metrohm Autolab B.V, Utrecht, Netherlands) controlled with GPES Multichannel 4.7 software package was the instrumentation employed. A complete voltammogram was recorded in a potential window from −1.5 to 1.8 V versus Ag/AgCl electrode at 50 $mV{\cdot}s^{-1}$ and a step potential of 10 mV. The experiments were performed at room temperature without stirring, using a typical three electrodes configuration formed by a combined platinum auxiliary and Ag/AgCl reference electrode, and an array of five different GECs as working electrodes. The sensor array was built in the laboratory following the standard methodology established in our research group [47], where first a gold connector is soldered to a copper disk. Then it is introduced into a PVC tube, and the empty cavity left on top of the copper disk is filled with the previous homogenized GEC paste. The ratio of resin, graphite powder, and modifier in the composite was 83:15:2 (w/w). Once the body of the electrode was filled, it was left for 48 h at 80 °C for the resin to harden. Finally, the surface was polished with sandpaper of different degrees to achieve a mirror-like finish.

As mentioned above, the array was formed by five different GEC electrodes, using as modifiers polypyrrole polymer and four kinds of NPs (Bi_2O_3, Sn_2O_3, Cu and ZnO). The selected electrodes were chosen from a wide range of modifiers available in the laboratory after the initial screening of their response against the target analytes. A scheme of the measuring procedure is depicted in figure 4.4.

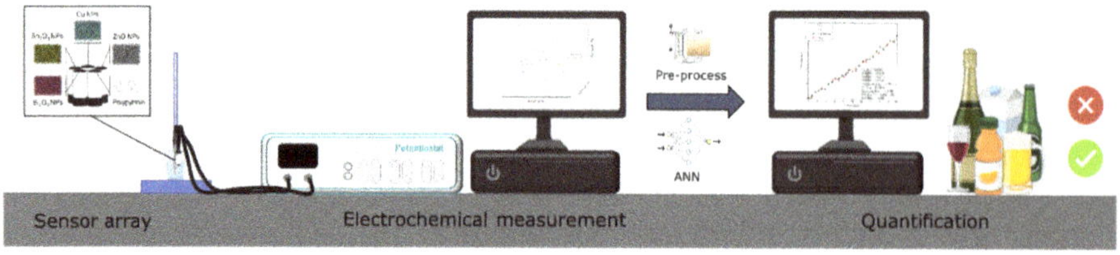

Figure 4.4. Schematic representation of the measuring methodology. The first step is the selection of the sensors, once the final array is chosen the electrochemical measurements to feed the chemometric model are carried out. Finally, the model is developed and used for the quantification in real samples.

For the analysis of the BAs mixtures, a set of samples based on a 3^3 factorial design was prepared. These samples are required to build the chemometric model (train set). Moreover, to assess the performance of the build model, another set of samples with concentrations randomly varying within the experimental domain was also prepared (test set). Under the above-described conditions, all those samples were analysed, obtaining five complete voltammograms for each of the samples.

4.4.2 Compression of voltammetric data

As already stated, before building the quantification model, and especially if ANNs are to be used, a pre-processing step to reduce the high dimensionality of the data is required. This is particularly critical with ANNs as it contributes to preventing the under-determination problem encountered with an oversized ANN model with excessively complex data, while at the same time it significantly reduces the computational requirements and modelling time. Moreover, in general, by avoiding redundancy in the input data, simpler models are generated, which reduces the risk of overfitting, with the consequent improvement in its performance and generalization ability. Furthermore, despite not being so critical for some methods such as PLS, even in such cases, the usage of a compression stage has been demonstrated that can improve the performance of the generated models [8, 48].

Over the next paragraphs, a brief introduction to the fundamentals of some of the more common algorithms employed for the compression of voltammetric data will be presented. The results obtained after applying the different pre-processing methods are summarized in table 4.2. In all the cases, parameters of the algorithms were selected to ensure that practically all the information contained in the original data is preserved, as demonstrated in the case of PCA, FFT and DWT.

Precisely, to quantify the reconstruction degree in the cases where the signals could be reconstructed from the obtained coefficients, two different metrics were used, the coefficient of determination (R^2) and a similarity factor named f_c that compares the area under both signals when superimposed ($f_c = (A \cap B)/(A \cup B)$, where A and B correspond to the two signals, figure 4.5) [49]. Both of them range from 0 to 1, increasing as the similarity between signals do, but the latter being more sensitive to small differences.

Table 4.2. Number of features used as input in the modelling stage and compression ratio for the different pre-processing techniques.

	Number of features	Compression ratio	Explained variance	R^2	f_c
PCA	5 × 4 PCs	99.4%	97.1%		
Integral	5 × 9 coeffs.	98.6%			
FFT	5 × 32 coeffs.	95.1%		0.999	0.969
DWT	5 × 47 coeffs.	92.8%		0.998	0.974
CI	51	98.4%			

CI, causal index; *DWT*, discrete wavelet transform; *FFT*, fast Fourier transform; *Integral*, windowed slicing integral method; *PCA*, principal component analysis.

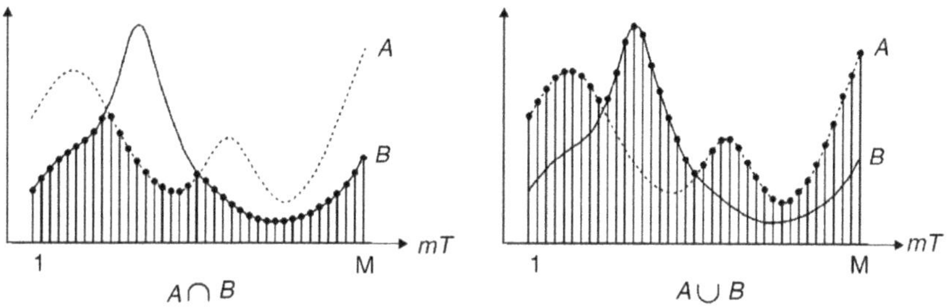

Figure 4.5. Schematic representation of the calculation of f_c index. Reproduced from [49], copyright (2006) with permission from Elsevier.

4.4.2.1 Principal component analysis

Probably one of the simplest and more efficient strategies for data compression (in terms of data reduction) is using PCA as feature extraction tool. PCA allows projecting the majority of the variability contained in the original variables onto a smaller number of orthogonal variables called principal components (PCs), with new coordinates called scores obtained after data transformation (figure 4.6(a)). Such transformation is done so that the first PC takes the direction that preserves as much of the variance of the data as possible, and the second one is then orthogonal to the previous one and captures as much of the remaining variance as possible, and so on.

Thus, by keeping only the first PCs, it is possible to describe the information in the original data set with considerably fewer variables than those originally present. In this manner, the accumulated variance can be used as a criterion to select the number of PCs to be kept for subsequent analysis.

This can be more clearly evidenced if taking the BAs data as an example. As can be seen from figure 4.6(b), where the increase of the accumulated variance (averaged for the different sensors) is plotted against the number of PCs taken, there is a sharp increase at the beginning, allowing to capture ca. 90% of the total variance in the original data with as few as the first two PCs. Nevertheless, to ensure that even better representation of the data is achieved, up to four PCs for each of the sensors were taken so that accumulated variance was above 97%.

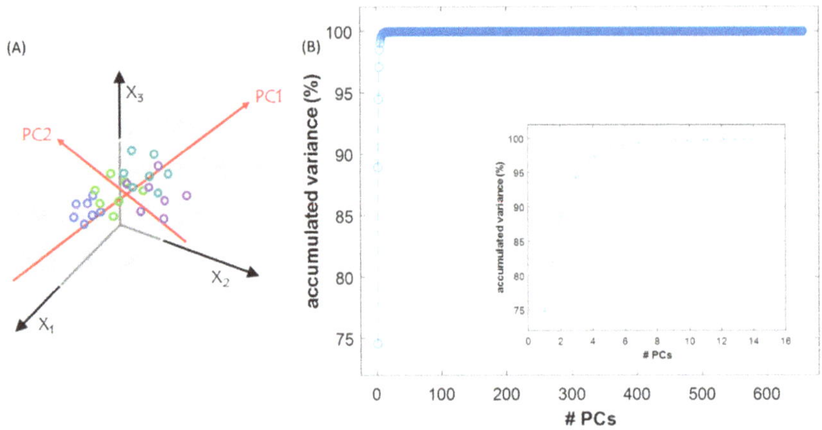

Figure 4.6. (a) Schematic representation of the PCA data transformation. (b) Percentage of explained variance by PCA depending on the number of PCs selected.

4.4.2.2 Windowed slicing integral method

Another simple and computationally efficient approach is to take the calculation of the area under sections of the voltammetric curve as the extracted feature [50]. The idea is to somehow capture information about the shape of the voltammogram; while its effectiveness lies in the fact that it is an approximation to the traditional way of working in cyclic voltammetry, in which oxidation or reduction peaks corresponding to each analyte of interest are identified and either the peak height (current maximum) or the area (total current) is calculated, parameters which are proportional to the concentration of the analyte. However, because in e-tongues we often have overlapping between the individual signals obtained by each analyte, it is not possible to perform a calibration line in a conventional way, and thus the area of each peak is replaced by the area within the voltammetric curve calculated from different sections.

To this aim, the voltammogram is divided into k sections and the area under each section is used as input (figure 4.7). By increasing the number of sections, better representation of the original data might be attained, but at the expense of the compression ratio. Thus, given that the *a priori* selection is arbitrary, this is a parameter that may be optimized during the modelling stage.

In the current case, after iterative variation of the number of sections for each of the different ANN topologies considered, over 4500 models were built, achieving the best performance for $k = 9$. Despite the simplicity of the method, and regardless of the performance that may be achieved with the related build model, it is evident that the optimization of this parameter requires a significant number of calculations in comparison to other approaches as it is not possible to carry an objective selection *a priori*. However, it must be reckoned that this optimization can be neglected at the expense of a slight decrease in the modelling performance, as generally the biggest variation occurs taking small k values, whereas similar performance is observed above a certain value (as shown in [50]). Therefore, taking a relatively large value (e.g., 5–10) is enough to ensure good signal representation and performance of the associated model.

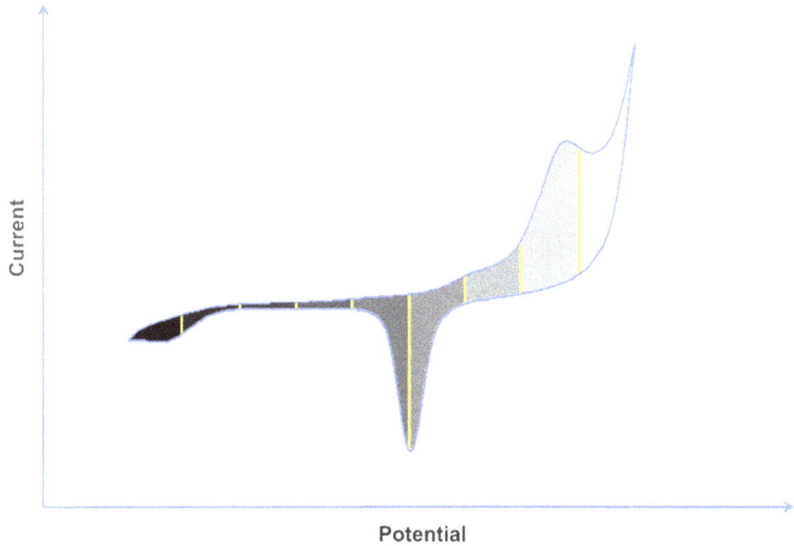

Figure 4.7. Schematic representation of the windowed slicing integral method when $k = 9$.

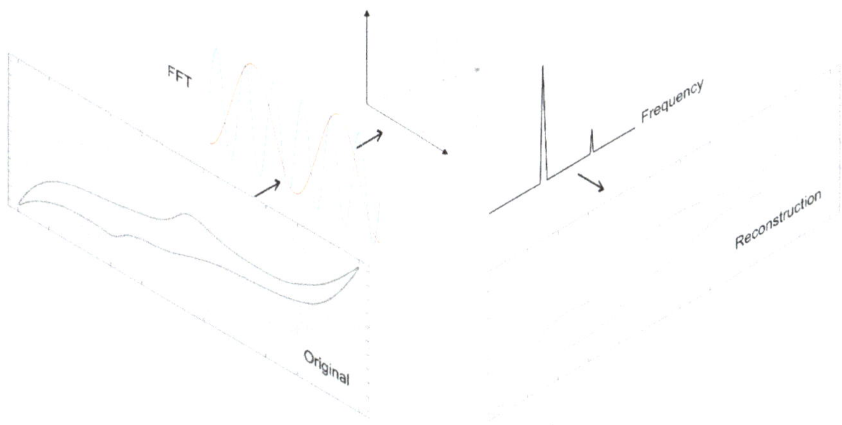

Figure 4.8. Schematic representation of the FT approach.

4.4.2.3 Fourier transforms

A more sophisticated approach is the usage of Fourier-based transforms (FTs), which allow the decomposition of a given signal from time domain into complex-valued function of frequency, whose magnitude represents their presence in the original signal (figure 4.8) [51]. Fourier transform is the term used to refer to different variants of transforms including fast Fourier transform (FFT) and discrete cosine transform (DCT), the two most used variants and the ones that have already been reported for the compression of voltammetric data [41, 52].

In other words, FTs allow us to decompose a given signal into components of different frequencies. The number of coefficients obtained after FT will be equal to

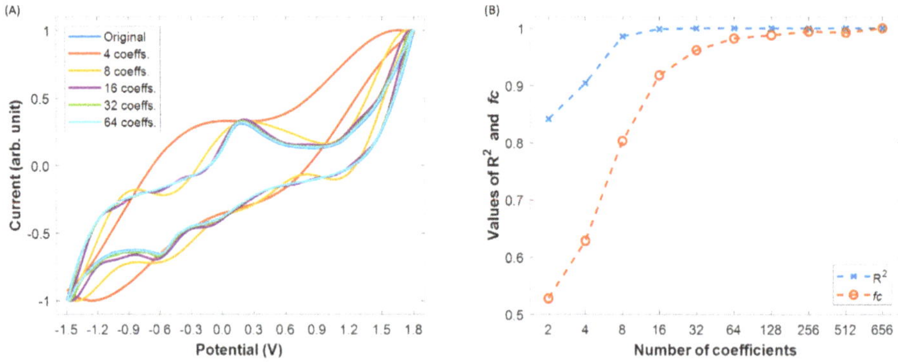

Figure 4.9. Selection of the optimal number of coefficients for signal compression with FFT. Changes in (a) the reconstruction of the voltammetric signal and (b) in the coefficient of determination (R^2, x) and f_c (○).

the number of points in the original signal, but as in the case of PCA, the first coefficients (the lower frequencies) are those preserving the most relevant information. Consequently, compression of the signal without significant loss of information can be attained by preserving only the first coefficients and discarding the rest; what is more, this also enables one to filter the noise present in the signals through the removal of the high frequency contributions, normally related to it.

For the selection of the optimal number of coefficients required, this can be evaluated from the comparison of the original signal with its approximation produced from the inverse transform of the preserved coefficients.

To demonstrate the potential of FFT, the variation of the reconstructed voltammetric signal when taking a different number of coefficients in comparison to the original signal for a given electrode is shown in figure 4.9(a). As can be observed, quite poor representation of the signal is obtained when a reduced number of coefficients is kept, but there is an exponential increase of the similarity between both signals as the number of coefficients is increased (figure 4.9(b)). For example, in this case, a quite good representation is obtained with as few as eight coefficients, and almost indistinguishable signals are obtained when taking up to 32 coefficients.

4.4.2.4 Discrete wavelet transform

Analogously to FTs, wavelet transforms (WTs) decompose the signal into a set of basic functions, obtained by dilations and translations of a unique function called mother wavelet [49]. The main advantage of WT over FT is its temporal resolution, as it captures both frequency and location of an event. In other words, if taking a non-stationary signal, while FTs will only output which frequencies are present, WT will also provide information on when they occur.

Among the different wavelets, a frequently used set of DWTs is the *Daubechies* wavelets. DWT is implemented using Mallat's pyramidal algorithm, which operates over a single discrete signal of length M by decomposing it into orthogonal subspaces of length ca. $M/2$ in each step (figure 4.10(a)), one space retains the

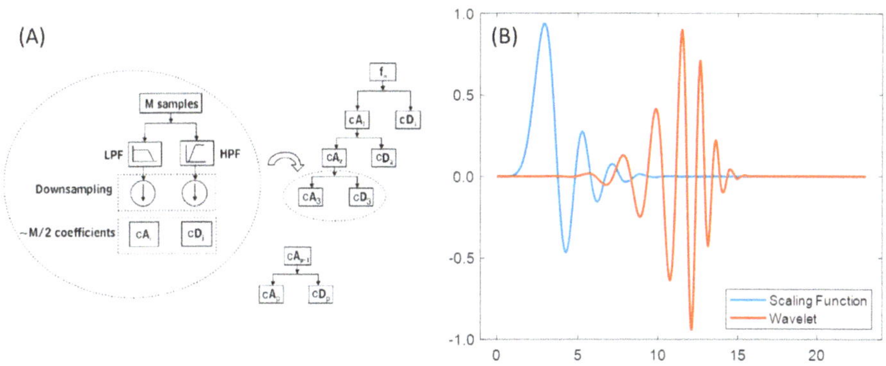

Figure 4.10. (a) Schema of the implementation of DWT by means of Mallat's pyramidal algorithm: (fn) original signal, (cA) approximation coefficients, (cD) detail coefficients, and subscripts denote decomposition level. Reproduced from [49], copyright (2006) with permission from Elsevier. (b) Representation of the Daubechies 4 scaling function and the wavelet used in the explained case.

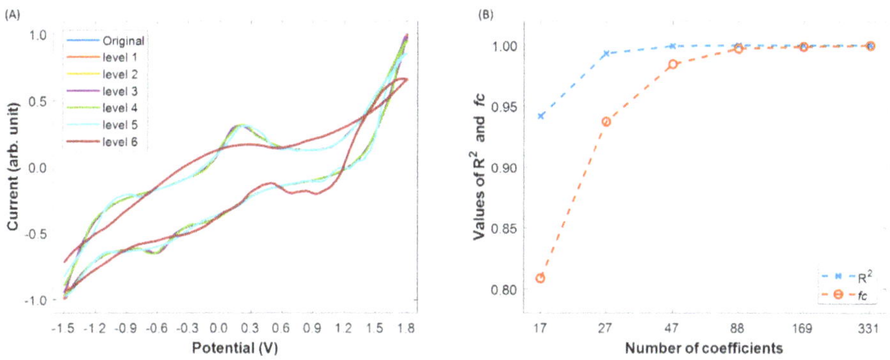

Figure 4.11. Selection of the optimal number of coefficients for signal compression with DWT. Changes in (a) the reconstruction of the voltammetric signal and (b) in the coefficient of determination (R^2, x) and f_c (○).

general traits of the signal, and the other retains the details [49, 53]. In this way, by increasing the decomposition level (i.e., repeating the decomposition process) k times, the degree of signal compression can be increased, but at the expense of signal reconstruction, as the accuracy is degraded at each step.

As happened with FT, WT can be applied for signal compression by keeping only the most significant coefficients, with the same added advantages that the signals might be de-noised during the transformation and that the inverse transform allows one to evaluate the optimal decomposition level to ensure good signal reconstruction.

The changes in the reconstruction of the signal when employing DWT with *Daubechies* mother function (figure 4.10(b)) and different decomposition levels are shown in figure 4.11. Unlike what was done with FFT, where the number of coefficients was increased, in this case, since Mallat's pyramidal algorithm is employed, the number of coefficients decrease as the decomposition level increases, showing a decay in the signal representation when the decomposition level was above 4 (47 coefficients).

4.4.2.5 *Feature selection: genetic algorithms and causal index*

Lastly, we have presented so far feature extraction methods, but another option is the use of feature selection approaches. The main difference between them is that while the latter consists in the selection of a smaller subset of the original variables, the former implies some sort of transformation of the original variables to generate new ones. A wide variety of criteria and methods for feature selection are reported in the literature, but among those, there are two that stand out: causal index (CI) and genetic algorithms (GAs).

CI is based on the usage of the selected modelling tool (generally ANNs) for variable selection [54]. The aim is to locate an optimal set of inputs that can successfully classify or predict the desired outputs by iteratively discarding those that barely contribute (or do not at all) to the prediction success; achieving in this way the dimensionality reduction (figure 4.12(a)). This is a two-step process, in which first a model is built, and then, the focus is directed towards the analysis of the weights that the model has assigned to each of the inputs, which allows the identification of the less-relevant features. These are then discarded (pruned), and this process is iteratively repeated until a near-optimal, small, set of inputs is identified or a significant decrease on the modelling performance is observed. Finally, optimization of any model parameters can proceed, and the actual model is built.

It has to be taken into account that there are two main drawbacks with this approach. The first one is that the minimal sets are not unique and will depend on the model configuration (e.g., if ANNs are used, the number of functions in the hidden layer or the transfer functions selected). The second one is that being an iterative process departing from the whole voltammograms, it is computationally intensive.

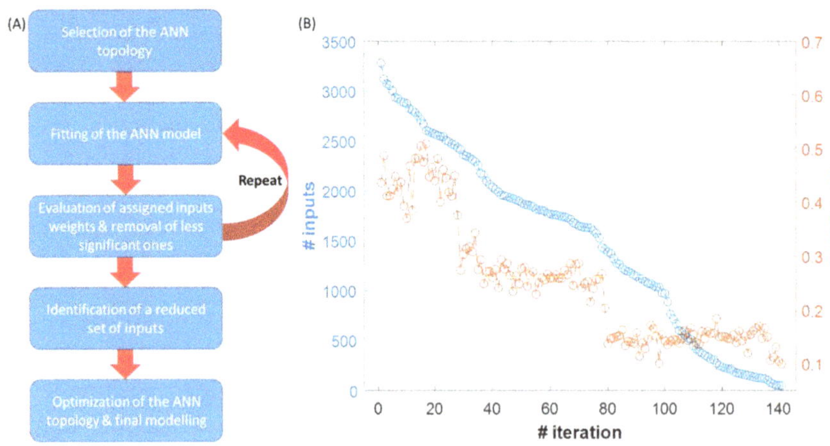

Figure 4.12. (a) Schematic representation of the steps involved in causal index: in each iteration a model is fitted, its RMSE for the testing subset is calculated, and the inputs to which the model has assigned lower weights are discarded. (b) Feature selection and reduction of the number of inputs by means of CI–ANN method.

As an example, the variation of the number of inputs after removal of the less-relevant features and the associated RMSE for each of the iterations is shown in figure 4.12(b). As can be seen, there is an almost steady decrease in the number of inputs after each iteration (although this number is not constant) and a decrease trend in the error (despite some fluctuation is observed). This iterative removal process is usually repeated until enough small sets of inputs are identified, all the relative contributions of the remaining features are too significant for those to be removed or a significant decrease in the modelling performance is obtained.

GAs correspond to one of the most advanced methods for feature selection, and are inspired by Darwin's theory of evolution and the idea of natural selection [55]. GAs are a heuristic optimization approach that involves five main steps: (i) coding of the variables, where each variable (gene) is binary coded into a vector (chromosome); (ii) initiation of the population, where the original chromosomes are perturbed to generate a large enough number of chromosomes; (iii) evaluation of the response, by generating a model for each of the chromosomes in order to obtain a numeric value describing the quality of the model (fitness value); (iv) reproduction, where a new population of chromosomes is created combining the original chromosomes, in what is considered the next generation, and where chromosomes with a higher fitness value have more chances to reproduce; and (v) mutation, to avoid, e.g., having genes that are never selected from the original chromosomes or being 'stuck' in specific conditions (in simpler words, to 'jump' to new regions of the experimental domain). Steps (iii) to (iv) are then repeated until a termination criterion is reached, which can either be a lack of improvement on the modelling, a maximum number of generations, etc. As with CI, the final sets are not unique, and the process can be reinitiated to obtain a new set of selected features.

4.4.3 ANN modelling

Upon compression of the voltammetric data with the desired algorithm, the next step is its modelling. In this case, this was achieved by means of ANNs, and the CI-ANN results will be presented as an example.

The first step when building an ANN model is the optimization of its topology (number of neurons in the hidden layer, transfer functions, etc). This selection consists in a trial-and-error process where several parameters are systematically tuned in order to find the configuration that leads to the best performance. To this aim, different models are built employing the data of the training subset, and its performance evaluated towards the data of the testing subset. This allows one to better assess their accuracy, as that data is not used at all during the modelling stage, which contributes to obtain more unbiased metrics as well as to detect possible over-fitted models.

Upon completion of the optimization, the selected ANN model had 51 neurons (corresponding to the selected features with the CI algorithm) in the input layer, five neurons and *tansig* transfer function in the hidden layer and three neurons (one for each of the analytes) and *purelin* transfer function in the output layer. Next, comparison graphs of predicted versus expected concentrations were built and the

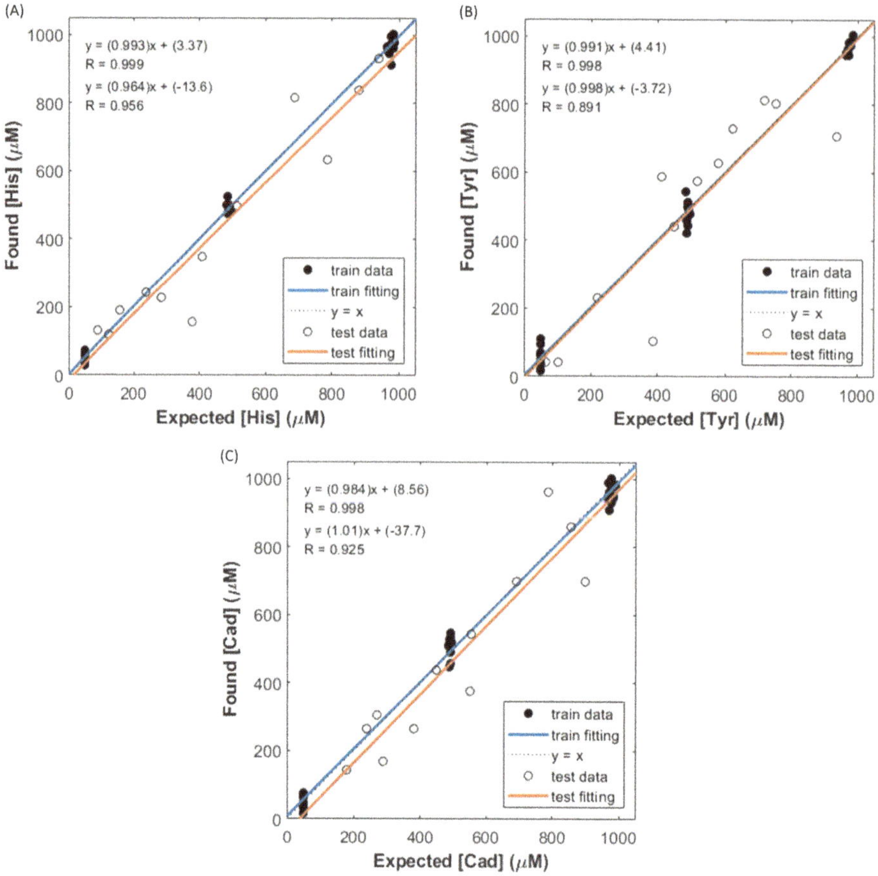

Figure 4.13. Modelling ability of the optimized CI-ANN. Comparison graphs of obtained versus expected concentrations of (a) histamine, (b) tyramine and (c) cadaverine.

linear regressions fitted (figure 4.13). As can be seen, a satisfactory trend was obtained in all cases, with regression lines close to the ideal ones (slope and correlation coefficients close to 1, and intercept values close to 0).

Finally, in order to illustrate the differences of the above discussed algorithms, different models were built for each of them, and the respective RMSEs were calculated to evaluate their performance (figure 4.14). At this stage, it has to be reckoned that these results merely present some general trends, given the optimal method could change depending on the study case; but the aim is not to determine the best algorithm, but to give an overview of them.

In general trends, we can see how all of them give similar performance, thus indicating that simple algorithms such as PCA or Integral are a good point to start with. Next, FFT and DWT provide similar and improved performance compared to the former, as could be expected from such more advanced signal processing methods. Lastly, although CI-ANN provided the best results, its main drawback is the considerably larger time required for the feature selection process. However,

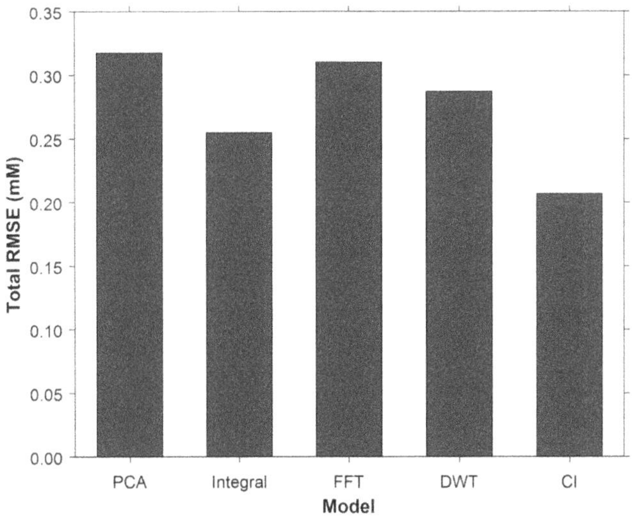

Figure 4.14. Obtained total RMSE values (expressed in mM) for the testing subset and the different models constructed.

this demonstrates the importance that the feature selection/extraction step has in the modelling, and how the removal of the less-significant features contributes to improve the model performance.

4.5 Conclusions and future trends

This chapter aimed to present the historical background of voltammetric e-tongues, the main stages of their development and the basic concepts so that it can serve as an introductory text for new researchers in the field. In this direction, section 4.2 intended to provide a general overview of the fundamentals and state of the art of voltammetric (bio)e-tongues by describing the more common choices in regards to the voltammetric (bio)sensor array and the chemometrics stage. Moreover, in section 4.4 special attention has been given in the presentation of some of the dimensional reduction methods that are commonly reported in the literature, given their importance in voltammetric e-tongue analysis. Finally, to better illustrate this, the detection of BAs by means of a voltammetric e-tongue was taken as an example to illustrate the differences between the several methods as well as the use of a supervised method such as ANNs.

Overall, voltammetric e-tongues represent a complementary alternative approach to potentiometric e-tongues, as voltammetric sensors provide information about the compounds present in the sample that can be oxidized or reduced, whereas potentiometric sensors provide information about the ionic composition of the samples (either cations or anions). Additionally, given the larger versatility of voltammetry depending on the specific technique chosen and the possibility to adjust its parameters conveniently, voltammetric e-tongues offer further advantages in regards to, e.g., sensitivity or the possibility for the simultaneous determination of several analytes from one voltammogram. Lastly, the main difference between both approaches is the high-dimensionality of the generated data, which has the advantage of providing richer data to be fed to the chemometric models, but at the same time is one of their disadvantages as it hinders the data treatment stage.

As future remarks, those can be divided into two different areas, which are actually related to the two main parts that define an e-tongue. On one side, we want to highlight the recent trends in the incorporation of catalysts and/or (bio)receptors aimed to improve the sensitivity and selectivity of the sensor array, which allows us to spawn new application fields. In this sense, some of the modifiers that have been already reported include the use of metallic NPs or other nanomaterials such as graphene or quantum dots (QDs), but also biological materials (which led to the concept of bioe-tongues). Regarding the latter, it has to be reckoned that up to now most of the reported systems were based on the usage of enzymes and, consequently, there is still a wide range of bioreceptors (such as antibodies or DNA) whose use has not yet been exploited. On the other side, that is the use of chemometric methods, the main challenge is the use of multi-way methods which permit one to better exploit the information of the original data. Lastly, special mention has to be given to the use of data fusion approaches, which seek to combine the measurements originated from sensors of very different nature.

Acknowledgments

The authors acknowledge the support of the Spanish Ministry of Science and Innovation (MCINN) through project PID2019-107102RB-C21. MBS thanks the support of the Government of Catalonia Secretariat for Universities and Research of the Ministry of Economy and Knowledge. MdV is grateful for the support from the program ICREA Academia.

References

[1] Skoog D A, West D M, Holler F J and Crouch S R 2014 *Fundamentals of Analytical Chemistry* (Belmont, CA: Cengage Learning)

[2] Persaud K and Dodd G 1982 Analysis of discrimination mechanisms in the mammalian olfactory system using a model nose *Nature* **299** 352–5

[3] del Valle M 2011 Bioinspired sensor systems *Sensors* **11** 10180–6

[4] Otto M and Thomas J D R 1985 Model studies on multiple channel analysis of free magnesium, calcium, sodium, and potassium at physiological concentration levels with ion-selective electrodes *Anal. Chem.* **57** 2647–51

[5] Vlasov Y and Legin A 1998 Non-selective chemical sensors in analytical chemistry: from 'electronic nose' to 'electronic tongue' *Fresenius J. Anal. Chem.* **361** 255–60

[6] Hayashi K, Yamanaka M, Toko K and Yamafuji K 1990 Multichannel taste sensor using lipid membranes *Sens. Actuators B Chem.* **2** 205–13

[7] Winquist F, Wide P and Lundström I 1997 An electronic tongue based on voltammetry *Anal. Chim. Acta* **357** 21–31

[8] Cetó X, Céspedes F and del Valle M 2013 Comparison of methods for the processing of voltammetric electronic tongues data *Microchim. Acta* **180** 319–30

[9] Tønning E *et al* 2005 Chemometric exploration of an amperometric biosensor array for fast determination of wastewater quality *Biosens. Bioelec.* **21** 608–17

[10] Bachmann T T and Schmid R D 1999 A disposable multielectrode biosensor for rapid simultaneous detection of the insecticides paraoxon and carbofuran at high resolution *Anal. Chim. Acta* **401** 95–103

[11] Nelson D L and Cox M M 2017 *Lehninger Principles of Biochemistry* (New York: W. H. Freeman)

[12] Gutiérrez J M, Haddi Z, Amari A, Bouchikhi B, Mimendia A, Cetó X and del Valle M 2013 Hybrid electronic tongue based on multisensor data fusion for discrimination of beers *Sens. Actuators B Chem.* **177** 989–96

[13] Kutyła-Olesiuk A, Zaborowski M, Prokaryn P and Ciosek P 2012 Monitoring of beer fermentation based on hybrid electronic tongue *Bioelectrochemistry* **87** 104–13

[14] Khaleghi B, Khamis A, Karray F O and Razavi S N 2013 Multisensor data fusion: a review of the state-of-the-art *Inform. Fusion* **14** 28–44

[15] Holmin S, Björefors F, Eriksson M, Krantz-Rülcker C and Winquist F 2002 Investigation of electrode materials as sensors in a voltammetric electronic tongue *Electroanalysis* **14** 839–47

[16] Gutiérrez J M, Moreno-Barón L, Pividori M I, Alegret S and del Valle M 2010 A voltammetric electronic tongue made of modified epoxy-graphite electrodes for the qualitative analysis of wine *Microchim. Acta* **169** 261–8

[17] Arrieta A A, Apetrei C, Rodriguez-Mendez M L and de Saja J A 2004 Voltammetric sensor array based on conducting polymer-modified electrodes for the discrimination of liquids *Electrochim. Acta* **49** 4543–51

[18] Paixão T R L C and Bertotti M 2009 Fabrication of disposable voltammetric electronic tongues by using Prussian Blue films electrodeposited onto CD-R gold surfaces and recognition of milk adulteration *Sens. Actuators B Chem.* **137** 266–73

[19] Winquist F 2008 Voltammetric electronic tongues—basic principles and applications *Microchim. Acta* **163** 3–10

[20] Medina-Plaza C, García-Hernández C, de Saja J A, Fernández-Escudero J A, Barajas E, Medrano G, García-Cabezón C, Martin-Pedrosa F and Rodriguez-Mendez M L 2015 The advantages of disposable screen-printed biosensors in a bioelectronic tongue for the analysis of grapes *LWT-Food Sci. Technol.* **62** 940–7

[21] Lange J and Wittmann C 2002 Enzyme sensor array for the determination of biogenic amines in food samples *Anal. Bioanal. Chem.* **372** 276–83

[22] Wang J, Kong S, Chen F, Chen W, Du L, Cai W, Huang L, Wu C and Zhang D-W 2019 A bioelectronic taste sensor based on bioengineered *Escherichia coli* cells combined with ITO-constructed electrochemical sensors *Anal. Chim. Acta* **1079** 73–8

[23] Qin C *et al* 2019 A bioinspired *in vitro* bioelectronic tongue with human T2R38 receptor for high-specificity detection of N-C = S-containing compounds *Talanta* **199** 131–9

[24] Hui G, Mi S, Chen Q and Chen X 2014 Sweet and bitter tastant discrimination from complex chemical mixtures using taste cell-based sensor *Sens. Actuators B Chem.* **192** 361–8

[25] Ebrahimi D, Chow E, Gooding J J and Hibbert D B 2008 Multi-analyte sensing: a chemometrics approach to understanding the merits of electrode arrays versus single electrodes *Analyst* **133** 1090–6

[26] Hayat A and Marty J L 2014 Disposable screen printed electrochemical sensors: tools for environmental monitoring *Sensors* **14** 10432

[27] Taleat Z, Khoshroo A and Mazloum-Ardakani M 2014 Screen-printed electrodes for biosensing: a review (2008–2013) *Microchim. Acta* **181** 865–91
[28] Sarma M, Romero N, Cetó X and del Valle M 2020 Optimization of sensors to be used in a voltammetric electronic tongue based on clustering metrics *Sensors* **20** 4798
[29] Cetó X, Apetrei C, del Valle M and Rodríguez-Méndez M L 2014 Evaluation of red wines antioxidant capacity by means of a voltammetric e-tongue with an optimized sensor array *Electrochim. Acta* **120** 180–6
[30] Giacometti J A, Shimizu F M, Carr O and Oliveira O N 2016 A guiding method to select and reduce the number of sensing units in electronic tongues 2016 IEEE Sensors pp 1–3
[31] Bro R 1997 PARAFAC tutorial and applications *Chemometr. Intell. Lab.* **38** 149–71
[32] Bro R 2006 Review on multiway analysis in chemistry— 2000–5 *Crit. Rev. Anal. Chem.* **36** 279–93
[33] Maggio R M, Muñoz de la Peña A and Olivieri A C 2011 Unfolded partial least-squares with residual quadrilinearization: a new multivariate algorithm for processing five-way data achieving the second-order advantage. Application to fourth-order excitation-emission-kinetic-pH fluorescence analytical data *Chemometr. Intell. Lab.* **109** 178–85
[34] Bro R and Smilde A K 2003 Centering and scaling in component analysis *J. Chemometr.* **17** 16–33
[35] Palit M, Tudu B, Bhattacharyya N, Dutta A, Dutta P K, Jana A, Bandyopadhyay R and Chatterjee A 2010 Comparison of multivariate preprocessing techniques as applied to electronic tongue based pattern classification for black tea *Anal. Chim. Acta* **675** 8–15
[36] Ciosek P and Wroblewski W 2007 Sensor arrays for liquid sensing—electronic tongue systems *Analyst* **132** 963–78
[37] Richards E, Bessant C and Saini S 2002 Multivariate data analysis in electroanalytical chemistry *Electroanalysis* **14** 1533–42
[38] Kumar N, Bansal A, Sarma G S and Rawal R K 2014 Chemometrics tools used in analytical chemistry: an overview *Talanta* **123** 186–99
[39] Bro R 1998 *Multi-way Analysis in the Food Industry. Models, Algorithms and Applications* (Amsterdam: University of Amsterdam)
[40] Despagne F and Massart D L 1998 Neural networks in multivariate calibration *Analyst* **123** 157R–78R
[41] Cetó X, Céspedes F and del Valle M 2012 BioElectronic tongue for the quantification of total polyphenol content in wine *Talanta* **99** 544–51
[42] Garcia-Hernandez C, Garcia-Cabezon C, Martin-Pedrosa F and Rodriguez-Mendez M L 2019 Analysis of musts and wines by means of a bio-electronic tongue based on tyrosinase and glucose oxidase using polypyrrole/gold nanoparticles as the electron mediator *Food Chem.* **289** 751–6
[43] Taylor S L and Eitenmiller R R 1986 Histamine food poisoning: toxicology and clinical aspects *CRC Crit. Rev. Toxicol.* **17** 91–128
[44] Food and Drug Administration 2020 *Fish and Fishery Products Hazards and Controls Guidance* (Gainesville: FL: FDA) pp 113–52
[45] Danquah A O, Benjakul S and Simpson B K 2012 *Food Biochemistry and Food Processing* (Oxford: Wiley-Blackwell) pp 820–32
[46] World Health Organization and Food and Agriculture Organization of the United Nations 2013 *Joint FAO/WHO expert meeting on the public health risks of histamine and other biogenic amines from fish and fishery products* meeting report World Health Organization

[47] Alegret S, Alonso J, Bartroli J, Céspedes F, Martínez-Fàbregas E and del Valle M 1996 Amperometric biosensors based on bulk-modified epoxy graphite biocomposites *Sens. Mater.* **8** 147–54
[48] Palacios-Santander J M, Cubillana-Aguilera L M, Cocchi M, Ulrici A, Naranjo-Rodríguez I, Seeber R and Hidalgo-Hidalgo de Cisneros J L 2008 Multicomponent analysis in the wavelet domain of highly overlapped electrochemical signals: Resolution of quaternary mixtures of chlorophenols using a peg-modified sonogel–carbon electrode *Chemometr. Intell. Lab.* **91** 110–20
[49] Moreno-Barón L, Cartas R, Merkoçi A, Alegret S, del Valle M, Leija L, Hernandez P R and Muñoz R 2006 Application of the wavelet transform coupled with artificial neural networks for quantification purposes in a voltammetric electronic tongue *Sens. Actuators B Chem.* **113** 487–99
[50] Cetó X, Céspedes F and del Valle M 2013 Assessment of individual polyphenol content in beer by means of a voltammetric bioelectronic tongue *Electroanalysis* **25** 68–76
[51] Rader C and Brenner N 1976 A new principle for fast Fourier transformation *IEEE T Acoust. Speech.* **24** 264–6
[52] Cetó X and Pérez S 2020 Voltammetric electronic tongue for vinegar fingerprinting *Talanta* **219** 121253
[53] Mallat S G 1989 A theory for multiresolution signal decomposition—the wavelet representation *IEEE Trans. Pattern Anal. Mach. Intell.* **11** 674–93
[54] Boger Z 2003 Selection of quasi-optimal inputs in chemometrics modeling by artificial neural network analysis *Anal. Chim. Acta* **490** 31–40
[55] Niazi A and Leardi R 2012 Genetic algorithms in chemometrics *J. Chemometr.* **26** 345–51
[56] Cetó X, González-Calabuig A and del Valle M 2015 Use of a bioelectronic tongue for the monitoring of the photodegradation of phenolic compounds *Electroanalysis* **27** 225–33
[57] Medina-Plaza C, de Saja J A and Rodriguez-Mendez M L 2014 Bioelectronic tongue based on lipidic nanostructured layers containing phenol oxidases and lutetium bisphthalocyanine for the analysis of grapes *Biosens. Bioelec.* **57** 276–83
[58] Cetó X, Capdevila J, Mínguez S and del Valle M 2014 Voltammetric bioelectronic tongue for the analysis of phenolic compounds in rosé cava wines *Food Res. Int.* **55** 455–61
[59] Mendes R K, Dantas M V C, Nogueira A B, Etchegaray A, Filgueiras P R and Poppi R J 2018 Simultaneous determination of different phenolic compounds using electrochemical biosensor and multivariate calibration *J. Braz. Chem. Soc.* **29** 482–9
[60] Salamanca-Neto C A R, Marcheafave G G, Scremin J, Barbosa E C M, Camargo P H C, Dekker R F H, Scarminio I S, Barbosa-Dekker A M and Sartori E R 2020 Chemometric-assisted construction of a biosensing device to measure chlorogenic acid content in brewed coffee beverages to discriminate quality *Food Chem.* **315** 126306
[61] Cipri A, Schulz C, Ludwig R, Gorton L and del Valle M 2016 A novel bio-electronic tongue using different cellobiose dehydrogenases to resolve mixtures of various sugars and interfering analytes *Biosens. Bioelec.* **79** 515–21
[62] Serrano N, Prieto-Simón B, Cetó X and del Valle M 2014 Array of peptide-modified electrodes for the simultaneous determination of Pb(II), Cd(II) and Zn(II) *Talanta* **125** 159–66
[63] Garcia-Cabezon C, Gobbi Teixeira G, Dias L G, Salvo-Comino C, García-Hernandez C, Rodriguez-Mendez M L and Martin-Pedrosa F 2020 Analysis of phenolic content in grape seeds and skins by means of a bio-electronic tongue *Sensors* **20** 4176

[64] Novakowski W, Bertotti M and Paixão T R L C 2011 Use of copper and gold electrodes as sensitive elements for fabrication of an electronic tongue: discrimination of wines and whiskies *Microchem. J.* **99** 145–51

[65] Herrera-Chacon A, González-Calabuig A, Campos I and del Valle M 2018 Bioelectronic tongue using MIP sensors for the resolution of volatile phenolic compounds *Sens. Actuators B Chem.* **258** 665–71

[66] Pérez-Ràfols C, Serrano N, Díaz-Cruz J M, Ariño C and Esteban M 2017 A screen-printed voltammetric electronic tongue for the analysis of complex mixtures of metal ions *Sens. Actuators B Chem.* **250** 393–401

[67] Campos I, Alcañiz M, Aguado D, Barat R, Ferrer J, Gil L, Marrakchi M, Martínez-Mañez R, Soto J and Vivancos J-L 2012 A voltammetric electronic tongue as tool for water quality monitoring in wastewater treatment plants *Water Res.* **46** 2605–14

Chapter 5

Voltammetric electronic tongues using sensors based on molecularly imprinted polymers

Anna Herrera-Chacon, Xavier Cetó and Manel del Valle

This chapter explores the use of molecularly imprinted polymers (MIPs) in voltammetric sensors arrays, combined with chemometric tools to generate MIP-based electronic tongues (e-tongues). MIPs are synthetic biomimetic receptors, sometimes referred to as the antibody's artificial counterpart, which can be used to significantly improve the electrochemical sensors' responses due to their high selectivity and preconcentration capabilities. Consequently, as happened through the incorporation of biosensors as part of the e-tongue sensor array, the combination of MIP-based sensors and chemometrics can unleash a new research area in the field of e-tongues. In order to demonstrate their potential, two applications carried out in our laboratories will be presented. The first case will cover the development and application of an MIP for the analysis of histamine (a biogenic amine) in wine samples, while the second case will demonstrate the application of an MIP-based e-tongue for the resolution of volatile phenolic compounds (concretely, 4-ethylphenol and 4-ethylguaiacol).

5.1 Introduction

The application of an electronic tongue (e-tongue) was reported for the first time in 1985, when Otto and Thomas attempted the determination of free metal ions with an array of non-specific ion-selective electrodes (ISEs) [1]. They demonstrated the advantages that could be obtained from the use of partial least squares regression (PLS) in comparison to ordinary least squares regression (OLS).

However, it was not until a few years later that the term and definition of e-tongue were coined [2]. An e-tongue is a biomimetic system applied to the analysis of liquid samples, which consists in the coupling of an array of sensors capable of giving a wide and complete response of the analysed species, plus a chemometric processing tool able to interpret and extract meaningful data from the complex readings (figure 5.1).

doi:10.1088/978-0-7503-3687-1ch5

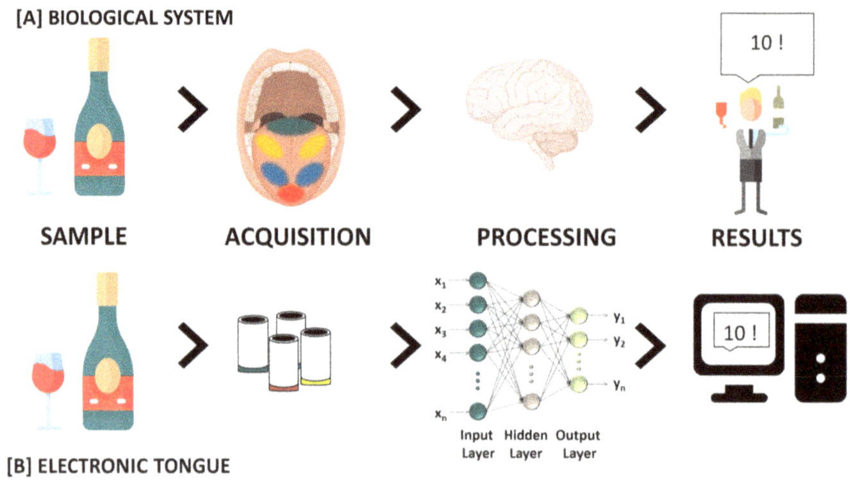

Figure 5.1. Schematic representation of the e-tongue biomimetic approach in comparison to the biological recognition process. Courtesy: Pixabay; © www.gograph.com/Peterhermesfurian; https://www.injurymap.com/free-human-anatomy-illustrations; Flaticon.com.

Initial studies involving the usage of e-tongues were mainly based on the usage of potentiometric sensors, and exploited the potential of chemometric methods to improve the performance of conventional ISEs in quantitative analysis [2]. However, soon after, the application of these systems in qualitative analysis for the recognition and discrimination of complex liquids started to emerge [3]. Such applications demonstrated not only the potential of e-tongues to derive meaning from complex or imprecise data, but also to accomplish novel applications not achievable by conventional approaches, such as mimicking the human taste perception.

Following the advance in e-tongue research, Winquist *et al* reported the application of a voltammetric e-tongue for the first time in 1997 [4]. The proposed system was composed by a double working electrode of gold and platinum, which with the aid of principal component analysis (PCA), allowed the discrimination of different beverages. However, what was really significant is the change in the measuring principle from potentiometry to voltammetry, which removed the limitation of detecting only charged compounds and opened a new range of applications. Furthermore, voltammetric measurements present several advantages such as their high sensitivity, versatility, robustness and tunability [4].

Over the next years, researchers in the field of e-tongues evaluated different types of sensors to further improve the performance and potential of such systems. In this manner, initial studies were based on the usage of different bare metals as working electrodes, but soon after, the usage of chemically modified electrodes was exploited to improve the performance of the former or to tackle the detection of different analytes.

More recently, Tønning *et al* proposed the term bioe-tongue to refer to the combination of one or several biosensors into the e-tongue sensor array, commonly sharing the same transduction principle to facilitate their compatibility [5]. Bioe-tongues exploit the higher selectivity of biosensors to address the lack of selectivity of e-tongues, and the capabilities of e-tongues to extract meaningful data from complex responses, to compensate biosensors' interference problems and/or to allow the multi-determination of several species. In this manner, a number of systems based on such an approach have been reported [6, 7], and despite the advantages that biosensors offer cannot be neglected, they might still present some disadvantages in comparison to 'conventional' e-tongues. These may be related to the potential higher cost of biosensors, their shorter long-term stability and inoperability out of physiological conditions, or the lack of a suitable biological recognition element suitable for any analyte(s) of interest. That being said, it must be remarked that the gain derived from the usage of biosensors surpasses those minor drawbacks, but evidences that there is still room for improvement.

The usage of synthetic receptors analogous in function to natural bioreceptors would be of great interest. The advantages derived from working with such artificial receptors are not only due to their affordability, but also their higher storage and chemical stability, which allows them to be used in harsher conditions (pH, temperature, solvents, etc). In this direction, among the different biomimetic (bio)receptors that have been used for the development of electrochemical sensors, there are two types that have demonstrated a potential above the rest: aptamers and molecularly imprinted polymers (MIPs); but with MIPs still being a cheaper and more versatile option in terms of operating conditions given their higher stability [8].

MIPs are biomimetic materials that are artificially synthesized to mimic molecular biorecognition, which is the basis of most biological processes such as ligand-receptor, substrate–enzyme, antigen–antibody interactions or the translation and transcription of the genetic code. The combination of these biomimetic materials with electrochemical techniques has allowed the development of advanced sensing devices, which significantly improve the performance of bare or catalyst-modified sensors, unleashing new applications [9–11]. However, despite the high selectivity exhibited by MIPs, these can still show some cross-response towards other molecules, especially with chemically analogous structures. Thus, the combination of MIPs with chemometrics opens the room for the development of what could be considered a new type of e-tongues based on their integration into the sensor array [12, 13].

In this sense, the current chapter will provide a brief overview of the concept of MIPs and their synthesis, focusing on their usage in electrochemical sensing. Finally, in order to demonstrate their potential, two applications carried out in our laboratories will also be presented. The first case will cover the development and application of an MIP for the analysis of histamine in wine samples, while the second case will demonstrate the application of an MIP-based e-tongue for the resolution of volatile phenolic compounds mixtures.

5.2 Molecularly imprinted polymers

5.2.1 A brief historical perspective of MIPs

The polymer imprinting technique has evolved a lot from the early concepts that appeared in the 1930s to what is known nowadays, and departs from the ideas of the 1970s when molecular imprinting to organic polymers was introduced. The first example of imprinting was reported by Polyakov in 1931 [14], who observed an unusual selective behaviour of silica particles depending on the different additives and solvents employed during its preparation [15]. At this time, the idea of antibody recognition was under debate by Brenil *et al* and Pauling [16, 17]. In 1949, Dickey reported the precipitation of silica gels in the presence of orange methyl dye, which resulted in a material that showed high selectivity towards the template dye among other different dyes [18]. It was in 1952 when Martin and Synge were Nobel laureated for a study that described the induction of selective footprints using methyl orange as an indicator ligand during the separation of amino acid derivatives [19]. These previous studies were the basis for the foundation of what was published in the 1970s by Wulff *et al* [20] and Klotz *et al* [21] describing that molecular imprinting could be achieved by introducing functional groups in association with the imprinted binding sites; analogous ideas were placed by Sagiv about the concept of silica imprinting [22]. However, it was not until the 1990s that the interest in exploiting MIPs started to grow exponentially, developing many different applications in several fields over different research areas [9, 23, 24]. From those, this chapter will focus on the usage of these receptors as a key element for the modification of electrochemical sensors and their application in e-tongues.

5.2.2 MIPs as recognition element

MIPs are plastic polymers that are designed and synthesized to contain tailor-made cavities selective in shape, size, and with chemical functional groups that favour their binding towards a (bio)molecule of interest (the template, which in analytical chemistry will be our target analyte) (figure 5.2). These biomimetic materials

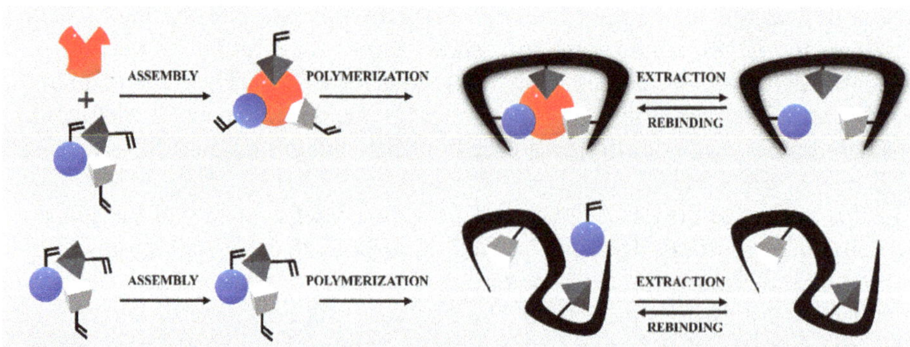

Figure 5.2. Schematic representation of the imprinting technique for the preparation of a (top) MIP and (bottom) a NIP.

emulate the host–guest functionality of their biological counterparts such as antibodies, enzymes or aptamers, which means that the aforementioned cavities provide the polymer with affinity and recognition properties versus the template. Indeed, recognition of the template by the MIP has been compared to Fischer's lock and key analogy used to describe the preferential interaction between an enzyme and its substrate [25].

As already stated, the synthetic nature of these materials allows for superb properties in terms of tunability (to suit almost any type of (bio)molecule), robustness, durability, low-cost, reusability and the possibility to work on a broader range of pH, temperature, etc.

Although MIPs are organic insulator materials, which have been mainly employed in chromatographic techniques, the above-mentioned characteristics have generated huge interest for their application in the sensing field, so that they may be used to improve the sensors' response or performance.

It is important to mention that when the MIP is synthesized, the non-imprinted polymer (NIP) should also be synthesized for comparison purposes. The NIP is obtained in the same conditions as the MIP, but without the addition of the template. Consequently, a polymer with non-specific and non-homogeneous cavities will be synthetized, which should demote its selectivity towards the target analyte. The aim is to demonstrate that there is actually an imprinting effect on the MIP and that its selectivity arises from the specific interaction in the generated cavities after the removal of the template. Moreover, by benchmarking the performance of the MIP against the NIP, the imprinting factor can be evaluated, which provides a measure of the ratio between the specific interactions in the generated cavities to the unspecific interactions with the polymeric material.

5.2.3 Synthesis of MIPs and integration

The imprinting technique has been developed for a diverse range of templates including ions, simple molecules, macromolecules, viruses, micro-organisms and micro- or nano-particles [26]. Several synthetic approaches have been reported since the imprinting technique gained attention in the early 1990s. As mentioned above, the imprinting technique is based on a pre-arrangement of the template and functional monomer(s) in an appropriate solvent, to create a complex through chemical interaction (the pre-polymerization complex). Then, this pre-arrangement is followed by a polymerization step where a polymer is knitted around this complex, commonly by the addition of a cross-linker and an initiator (in an inert atmosphere). Finally, the template is removed to generate the cavities tailored to the template molecule, which may be done in a simpler or more complex way depending on the nature of the previous interactions.

There are several synthetic approaches reported in the literature for polymer imprinting which, depending on the final application, will lead us to the choice of the preferable one as each has its advantages and disadvantages. Similarly, there are different approaches for the immobilization of MIPs onto the electrode, which is also a crucial step as this will have a direct impact on the sensor's response. For this reason, both aspects will be discussed hereunder.

As for the imprinting, Wulff and Sharan proposed the covalent approach [21, 27] that is based on the derivatization of the template to obtain a covalent binding with the functional monomer(s). The process creates precisely tailor-made cavities, but the removal of the template requires harsher conditions such as the use of concentrated acid or bases, among other processes, which still might not be enough for the total template removal. Moreover, this creates a material with slow kinetics. Consequently, this approach is not very suitable for sensing purposes.

Another approach was disseminated and promoted by Mosbach's laboratory in the 1980s, which is known as non-covalent imprinting [28, 29]. Unlike the previous one, the template-functional monomer(s) complex and the polymerization step are governed by weak interactions, e.g., hydrogen bonding, π-stacking, van der Waals or electrostatic interactions, between others. These weaker interactions lead to the creation of heterogenous tailor-made cavities (less specific than those generated with covalent imprinting), but at the expense of better rebinding kinetics and easier template removal.

These two approaches represent the two main milestones in the dissemination of the imprinting technique, which skyrocketed their application. From those, the possibility of obtaining imprinted polymers with molecular recognition has been feasible with different polymerization techniques such as bulk, precipitation, suspension, emulsion, core–shell or through films [12].

For example, a common approach for the synthesis of MIPs is the use of non-covalent imprinting with free-radical polymerization and using a thermal radical initiator (such as azobisisobutyronitrile, AIBN), which is the driving force to react an acrylic-based monomer (that is previously attached to the template) with ethylene glycol dimethacrylate (EGDMA) as cross-linker in an appropriate solvent under oxygen free conditions [30, 31].

Next, the synthesized MIP particles have to be immobilized onto the electrode if an electrochemical sensor is to be obtained. In fact, the immobilization of these artificial receptors onto the electrode surface is one of the key steps due to the insulator nature of MIPs that might hinder obtaining an adequate electrochemical response. The aim is to obtain a thin robust layer, without steric hindrance or fouling effects. The difficulty in doing so is one of the reasons that has been related with the slow development of MIP-based sensors [23]. However, nowadays, a plethora of options are reported in the literature, the main ones being: the incorporation of the MIP particles into a conductive paste electrode by mixing before its preparation; the immobilization of a previously prepared MIP or MIP-based composite with conductive nanomaterials onto the transducer surface (usually by drop-, spray- or spin-coating); or the entrapment of the MIP particles through electropolymerization [10].

More recently, electropolymerization has arisen as a simple one-step approach to directly obtain MIP films onto the electrode surface [9, 32]. For obvious reasons, this option has been widely employed in the field of electrochemistry. The main advantages of electropolymerization in comparison to the previously described methods are the relative simplicity of the polymerization process that involves both the synthesis and integration with the transducer. Moreover, there is the possibility of easily tuning and controlling the polymer growth and morphology by changing the electrochemical parameters, and additionally, there is also the ease of removal of the template molecule via oxidation/reduction processes [32].

5.3 MIP-based electrochemical sensor arrays

In order to illustrate the potential of MIPs in combination with voltammetric sensors and the benefits that could be attained when they are incorporated into the sensor arrays of e-tongues, two applications carried out in our laboratories are going to be presented hereafter. We will cover in depth the main steps to be considered when developing an MIP-based e-tongue, going from the MIP synthesis and characterization to the final application involving qualitative and quantitative tasks and the use of chemometric methods.

The first case is the selective determination of histamine in wine samples, evaluating the response to possible interferents using PCA and comparing the MIP-based approach against a reference method [33]. In the second case, the development and application of an MIP-based e-tongue for the simultaneous determination of two volatile phenols (VPs; namely, 4-ethylphenol (4-EP) and 4-ethylguaiacol (4-EG)) will be also demonstrated [34].

For the preparation of the MIP-based sensor arrays, a non-covalent imprinting approach was chosen, obtaining polymer microparticles (beads). This strategy was chosen as it allowed the decoupling of the synthesis and the immobilization process onto the electrode surface, thus permitting the optimization of each step separately.

For the immobilization of the obtained polymer beads, the sol–gel method was used, which allows the incorporation of different materials into a formed inorganic framework [35]. In this procedure, an alkoxide precursor (commonly tetraethyl orthosilicate, TEOS) is first hydrolysed (sol) and then polycondensated before drying (gel). This procedure is widely used in the modification of chemical sensors due to the possibility to obtain thin porous film layers, with their porosity adapted to the application.

5.3.1 MIP synthesis

Histamine MIP in beads form was synthesized following the more common technique reported in the literature, that is, non-covalent imprinting with methacrylic acid (MAA) as the functional monomer. Briefly, the synthesis starts by mixing histamine dihydrochloride (the template) with MAA (the functional monomer) in ethanol to allow the formation of the pre-polymerization complex, followed by the addition of ethylene dimethacrylate (EGDMA, the cross-linker) and 2.2'azobis(2,4-dimethylvaleronitrile) (AIVN, the initiator) and immediately purging with nitrogen. The polymerization was done in a thermostatically controlled water bath under magnetic stirring. Finally, the extraction of the template was done with a Soxhlet reflux system using a methanol:acetic acid (9:1) solution, after which the polymer particles were dried.

In the case of MIPs for VPs, divinylbenzene (DVB) was used as the functional monomer instead of MAA in order to promote hydrogen bonds and π–π stacking interactions, as otherwise the only other relevant feature of such molecules to be used for its imprinting is an alcohol group. This was the only difference in comparison to the above-mentioned protocol, which had very limited imprinting effect. In this manner, the solution containing the template (either 4-EP or 4-EG) and DVB (the monomer) was first stirred at low temperature. Then, EGDMA and AIVN were added into the solution, followed by the removal of oxygen through nitrogen purging. Finally, the polymerization and the template removal were done as already described.

5.3.2 Preparation of the sensors arrays

Graphite-epoxy composites (GECs) were employed as the transducer element of the sensing platform [33, 34]. For its preparation, first, a 5 mm copper disc is soldered onto an electrical connector, and introduced into a 6 mm inner diameter PVC tube. Next, an epoxy resin is mixed with graphite powder and allocated into the tube, which is cured at 40 °C for 48 h, providing afterwards a conductive and rigid material. Finally, those are polished with sandpaper in order to obtain a mirror-like surface.

In both study cases, the same immobilization procedure was performed. Sol–gel was prepared by mixing vigorously TEOS, ethanol, water and hydrochloric acid, then rested to reach the syneresis stage, and finally mixed with graphite and the polymer (either the MIP or the NIP). Lastly, the resulting solution was spin-coated onto the electrode surface and dried overnight at 4 °C (figure 5.3).

In this manner, for the determination of histamine, the array of sensors was formed by three working GEC electrodes: one modified with the histamine MIP, one modified with the NIP and a bare electrode. Similarly, for the determination of the VPs, the sensor array included five GEC electrodes: one bare electrode plus four of them modified with 4-EP-MIP, 4-EG-MIP, NIP and the sol–gel (only with graphite, but without MIP/NIP particles).

5.3.3 MIP characterization

Morphological characterization of the synthesized MIPs was done by scanning electron microscopy (SEM), both before and after the modification of the electrodes.

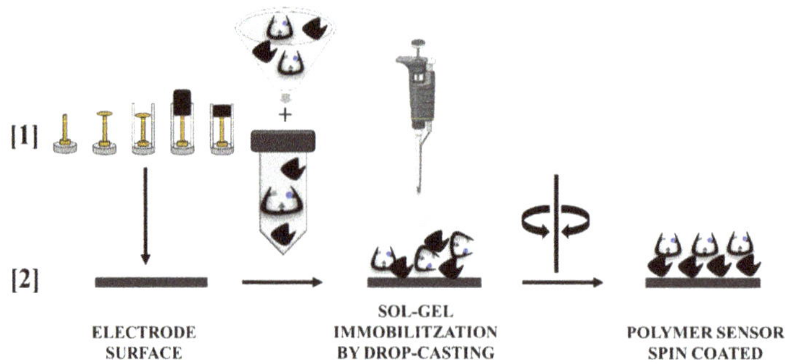

Figure 5.3. [1] Schematic of the preparation of the GECs. [2] Sol–gel immobilization process by drop-casting or spin coating.

In this manner, the shape and average size of the particles were inferred from the images of the synthesized MIP particles, whereas images after their immobilization allowed one to confirm their fixation and proper distribution along the electrode surface. In addition to this, confocal microscopy was used to characterize and compare the complexing ability of the MIP and NIP particles.

5.3.4 Voltammetric analysis

Differential pulse voltammetry (DPV) measurements were carried out at room temperature under quiescent conditions in a multichannel configuration. Prior to proceeding with the actual measurement, an enrichment step was performed in which the sensors were dipped in the solution, but without applying any polarization potential. The aim of this step was to allow enough time for the rebinding of the analyte into the MIP cavities. Upon completion of the measurement, no physical regeneration of the electrodes was required, but only the use of an electrochemical cleaning stage between each measurement in order to prevent any accumulative effect. Such a simple step was enough to minimize any fouling of the electrode or MIP and recover the original baseline.

5.3.5 Chemometric analysis

Chemometric analysis was done in Matlab R2016b using its Statistical, Wavelet, and Neural Network toolboxes by specific routines written by the authors [33, 34]. Briefly, principal component analysis (PCA) was used to identify initial patterns in the data and to assess the potential of the sensor array to discriminate between the different compounds. In contrast, artificial neural networks (ANNs) were used to build the quantitative model that allowed the individual quantification of the different compounds. Discrete wavelet transform (DWT) was used to reduce the large dimensionality and complexity of the generated data, which allows gaining advantages in training time, avoiding redundancy in input data, and obtaining a model with better generalization ability [36].

5.3.6 Analysis of histamine in wine samples

Histamine is a nitrogenous compound belonging to the class of biogenic amines (BAs), formed mainly by microbial decarboxylation of their precursor amino acids. Thus, they can be related to inadequate handling or storage practices, and actually its level is commonly used as an indicator of improper preservation or poor sanitation [37]. Moreover, although low concentrations of BAs such as histamine and tyramine may be beneficial (e.g., histamine is involved in local immune responses), ingestion of higher concentrations can be detrimental to health, especially in the presence of ethanol [37]. Consequently, the development of rapid and low-cost methods for the determination of histamine is of utmost importance.

Upon completion of the synthesis of the MIP and NIP particles as above described, the next step was its characterization by SEM imaging (figure 5.4). Non-regular spherical-shaped beads were obtained, without significant observable differences between both polymers at first sight. A more detailed analysis revealed an

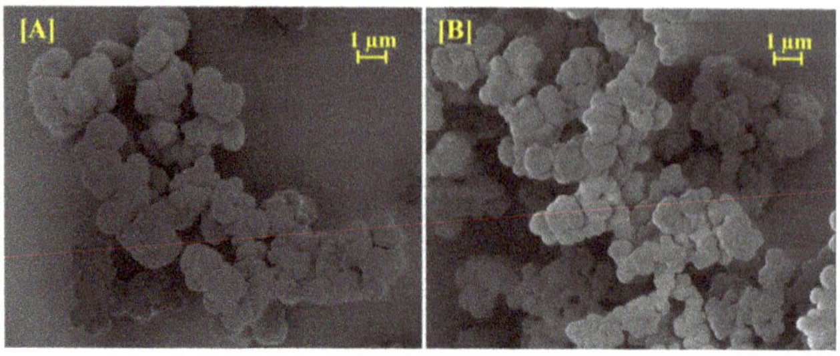

Figure 5.4. SEM images of the (a) MIP and (b) NIP beads.

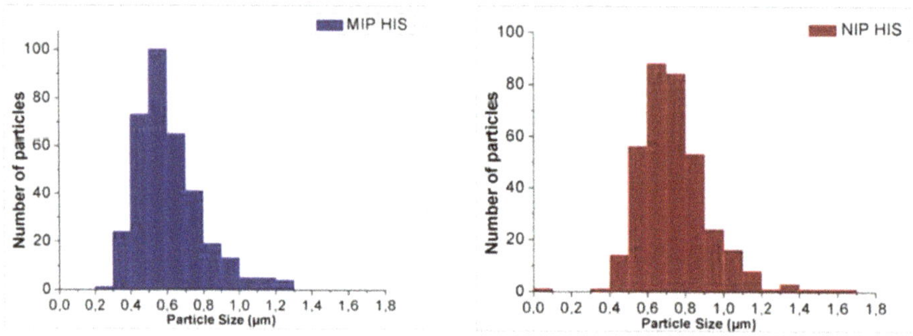

Figure 5.5. Histograms corresponding to the measured bead size from the SEM images for the (a) MIP and (b) NIP beads. Reproduced from [33], copyright (2020) with permission from Elsevier.

Figure 5.6. SEM images of the (a) MIP beads and (b) NIP beads immobilized via sol–gel and (c) bare graphite–epoxy composite (GEC) electrode. Reproduced from [33], copyright (2020) with permission from Elsevier.

average size of 0.61±0.18 μm and 0.74±0.19 μm for the MIP and NIP particles, respectively (figure 5.5).

Next, to confirm the immobilization of the particles on top of the electrode by the sol–gel method, SEM imaging was repeated (figure 5.6). From the comparison of (c) with (b) and (a), it is clear that the surface has been modified. While the former seems relatively flat and homogeneous, the latter two seem to have some large particles that can be attributed to the MIP and graphite. Indeed, by examining the images in more detail, some spherical shapes can still be seen (although not straightforward as they are now covered with sol–gel) that resemble the particle aggregates shown in figure 5.4. Thus, confirming the suitability of the approach to integrate the particles along with the transducer.

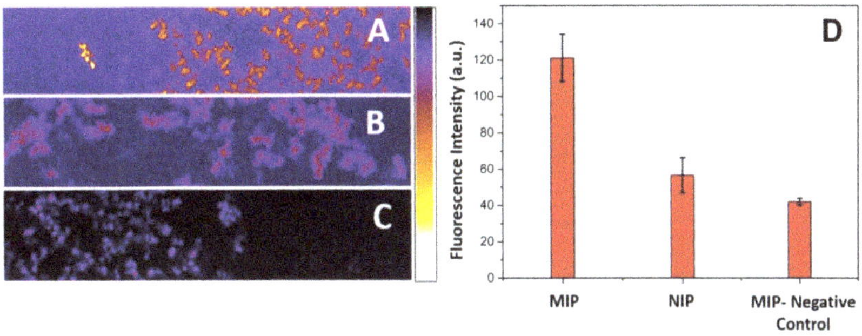

Figure 5.7. Confocal microscopy of (a) MIP beads and (b) NIP beads after incubation with histamine and reaction with OPA. (c) Background fluorescence of MIPs without the incubation with histamine as negative control. (d) Fluorescence intensity (a.u.). Reproduced from [33], copyright (2020) with permission from Elsevier.

As the only difference between the MIP and NIP are the cavities generated during the imprinting of histamine, it could be expected that no significant morphological differences are observed from the SEM images. Thus, to characterize the differences in affinity towards the template between both, confocal microscopy was used. To this aim, the complex formed upon reaction of *o*-phtaldialdehyde (OPA) with histamine was used as the fluorescent dye. OPA reacts specifically with primary amines, yielding a maximum fluorescence at 455 nm upon excitation at 340 nm [38].

In this manner, the MIP and NIP particles were incubated with histamine to allow the binding, followed by OPA addition generating the fluorescent OPA-histamine complex *in situ*, which allowed visualization of the interaction of the polymer with histamine. As shown in figure 5.7, the MIP exhibited a much higher intensity than the NIP, confirming the imprinting of histamine in the polymeric material. That is, the capture is a result of the interaction at the specific cavities generated and not due to unspecific interaction with the polymeric material, although the latter will always be present to some extent. Moreover, to further confirm the results of this assay, a negative control experiment in which no histamine was added was also run (figure 5.7(c)), showing just some residual fluorescence. Lastly, the numerical comparison of the fluorescence intensities between the MIP and NIP, revealed 2.15 times higher readings for the MIP, a significant value that could be taken as the imprinting factor.

The next step was to optimize the experimental conditions under which the electrochemical measurements are to be carried out to maximize the sensor response. Although the details about the optimization will not be discussed for brevity, its importance has to be highlighted as such parameters will have a significant impact on the final result. Those include the measuring buffer and its pH (given the basic nature of histamine with two different pKa), the incubation time to allow the rebinding of histamine into the MIP cavities, the cleaning of the electrodes upon measurement, etc.

Under the optimized conditions, the analytical performance of the developed MIP-based sensor was characterized in terms of linearity, sensitivity, limit of

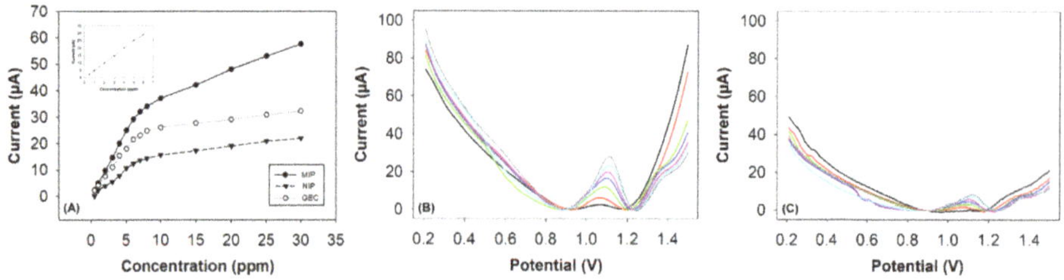

Figure 5.8. Adsorptive DPV response of the developed sensors against histamine in phosphate buffer at pH 7.0. (a) Generic calibration range examined with the inset of the low concentration linear range; (b) obtained voltammograms for the MIP sensors; (c) obtained voltammograms for the NIP sensors. Reproduced from [33], copyright (2020) with permission from Elsevier.

detection (LOD), repeatability, etc. From the calibration plot (figure 5.8(a)) the sensitivity in the linear range from 0.5 to 6.0 ppm (2.7 to 32.9 μM, $R^2 = 0.9992$) was calculated as 4.95 μA/ppm (40.90 μA/μM), with LOD and LOQ values of 0.19 ppm (1.0 μM) and 0.62 ppm (3.4 μM), respectively. Repeatability (intra day) was calculated using the same electrode for twelve repetitive measurements of a 5 ppm (27.4 μM) histamine solution, whereas reproducibility (inter-day) was estimated on three different days from the variation of the sensitivity of the same electrode. Repeatability and reproducibility values, expressed as RSD %, were 4.65 and 2.61, respectively.

In the same manner, the calibration plot for the NIP-based sensor and a bare GEC electrode were built for comparison purposes (figure 5.8(a)). The voltammetric responses of the NIP were clearly smaller (figure 5.8(c)), which was attributed to the lack of binding sites and the blockage of the surface by the NIP particles and the sol–gel (both non-conducting materials). Again, from the comparison of the sensitivities of the MIP to NIP, a factor of 2.27 was obtained; a value in agreement with the previous factor calculated from the confocal imaging, confirming an imprinting factor of ca. 2.2. For the GEC electrode, the recorded currents were between those of the MIP and the NIP, which agrees with the hypothesis that the MIP actually provides a pre-concentration effect, whereas the NIP acts oppositely.

The ability of the MIP to discriminate histamine from other compounds was evaluated by PCA from the calibration plots built for potential interfering compounds in the determination of histamine present in wine. The compounds considered were tyramine (another common BA), gallic acid (one of the major phenolic compounds found in wine), and *p*-coumaric acid (another common phenolic compound and relatively analogous to tyramine).

The obtained scores plot is shown in figure 5.9, from which it can be seen how clear clusters are obtained for each of the considered compounds. On the left of the plot, there are the clusters corresponding to tyramine and *p*-coumaric acid, with similar values for PC1 and being distinguished only by PC2; a fact that can be explained due to the similarities between both molecules. On the top of the plot, there is the cluster corresponding to gallic acid, whereas the cluster corresponding to histamine appears on the right side. Especially significant is the fact that the

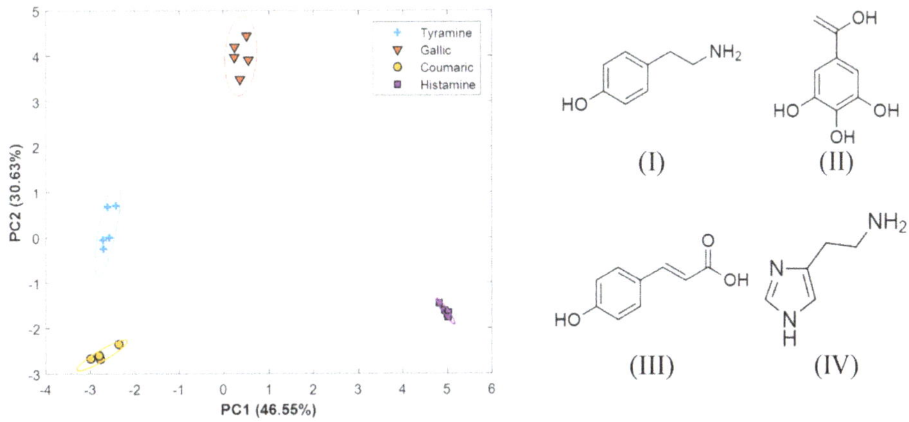

Figure 5.9. (Left) scores plot obtained from the voltammetric responses of the MIP-modified sensor for different concentration solutions of (+) tyramine, (▾) gallic acid, (●) *p*-coumaric acid and (■) histamine. Ellipses plotted correspond to 95% confidence limits for each of the clusters. (Right) chemical structure of the different compounds: (I) tyramine, (II) gallic acid, (III) *p*-coumaric acid and (IV) histamine.

Table 5.1. Results obtained for the determination of histamine (expressed in ppm) for two wine samples using the developed MIP sensor, in comparison with the reference fluorimetric method. Confidence intervals calculated at 95% confidence level.

	Fluorimetric	Voltammetric
Wine 1	0.037 ± 0.004	0.042 ± 0.010
Wine 2	0.287 ± 0.019	0.241 ± 0.025

latter is the cluster that is further from any other, as it has to be kept in mind that PC1 is the PC that holds the largest variance, thus, meaning that histamine samples are identified as much more different from any other of the compounds by the MIP.

Finally, after completing the morphological, affinity, and analytical characterization of the developed MIP, it was applied for the determination of histamine in two different wine samples. To validate the results obtained with the voltammetric approach, the same samples were analysed with the fluorimetric reference method based on the reaction of histamine with OPA [39]. The obtained values are summarized in table 5.1, where a good agreement between both methods can be seen. Thus, demonstrating the potential of the MIP-based approach to carry out the selective determination of histamine in such a complex matrix as it is the case of wine. In this regard, more significant is the performance of the MIP-based sensor if taking into account that the reference method requires an extraction step to minimize the matrix effect, whereas any pre-treatment stage was not required for the former.

5.3.7 Analysis of volatile phenols

The next application that will be presented herein deals with the determination of 4-ethylphenol (4-EP) and 4-ethylguaiacol (4-EG), two of the most common VPs found in wine [40]. VPs are produced by microbial decarboxylation of phenolic acids present in wine (such as ferulic and *p*-coumaric acids) during the fermentation, and are usually associated with *Brettanomyces* infection [37]. Depending on the concentration of VPs in wine, those can contribute positively or negatively to its aroma. However, their sensory threshold may vary depending on the wine, e.g., values above 368–425 ppb for 4-EP have been already reported to have a negative impact on the quality of many wines [41, 42].

The common approach for its determination is based on headspace-solid phase microextraction (HS-SPME) combined with gas chromatography-mass spectrometry (GC-MS). However, this analysis turns out to be expensive, time-consuming and requires high-cost equipment and qualified personnel. Alternatively, its simultaneous multi-determination by making use of a MIP-based e-tongue is proposed herein, in what represents a common application of e-tongues, but with the added advantage derived from the incorporation of MIPs.

Similar to what has been discussed in the previous section, upon the synthesis of the MIPs, those were characterized both by SEM as well as electrochemically. However, for brevity, we will focus on the part relevant to the e-tongue application.

An important feature of e-tongues derived from their definition is that sensors should show cross-response features. Thus, cross-calibration curves for both MIPs and the NIP were built (table 5.2). As already seen with the previous application, the NIP-modified sensor is the one giving the smallest response, with each MIP showing the highest response towards the respective template (as it acts as a pre-concentration element), but although they are aimed to be highly selective, still showing a significant degree of cross-reactivity, which is highly desirable for developing an e-tongue application.

After this initial confirmation, the next step was to assess if the sensors were able to discriminate the considered VPs from other compounds that may be present in wine, such as phenolic compounds (from which VPs are formed). To this aim, 25 ppm stock solutions of gallic acid and quercetin, as well as 3 ppm solutions of 4-EP and 4-EG, were measured with the proposed sensor array (figure 5.10(a)), and the obtained responses submitted to PCA. The obtained scores plot is shown in figure 5.10(b), where separate clusters are obtained for each of the considered compounds.

Table 5.2. Sensitivity values (μA/μM) derived from the cross-calibration curves for 4-EP and 4-EG for the MIPs, the NIP and bare sol–gel electrodes.

Compound	4-EP MIP	4-EG MIP	NIP	Sol–gel
4-EP	33.0	29.9	11.5	28.2
4-EG	33.0	61.0	10.5	45.5

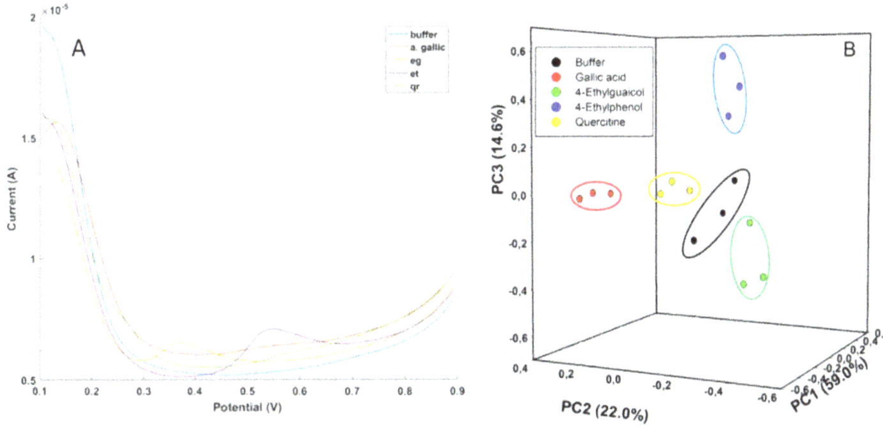

Figure 5.10. (a) DPV signals obtained for the considered compounds with the MIP (solid line) and GEC (dashed line) sensors and (b) scores plot for the three first principal components for the analytes 4-EP, 4-EG, and two polyphenols present in wine, quercetin and gallic acid. Reproduced from [34], copyright (2018) with permission from Elsevier.

Finally, the last step was the simultaneous quantification of VPs mixtures. The set of samples required to build the chemometric model were prepared based on a 4^2 full factorial design within the range 3 to 20 ppm (16 samples, training subset), while an extra subset of nine samples (testing subset) with concentrations within the experimental domain defined by the former subset of samples were also prepared.

For the modelling of the data, ANNs were chosen as the chemometric method. Consequently, prior to the modelling stage, a data compression stage was required, given the high complexity and dimensionality of the input information. This step contributes to preventing the under-determination problem encountered with an oversized ANN with excessively complex data, while at the same time it reduces the computational requirements and modelling time significantly. Moreover, in general, by avoiding redundancy in the input data, simpler models are generated, which reduces the risk of overfitting, with the consequent improvement in its performance and generalization ability. In this case, this was achieved by means of DWT, employing the mother function Daubechies 3 and a third decomposition level, achieving a compression ratio of ca. 85.4% without any loss of significant information.

After optimization of the ANN architecture by iteratively changing the different model parameters, the ANN model had 144 neurons (corresponding to the DWT coefficients) in the input layer, four neurons and *tansig* transfer function in the hidden layer, and two neurons (one for each of the analytes) and *purelin* transfer function in the output layer.

Next, the comparison graphs of predicted versus expected concentrations were built and the linear regressions fitted (figure 5.11). As can be seen, a satisfactory trend was obtained in all the cases, with regression lines close to the ideal ones (slope and correlation coefficients close to 1, and intercept values close to 0).

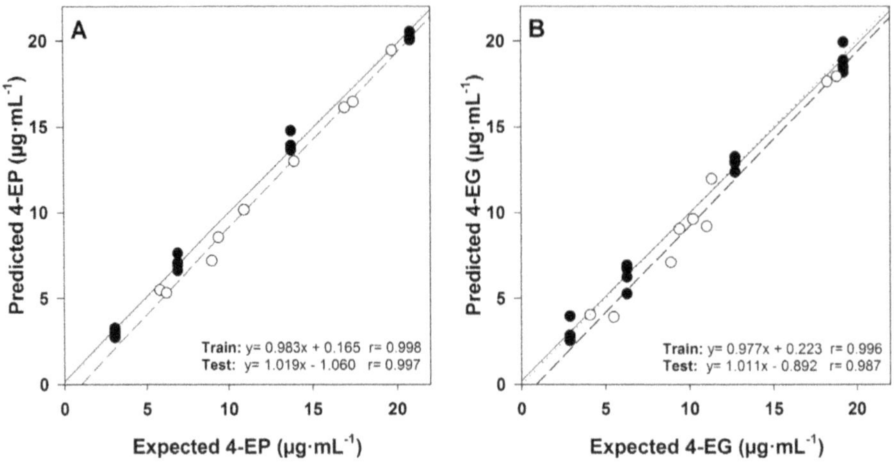

Figure 5.11. Comparison graphs of obtained versus expected concentrations for (a) 4-EP and (b) 4-EG, both for training (●, solid line) and testing subsets (○, dashed line). The dotted line corresponds to ideal comparison line ($y = x$). Reproduced from [34], copyright (2018) with permission from Elsevier.

5.4 Conclusions

This chapter aimed to present the potential of MIPs as biological receptors counterparts for the development of electrochemical MIP-based sensors that may be later incorporated into e-tongues. As such, MIPs can counterbalance some of the biosensors' drawbacks such as their potential higher cost, their long-term stability and inoperability out of physiological conditions, allowing them to be used in harsher conditions (pH, temperature, solvents, etc). However, despite the high selectivity that MIPs exhibit, those can still show some cross-response towards other molecules, as has been demonstrated herein. Thus, the combination of MIPs with chemometric methods opens the room for the development of what could be considered a new type of e-tongues based on its usage.

In this direction, e-tongue approach based on the usage of a sensor array may resolve MIP selectivity issues, a point normally tackled by chromatographic approaches, in a more direct and simple way. Interesting advantages of this coupling can be the gain in sensitivity and selectivity (derived from the MIP ability to bind and preconcetrate the target analyte), plus providing a generic approach for resolving mixtures of related compounds, with the unique condition of being electroactive (unless indirect measurements such as the usage of a redox probe to follow the pore blockage are considered).

In order to illustrate these ideas, two different applications carried out in our laboratories were presented and used to cover the main steps to be followed when working with MIPs.

Acknowledgments

The authors acknowledge the support of the Spanish Ministry of Science and Innovation (MCINN) through project PID2019-107102RB-C21. AHC is grateful for the support of the project CTQ-2016–80170-P. MdV is grateful for the support from the program ICREA Academia.

References

[1] Otto M and Thomas J D R 1985 Model studies on multiple channel analysis of free magnesium, calcium, sodium, and potassium at physiological concentration levels with ion-selective electrodes *Anal. Chem.* **57** 2647–51

[2] Vlasov Y and Legin A 1998 Non-selective chemical sensors in analytical chemistry: from 'electronic nose' to 'electronic tongue' *Fresenius J. Anal. Chem.* **361** 255–60

[3] Hayashi K, Yamanaka M, Toko K and Yamafuji K 1990 Multichannel taste sensor using lipid membranes *Sens. Actuators B Chem.* **2** 205–13

[4] Winquist F, Wide P and Lundström I 1997 An electronic tongue based on voltammetry *Anal. Chim. Acta* **357** 21–31

[5] Tønning E *et al* 2005 Chemometric exploration of an amperometric biosensor array for fast determination of wastewater quality *Biosens. Bioelec.* **21** 608–17

[6] Cetó X, Voelcker N H and Prieto-Simón B 2016 Bioelectronic tongues: new trends and applications in water and food analysis *Biosens. Bioelec.* **79** 608–26

[7] del Valle M 2017 *Trends in Bioelectroanalysis* (Cham: Springer International Publishing) pp 143–202

[8] Canfarotta F, Cecchini A and Piletsky S 2018 *Molecularly Imprinted Polymers for Analytical Chemistry Applications* (London: The Royal Society of Chemistry) pp 1–27

[9] Malitesta C, Mazzotta E, Picca R, Poma A, Chianella I and Piletsky S 2012 MIP sensors—the electrochemical approach *Anal. Bioanal. Chem.* **402** 1827–46

[10] Rebelo P, Costa-Rama E, Seguro I, Pacheco J G, Nouws H P A, Cordeiro M N D S and Delerue-Matos C 2021 Molecularly imprinted polymer-based electrochemical sensors for environmental analysis *Biosens. Bioelec.* **172** 112719

[11] Feroz M and Vadgama P 2020 Molecular imprinted polymer modified electrochemical sensors for small drug analysis: progress to practical application *Electroanalysis* **32** 2361–86

[12] Herrera-Chacon A, Cetó X and del Valle M 2021 Molecularly imprinted polymers towards electrochemical sensors and electronic tongues *Anal. Bioanal. Chem.* **413** 6117–40

[13] Huynh T-P and Kutner W 2015 Molecularly imprinted polymers as recognition materials for electronic tongues *Biosens. Bioelec.* **74** 856–64

[14] Polyakov M 1931 Adsorption properties and structure of silica gel *Zhur. Fiz. Khim* **2** 799–805

[15] Kuznetsova T F and Eremenko S I 2002 Adsorption properties of silica gel prepared from a decationized *Sol Colloid J.* **64** 451–6

[16] Breinl F and Haurowitz F 1930 Chemical examinations on the precipitate from haemoglobin and anti-haemoglobin serum and comments on the nature of antibodies *H.-S Z. Physiol. Chem.* **192** 45–57

[17] Pauling L 1940 A theory of the structure and process of formation of antibodies *J. Am. Chem. Soc.* **62** 2643–57

[18] Dickey F H 1949 The preparation of specific adsorbents *Proc. Natl. Acad. Sci.* **35** 227–9

[19] Nobel lectures in chemistry 1942—1962

[20] Takagishi T and Klotz I M 1972 Macromolecule-small molecule interactions; introduction of additional binding sites in polyethyleneimine by disulfide cross–linkages *Biopolymers* **11** 483–91

[21] Wulff G 1972 The use of polymers with enzyme-analogous structures for the resolution of racemates *Angew. Chem. Int. Edit* **11** 341
[22] Sagiv J 1979 Organized monolayers by adsorption. III. Irreversible adsorption and memory effects in skeletonized silane monolayers *Isr. J. Chem.* **18** 346–53
[23] Blanco-López M C, Lobo-Castañón M J, Miranda-Ordieres A J and Tuñón-Blanco P 2004 Electrochemical sensors based on molecularly imprinted polymers *TrAC-Trend. Anal. Chem.* **23** 36–48
[24] Haupt K and Mosbach K 2000 Molecularly imprinted polymers and their use in biomimetic sensors *Chem. Rev.* **100** 2495–504
[25] Fischer E 1894 Einfluss der Configuration auf die Wirkung der Enzyme *Ber. Dtsch. Chem. Ges.* **27** 2985–93
[26] 2001 *Molecularly Imprinted Polymers. Man-made Mimics of Antibodies and their Applications in Analytical Chemistry* vol 23 (Amsterdam: Elsevier)
[27] Wulff G, Vesper W, Grobe-Einsler R and Sarhan A 1977 Enzyme-analogue built polymers, 4. On the synthesis of polymers containing chiral cavities and their use for the resolution of racemates *Makromol. Chem.* **178** 2799–816
[28] Arshady R and Mosbach K 1981 Synthesis of substrate-selective polymers by host-guest polymerization *Makromolekul. Chem.* **182** 687–92
[29] Norrlöw O, Glad M and Mosbach K 1984 Acrylic polymer preparations containing recognition sites obtained by imprinting with substrates *J. Chromatogr.* A **299** 29–41
[30] Cormack P A G and Elorza A Z 2004 Molecularly imprinted polymers: synthesis and characterisation *J. Chromatogr.* B **804** 173–82
[31] Beyazit S, Tse Sum Bui B, Haupt K and Gonzato C 2016 Molecularly imprinted polymer nanomaterials and nanocomposites by controlled/living radical polymerization *Prog. Polym. Sci.* **62** 1–21
[32] Sharma P, Pietrzyk-Le A, D'Souza F and Kutner W 2012 Electrochemically synthesized polymers in molecular imprinting for chemical sensing *Anal. Bioanal. Chem.* **402** 3177–204
[33] Herrera-Chacon A, Dinç-Zor Ş and del Valle M 2020 Integrating molecularly imprinted polymer beads in graphite-epoxy electrodes for the voltammetric biosensing of histamine in wines *Talanta* **208** 120348
[34] Herrera-Chacon A, González-Calabuig A, Campos I and del Valle M 2018 Bioelectronic tongue using MIP sensors for the resolution of volatile phenolic compounds *Sens. Actuators B Chem.* **258** 665–71
[35] Mujahid A, Lieberzeit P A and Dickert F L 2010 Chemical sensors based on molecularly imprinted sol-gel *Materials* **3** 2196–217
[36] Cetó X, Céspedes F and del Valle M 2013 Comparison of methods for the processing of voltammetric electronic tongues data *Microchim. Acta* **180** 319–30
[37] Buglass A J, Caven-Quantrill D J, Fuller N J, Hudson J A, Lee C G, Lee S H and McKay M 2011 *Handbook of Alcoholic Beverages: Technical, Analytical and Nutritional Aspects* (Hoboken, NJ: Wiley)
[38] Roth M 1971 Fluorescence reaction for amino acids *Anal. Chem.* **43** 880–2
[39] AOAC International 2012 Histamine in seafood. *Fluorometric Method* (Gaithersburg, MD) pp 35.1.2
[40] Chatonnet P, Dubourdie D, Boidron J-N and Pons M 1992 The origin of ethylphenols in wines *J. Sci. Food Agric.* **60** 165–78

[41] Chatonnet P, Dubourdieu D, Boidron J-N and Lavigne V 1993 Synthesis of volatile phenols by Saccharomyces cerevisiae in wines *J. Sci. Food Agr.* **62** 191–202
[42] Curtin C, Bramley B, Cowey G, Holdstock M, Kennedy E, Lattey K, Coulter A, Henschke P, Francis L and Godden P 2007 Sensory perceptions of 'Brett' and relationship to consumer preference *Proc. of the Thirteenth Australian Wine Industry Technical Conference*; K Beames *et al* (Adelaide: Australian Wine Industry Technical Conf. Inc) pp 207–11

Chapter 6

Fundamentals and applications of impedimetric electronic tongues

Murilo H M Facure, Rafaela S Andre, Luiza A Mercante, Lucas Fugikawa-Santos, Osvaldo N Oliveira Jr and Daniel S Correa

In this book chapter, we discuss the fundamentals of impedimetric electronic tongues (e-tongues), starting with a brief introduction to the principles of impedance spectroscopy and how this technique can be used in e-tongues for various applications. We also elaborate upon the types of interdigitated electrodes (IDEs) employed in the sensing units of e-tongues, their fabrication process, geometric characteristics, and experimental conditions that may affect sensing performance. Next, we explore the vast number of materials employed to modify IDEs, including conducting polymers, polyelectrolytes, inorganic nanoparticles, carbon-based, and inorganic 2D nanomaterials in order to improve e-tongue performance. Finally, we provide recent examples of the use of impedimetric e-tongues for the analysis of the basic tastes, as well as complex matrices as wine, juice, and gluten-containing food. Emphasis is also given to e-tongues employed to detect trace levels of water contaminants like pesticides, heavy metals, and antibiotics.

6.1 Introduction to impedimetric e-tongues

There is a growing concern about the harmful effects of various pollutants on human health and the environment. Heavy metals, pesticides, hormones, and antibiotics released into the environment without proper treatment or control are examples of such pollutants [1–4] that act contaminating water resources, including groundwater, rivers, and water supplies to the population [5–7]. Food contaminants, such as toxins and microorganisms, are another concern once they pose risks even at very low concentrations [8–11]. Analytical techniques providing fast, reliable results with high sensitivity and low detection limits are required for water and food quality control, but the most employed methods are time-consuming, expensive, and require

doi:10.1088/978-0-7503-3687-1ch6

trained personnel [12, 13]. In this scenario, chemical sensors may represent a solution in assisting or even replacing standard techniques [14–16]. They may be biosensors, which are highly specific for target analytes, or sensors based on global selectivity concepts, such as the electronic noses (e-noses) and electronic tongues (e-tongues). The latter can mimic the gustatory system, being capable of analyzing complex samples and discriminating them even when the concentrations of the analytes (contaminants or not) are down to parts per billion (ppb) or parts per trillion (ppt) [3, 13, 17, 18].

E-tongues may operate under distinct transduction mechanisms, such as voltammetry [19–21], amperometry [22–24], potentiometry [25–28], and impedance spectroscopy [29–32]. This book chapter focuses on impedimetric e-tongues whose methods of fabrication, measuring procedures, and data analysis are illustrated in figure 6.1. A few (nano)materials used in e-tongues are shown in figure 6.1(a), but there is no limit to which material can be used. However, the choice of suitable materials to modify interdigitated electrodes (IDEs) is crucial, as they affect the cross-sensitivity and global selectivity of the device [15, 32, 33]. Different types of IDEs can be fabricated and then modified using several techniques, as schematized in figure 6.1(b) [34–36]. The array of sensing units is connected to an impedance analyzer and a computer to perform the analyses of the liquid and collect the electrical responses (figure 6.1(c)). Due to the large amounts of data, statistical and computational techniques are employed to establish a fingerprint of the specific liquid sample [17]. These techniques include multidimensional projection methods such as principal components analysis (PCA) [17, 37, 38], neural networks, and other machine learning methods [39–42]. A few examples of e-tongue applications are illustrated in figure 6.1(d), e.g., food and taste analysis, environmental monitoring, and quality control.

This book chapter will address the fundamental aspects and applications of impedimetric e-tongues, with sections divided as follows: impedance spectroscopy principles; IDEs: types, materials, and fabrication techniques; materials employed to modify IDEs; and applications of impedimetric e-tongues.

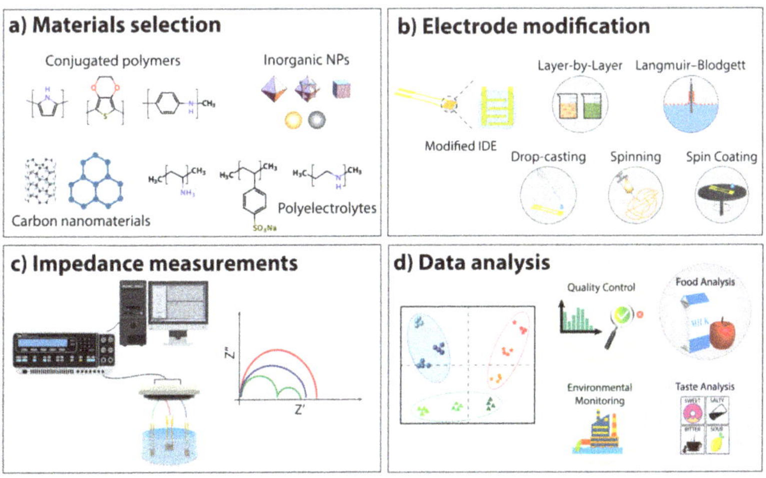

Figure 6.1. Scheme of the sequence of typical steps to develop an impedimetric e-tongue.

6.2 Impedance spectroscopy principles

In an impedance spectroscopy (IS) experiment, the electrical response (usually current) to a known signal stimulus (voltage) is obtained as a function of the excitation frequency. When a harmonic excitation voltage with fixed angular frequency ω, $V(t) = V_0\sin(\omega t)$, is applied to two electrodes separated by a material medium, the electrical current oscillates at the same angular frequency, but it can be phase-shifted by a phase angle δ, being represented by $I(t) = I_o\sin(\omega t - \delta)$. The current may be separated into two contributions: one responding in phase to the stimulus and related to processes such as charge drift and redox reactions, while the phase-shifted component is normally associated with dipolar orientation effects, space charge build-up, or polarization due to local charges [43]. If the response signal is shifted by $\pm \pi/2$ relative to the excitation stimulus (which is proportional to $\sin \omega t$), it is said to be *in quadrature* and proportional to $\cos \omega t$, since $\sin(\omega t \pm \pi/2) = \pm \cos\omega t$. Therefore, in the general case, the electrical current can be represented by a linear combination of terms proportional to $\sin\omega t$ (in-phase) and $\cos\omega t$ (in quadrature).

A convenient way to describe the relationship between the current response and the excitation voltage is to use complex representation, with the excitation voltage written as:

$$V^*(\mathrm{t}) = V_0 e^{i\omega t} = V_0[\cos(\omega t) + i \sin(\omega t)] \tag{6.1}$$

where $i = \sqrt{-1}$ is the imaginary number and V_0 is a purely real term.

The complex current, delayed by a phase angle δ, is:

$$I^*(t) = I_0 e^{i(\omega t - \delta)} = I_0 \frac{e^{i\omega t}}{e^{i\delta}} \tag{6.2}$$

Combining equations (6.1) and (6.2) allows us to define a complex function Z^* from the relationship:

$$V^*(t) = Z^* I^*(t) \tag{6.3a}$$

where

$$Z^* = \frac{V_0}{I_0} e^{i\delta} = \frac{V_0}{I_0}(\cos \delta + i \sin \delta) \tag{6.3b}$$

Note that Z^* does not depend on t, and two other time-independent complex functions V^* and I^* can be defined:

$$V^* = Z^* I^* \tag{6.4}$$

Equations (6.3a) and (6.4) are the equivalent of Ohm's law in alternating current systems, where the *complex impedance* Z^* is dependent on the angular frequency ω and phase-shift angle δ. The complex impedance can be represented by:

$$Z^* = Z' + iZ'' \tag{6.5}$$

where Z' and Z'' are, respectively, the real and the imaginary part of Z^*. Comparing equation (6.5) to the last term of equation (6.3b), one can define the phase shift angle as:

$$\delta = \tan^{-1}\left(\frac{Z''}{Z'}\right) \tag{6.6}$$

where

$$Z' = |Z|\cos\delta \tag{6.7a}$$

$$Z'' = |Z|\sin\delta \tag{6.7b}$$

and

$$|Z| = \sqrt{Z'^2 + Z''^2} = V_0/I_0 \tag{6.7c}$$

where $|Z|$ is the modulus (or the amplitude) of the complex impedance. The complex impedance can be determined by measuring the amplitude and phase shift of the current (I_0 and δ, respectively) for known values of the amplitude and frequency of the excitation voltage (V_0 and ω, respectively).

The current of an ideal resistor (of resistance R) is always in phase with the excitation signal ($\delta = 0$), and the complex impedance has only the real part, Z', which is equal to R. On the other hand, for an ideal capacitor with capacitance C, the current is the displacement current $I(t) = C\,dV/dt$. By differentiating the second term of equation (6.1) and combining to equation (6.3a), the complex impedance is given only by the imaginary part, $Z'' = -1/\omega C$, and the response is in quadrature ($\delta = -\pi/2$) with the electrical stimulus.

A single, ideal circuit element is normally insufficient to explain the electric response with accuracy. The impedance is better represented by a combination (association) of discrete circuit elements, whose electric response can be described by an *equivalent impedance* (Z_{eq}^*). Circuit elements can be associated in *parallel* (if all elements are subject to the same voltage) or *in series* (if the same current flows through all elements). The equivalent impedance from the association of n complex impedances Z_i^* is:

$$\left(Z_{eq}^*\right)^{-1} = \sum_i^n \frac{1}{Z_i^*}. \text{ (Parallel association)} \tag{6.8}$$

and

$$Z_{eq}^* = \sum_i^n Z_i^*. \text{ (Series association)} \tag{6.9}$$

The analytic expression for the equivalent complex impedance can be complicated depending on the complexity of the model, because the real and imaginary parts of each function cannot be combined and must be computed separately. Numerical simulations or fittings of the equivalent impedance are used instead of the analytic expressions in these cases [44]. The impedance determines the amplitude of the electrical current: the higher it is, the lower the current is. Hence, the impedance is the a.c. analogous of the electrical resistance R in a d.c. circuit. Similarly, the *electrical conductance* G is defined as the inverse of the resistance ($G = 1/R$). The *complex admittance* can be defined as:

$$Y^{*}(\omega) = \frac{1}{Z^{*}(\omega)} = Y' + iY'' \tag{6.10}$$

where $Y' = Z'/|Z|^2$ and $Y'' = -Z''/|Z|^2$ are the real and the imaginary parts of the admittance, respectively. For a single ideal resistor with resistance R, $Y' = G = 1/\mathrm{R}$. The complex admittance is equivalent to the *complex conductivity* $\sigma^{*}(\omega) = \sigma' + i\sigma''$, differing only by geometrical characteristics (e.g., separation and area of the electrodes). Another useful a.c. electrical response function is the *complex capacitance*, defined as:

$$C^{*}(\omega) = \frac{Y^{*}(\omega)}{i\omega} = C - iC'' \tag{6.11}$$

where $C' = Y''/\omega$ and $C'' = Y'/\omega$. Equation (6.11) is derived from the relationship between the complex permittivity (ε^{*}) and the complex conductivity of a linear medium: $\sigma^{*} = i\omega\varepsilon^{*}$. The complex permittivity also allows the determination of the complex dielectric constant ($\kappa^{*} = \varepsilon^{*}/\varepsilon_0$). These three complex dielectric functions (C^{*}, ε^{*} and κ^{*}) are equivalent and have the same functional dependence. They differ from each other only by the multiplication of (real) constant values. Some authors also prefer to use the *electrical modulus* ($M^{*} = 1/C^{*}$) as a complex dielectric function [45].

The general designation for any complex electric response function (Z^{*}, Y^{*}, C^{*}, M^{*}, etc) is *complex immittance.* As inferred from equations (6.10) and (6.11), they are all correlated and can be calculated from each other. The choice for a particular immittance function, therefore, depends on the characteristics to be emphasized. Conductive pathways and electrochemical reactions are more likely to be modeled by resistive/conductive functions, which make it more convenient to analyze the results by using Z^{*} and Y^{*}. Polarization effects and space charge distributions in the material, conversely, can be better represented by dielectric response functions such as C^{*} and M^{*}.

The association of circuit elements with an equivalent complex impedance that resembles the results from a measurement is useful to understand the physical processes, even though it is not realistic. The simplest representation of a homogeneous material with inert electrodes is the parallel circuit with a resistor R_b (representing the bulk resistance) and a capacitor C_g (representing the geometrical capacitance). Figure 6.2(a) shows the circuit for which the complex components of the equivalent impedance are:

$$Z' = \frac{R_b}{(1 + (\omega.\ \tau)^2)} \tag{6.12a}$$

and

$$Z'' = \frac{\omega.\ \tau.\ R_b}{(1 + (\omega.\ \tau)^2)} \tag{6.12b}$$

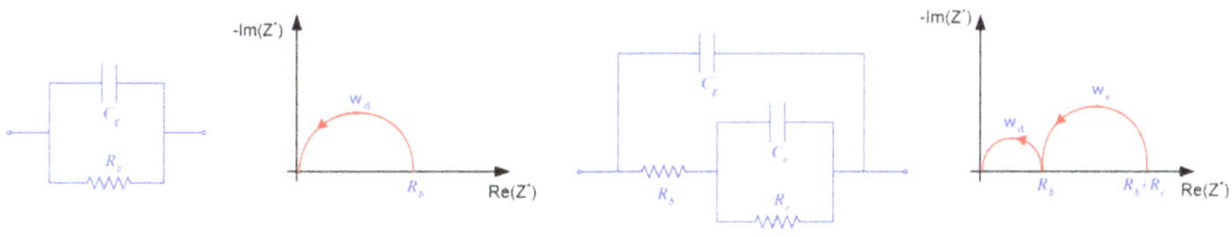

(a) (b) (c) (d)

Figure 6.2. (a) Single RC circuit from the parallel association of the material bulk resistance (R_b) and the geometrical capacitance (C_g), (b) Z complex plane plot for the response from a single RC circuit, where $\omega_d = 1/R_bC_g$, (c) equivalent circuit for a system including electrode reaction resistance (R_e) and diffusive double-layer capacitance (C_e) effects, and (d) Z complex plane plot for the response from the circuit in (c), where $\omega_e = 1/R_eC_e$.

where $\tau_d = R_bC_g = \varepsilon/\sigma$ is the dielectric relaxation time of the material. The complex impedance data may be presented in a *complex plane* as in figure 6.2(b), where the negative imaginary part of Z^* is plotted against the real part ($-Im(Z^*)$ versus $Re(Z^*)$). For the equivalent circuit in figure 6.2(a), the Z-plane displays a semicircle with diameter R_b and apex at the condition $\omega_d\tau_d = 1$ (figure 6.2(b)). The frequency, which does not appear explicitly in the Z-plane, increases in the counterclockwise direction, as indicated by the arrow in figure 6.2(b). The low-frequency limit for the real part, $Z'(\omega\to 0)$, is equal to the bulk resistance R_b at the point where the semicircle crosses the real axis.

The first attempt to depart from the ideal case of a homogeneous material with inert electrodes is to consider processes in the vicinity of the electrodes. These processes may include heterogeneous reactions or adsorption of specific species [46, 47]. They introduce a local resistance that differs from the bulk resistance and gives rise to delayed electric responses with distinct characteristic times. The simplest modification in the single RC circuit of figure 6.2(a) to account for interface effects is to add a parallel association of a resistance R_e and a capacitance C_e. These components may represent a reaction resistance and a diffusive double-layer capacitance in the polarization region near the electrode, respectively. The circuit elements due to the bulk resistance R_b and geometrical capacitance C_g are associated with R_e and C_e as shown in figure 6.2(c), whose Z-plane plot is illustrated in figure 6.2(d). Two semicircles appear when the dielectric relaxation time ($\tau_d = 1/\omega_d = R_bC_g$) is considerably smaller than the relaxation time associated with the electrode ($\tau_e = 1/\omega_e = R_eC_e$). The real part of the impedance tends to R_b in the high-frequency limit ($\omega\to\infty$) and to R_b+R_e in the low-frequency limit ($\omega\to 0$).

Another process to consider in impedance experiments is the dielectric relaxation arising from the electric polarization of charges/dipoles. When a dielectric (insulating) material is subjected to an external electric field, two major effects occur: (i) almost instantaneous polarization due to electrons displacement with respect to the nuclei, thus leading to the high-frequency ($\omega\to\infty$) dielectric constant (κ_∞) and geometric capacitance (C_g or C_∞), and (ii) orientation of electric dipoles, responsible for the static dielectric constant (ε_s) and static (or low-frequency) capacitance (C_s) for the low-frequency regime ($\omega\to 0$). In the ideal case of a homogeneous material,

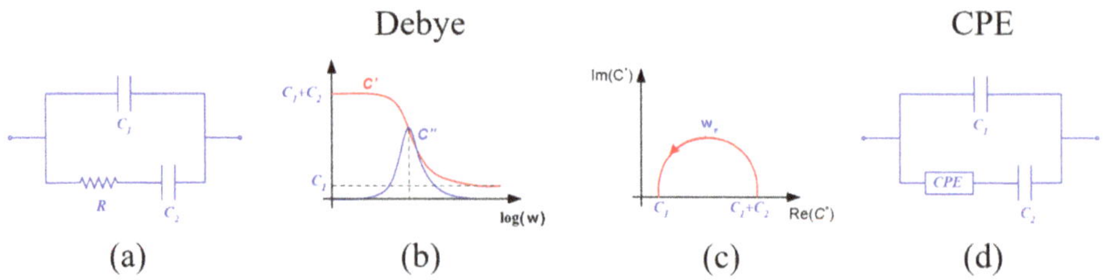

Figure 6.3. (a) Equivalent circuit to model a single relaxation time system, (b) capacitance Bode plot, (c) complex plane for a system which follows Debye relaxation, and (d) equivalent circuit comprising a constant phase element (CPE).

all the dipoles respond equally to the electric field, and the electrical response can be modeled with a single relaxation time in the equivalent circuit of figure 6.3(a). For high frequencies ($\omega \rightarrow \infty$), the impedance of the lower branch of the circuit in figure 6.3(a) is too high and the equivalent capacitance is $C_\infty = C_g = C_1$. In the low-frequency regime, the voltage drops over the resistance R (which can be associated with the energy dissipated to orientate the electric dipoles in a material medium) is null, and the circuit responds as the parallel association of two capacitors ($C_s = C_1 + C_2 = C_\infty + C_2$).

The frequency-dependent complex capacitance of the equivalent circuit from figure 6.3(a) is:

$$C^*(\omega) - C_\infty = \frac{C_s - C_\infty}{1 + (\omega\tau)^2} - i\frac{\omega\tau(C_s - C_\infty)}{1 + (\omega\tau)^2} \tag{6.13}$$

Equation (6.13) is the expression for *Debye relaxation* [44], characterized by a single relaxation time τ, as initially assumed. The graphics for the frequency response of the real and imaginary parts of $C^*(\omega)$ (Bode plots) and the capacitance complex plane plot are shown in figure 6.3(b) and (c). The imaginary part of the capacitance (C'') has a peak at the frequency that determines the relaxation time ($\omega_0 = 1/\tau_d$). As inferred from equation (6.11), C'' responds in phase with the excitation signal and is known as *dielectric loss* because it is related to the energy dissipated due to the dipolar orientation in the dielectric.

In most real situations, the material does not have a single relaxation time, as its microscopic properties depart from the average (or most probable value). The Debye relaxation model may then be corrected by assuming that the dielectric relaxation time follows a continuous distribution. Cole and Cole [48] proposed a symmetric distribution of relaxation times to describe the deformation of the semi-arcs formed by the dielectric complex plane (C'' versus C') for various materials. The Cole–Cole dielectric response function is:

$$C^*(\omega) - C_\infty = \frac{C_s - C_\infty}{[1 + (i\omega\tau)^{1-\alpha}]}, \quad (0 < \alpha < 1) \tag{6.14}$$

which is reduced to the Debye relaxation, equation (6.13), when $\alpha = 0$.

Another often used dielectric response function contains an asymmetric distribution of relaxation times, as proposed by Davidson and Cole [49, 50]:

$$C^*(\omega) - C_\infty = \frac{C_s - C_\infty}{[1 + (i\omega\tau)^\beta]}, \quad (0 < \beta < 1) \tag{6.15}$$

which is also reduced to the Debye relaxation if $\beta = 1$.

Deviations from the Debye relaxation, equation (6.13), can also be expressed by using an equivalent circuit similar to that of figure 6.3(a), but with the resistor R replaced by a *constant phase element* (CPE), as shown in figure 6.3(d). CPE follows an empirical impedance function represented by:

$$Z^*_{\mathrm{CPE}}(\omega) = \frac{1}{Q_0(i\omega)^n}, \quad (0 < n < 1) \tag{6.16}$$

For $n = 1$, CPE is an ideal capacitor of capacitance Q_0, whereas it is an ideal resistor of resistance $1/Q_0$ for $n = 0$. For any other values of n, CPE is interpreted as an element with a combined response of capacitive and resistive contributions to the electric current. An important distributed element is the *Warburg diffusion element* (Z_W) [51], which can be defined as a CPE with a constant phase angle of $\pi/4$, independent of the frequency, and amplitude proportional to the square root of the frequency, as follows:

$$Z_W = \frac{A_W}{\sqrt{\omega}} + \frac{A_W}{i\sqrt{\omega}} \tag{6.17}$$

where A_W is the Warburg coefficient.

The Warburg element is appropriate to describe systems involving diffusion effects, assuming a semi-infinite linear diffusion in large planar electrodes [52].

Finding a complex dielectric function to represent the experimental results from an impedance measurement is not a simple task; in some cases, it is not even possible to achieve. In sensing applications, IS results are employed to obtain a set of parameters for further data analysis. A common approach is to use a parallel (or series) frequency-dependent RC association for the electric response function [53, 54]. The pair of functions ($R(\omega)$ and $C(\omega)$) is equivalent to any pair of frequency-dependent functions formed by the real and imaginary parts of the complex response functions (Z^*, Y^*, C^*, etc) if the type of association (parallel or series) is established. Albeit inadequate for equivalent circuit modeling, another strategy for sensing applications [55, 56] is to consider the low-frequency (usually at 1 kHz) values from the real parts of Z^* and C^* as output parameters. If the experiment has already achieved the steady-state condition at the measuring frequency, $Re(Z^*)$ and $Re(C^*)$ are representative for both contributions (in-phase and in quadrature) to the electric current response and are sufficient to provide useful information from the processes at the material/electrode interfaces.

6.3 Interdigitated electrodes (IDEs): types, materials and fabrication techniques

Impedimetric sensors have been made with IDEs as transducer units [57, 58]. In impedimetric e-tongues, an array of IDEs (sensing units) is used for liquid analyses to obtain a global response or a recognition pattern that is like a fingerprint of the sample [17, 59]. This strategy has been adopted to analyze complex samples (in terms of chemical composition), combining global selectivity and cross-sensitivity concepts, mimicking the human gustatory system [59–61]. IDEs in e-tongues allow for using different materials as sensitive layers and increase the change in capacitance due to the increased effective area [30, 37, 62]. An IDE comprises a pair of metal strips as electrodes, each of which is composed of multiple interconnected electrode stripes, usually referred to as fingers or digits. They form an interdigitated electrode array deposited onto planar and insulating substrates, as depicted in figure 6.4(a) [57]. The spacing (s) between digits can vary from dozens of micrometers to tens of nanometers. The length, width (w), and number of digits can also be varied [35, 63, 64]. These parameters define the electrode geometry and electrode area, which is relevant for sensitivity [65]. Also, the interdigitated configuration eliminates the need for reference electrodes since the current flowing between a pair of IDEs is approximately zero under an open circuit (without sensing material deposition) [66, 67]. The electric field penetration is proportional to the distance between electrode digits ($w + s$) and depends on the frequency, with the current flowing close to the surface as depicted in figure 6.4(b) [57]. Additional advantages in using IDEs include the faster achievement of the steady-state conditions than with larger electrodes, increased signal-to-noise ratio, and ease of reading the electrical change via impedance measurements [58, 68]. The material used to coat the IDE is critical for impedance measurements and sensitivity [68, 69]. The impedance depends on the dielectric properties of the modified IDE and the target medium conductivity (liquid or gas). It also depends on the double-layer formed at the interface between the coating layer and the medium. The choice of materials and their influence on the sensitivity of e-tongues is discussed in detail in section 6.4.

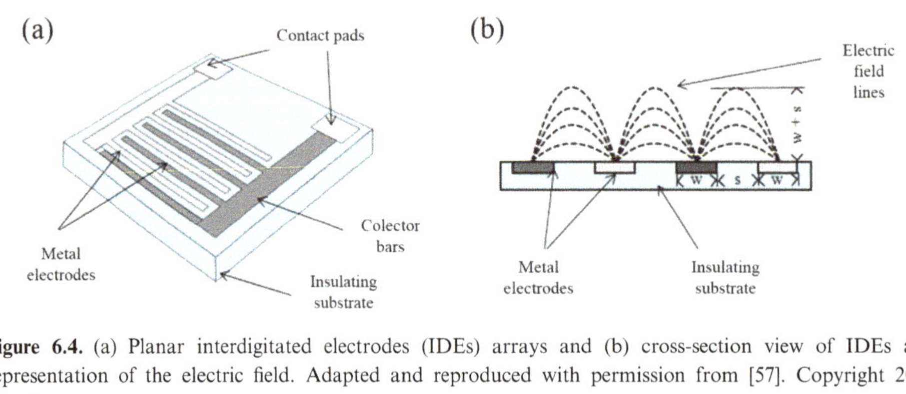

Figure 6.4. (a) Planar interdigitated electrodes (IDEs) arrays and (b) cross-section view of IDEs and representation of the electric field. Adapted and reproduced with permission from [57]. Copyright 2019 Elsevier.

6.3.1 Materials and IDE design

Materials to produce IDEs are limited by the fabrication technique and may affect the target application [34, 70]. The most used materials are gold (Au), chromium (Cr), platinum (Pt), aluminum (Al), silver (Ag), and graphite (C) [71]. The choice of the active electrode material is based on a combination of characteristics for an intended application and their suitability to the fabrication process and cost. For example, Au IDEs have a significantly higher signal-to-noise ratio than Pt IDEs, resulting in superior performance [65]. However, the latter have superior performance for immunosensor applications [68]. The materials for electrode fabrication are shaped using thin and thick-film techniques such as photolithography [72] and screen-printing [73]. Novel materials with improved performance have been employed in IDEs, as is the case of single-walled carbon nanotubes and graphene [73–75].

The IDE parameters and design are also relevant for sensitivity and performance [35, 36, 76], with sensitivity being dependent on electrode height, width and spacing of the digits, and material composition. Decreasing electrode dimensions down to a few micrometers results in higher sensitivity [66, 77]. The digit width and spacing are important to improve the signal-to-noise ratio, with the signal increasing with decreased spacing and narrower digits. The active area increases with the number of digits, but so do the signal and background noise, and therefore the signal-to-noise ratio is kept constant. In an alternative design, referred to as three-dimensional interdigitated electrode arrays (3D-IDEA), the electrode digits are separated with insulating barriers to enhance the sensitivity by boosting the effect from surface conductivity on impedance measurements [78–80]. In conclusion, IDE parameters can be designed and optimized with appropriate sizes to improve sensitivity for specific analytes.

IDEs made with functional materials at low cost may be used directly as sensors [74], as in the example of graphene IDEs applied as taste sensors [73]. The performance of the latter is attributed not only to graphene but also to the designed IDE parameters, which resulted in sensor responses capable of discriminating food constituents with a response time of 2 s and no dependence on temperature or humidity. However, reactive values at low frequency for some of the tested chemicals were close, indicating a limitation of operation [73]. The work illustrates the importance of IDE materials and design, especially for analyzing complex samples.

6.3.2 Fabrication techniques

The standard procedure to fabricate IDEs is the deposition of metallic electrodes on insulating substrates, but carbon-based materials have also been used [81]. The fabrication of IDEs requires steps for interdigit patterning and methods to reproduce the pattern designs at the microscale with a suitable resolution. These steps are added to the fabrication of conventional [82] and macroscale electrodes [83, 84], being essential for the IDE properties and performance [70]. With recent advances, these methodologies evolved towards low-cost and large-scale production with virtually no restrictions in materials and substrates, including flexible substrates [85]. The most used methods are summarized here, providing details on the methodology and main features that allow for a variety of IDEs applicable to e-tongues.

6.3.2.1 Optical lithography

Optical lithography (also known as photolithography) is the most traditional technique for micropatterning surfaces with high resolution [77, 86]. Combined with physical deposition methods, such as sputtering and electron-beam evaporation, photolithography enables fast fabrication of optoelectronic devices at a large scale with relatively low-cost [87]. The method is traditionally conducted by spin-coating a layer of photoresist onto the substrate, usually a glass or Si/SiO_2 slide. A patterned mask is placed on top of the substrate to expose the photoresist to UV light selectively. Since the photoresist is a light-sensitive organic compound, the light-exposed region will become soluble or insoluble, being classified as positive or negative photoresists [88], respectively. Typical positive photoresists are diazonaphthoquinone (DNQ) mixed with phenol-formaldehyde resin [89, 90], while epoxy-based polymers [91, 92] are negative photoresists. After UV exposure, the electrode material is deposited, and a solvent (the developer) is applied to remove the excess of photoresist, resulting in interdigitated micropatterning [93]. This removal process of the excess photoresist is known as lift-off, being followed by the IDE coating with the sensitive material according to the intended application [70]. The original technique has evolved to reach nanoscale resolution with techniques based on evanescent near-field optical lithography, near-field photolithography [86], light coupling and guiding through elastomeric masks [86], surface plasmon polariton [94], and direct writing methods such as e-beam lithography (EBL) [94, 95]. The first semiconductor nanocrystal patterned at the nanoscale with good electrical conductivity and no cracks was fabricated by EBL in 2012 [95]. Nanoscale patterns of PbS, PbSe, and CdSe cores and Zn0.5Cd0.5Se–Zn0.5Cd0.5S core–shell nanocrystals were obtained with dimensions as small as 30 nm. The pattern fabrication steps and micrographic images are depicted in figure 6.5(a).

6.3.2.2 Etch patterning

Etching techniques allow the fabrication of IDEs directly on the sensitive material to achieve high-performance electrodes [70, 85]. The method consists of etching away the undesired material, constructing the pattern directly on the sensitive layer, followed by a coating layer of electrolyte gel. There are dry and wet etching methods for IDE design [87, 98]. Dry etching techniques comprise plasma and laser etching methods, with the former being more common and cost-effective [77, 99]. In plasma etching the film coated with the electrode material is irradiated in a plasma chamber, exposing the entire film to the radiation [96]. Then, a thin layer of the electrode material is deposited on the sensitive material using physical deposition techniques and patterning masks. Next, the film-coated with the electrode material is irradiated in a plasma chamber, etching the exposed region away and forming IDE patterns. Several combinations can be used in electrodes such as carbon nanotubes [100], gold [96],

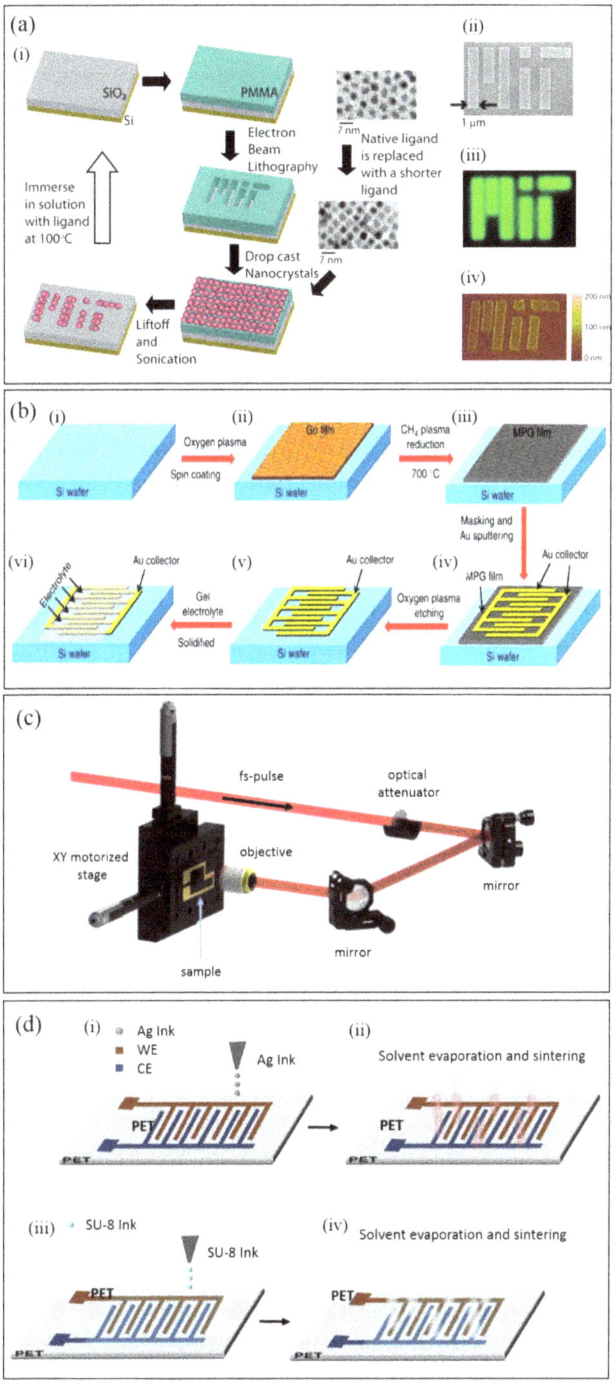

Figure 6.5. (a) (i) Overview of the photolithographic process: the patterns are positioned on the surface of the substrate with nanoscale precision, the film electrical conductivity is reached by annealing or exchanging the capping ligand for a smaller molecule. At the right side, transmission electron micrographs of the nanocrystal

films before and after cap exchange, (ii) scanning electron micrograph, (iii) fluorescence (actual color), and (iv) AFM images. Adapted and reproduced with permission from [95]. Copyright 2012 American Chemical Society. (b) Schematic illustration of IDE fabrication by plasma etching onto a silicon (Si) wafer. The fabrication process includes (i) oxygen plasma surface treatment of silicon, spin-coating deposition of the GO on surface-modified silicon, (ii) CH_4 plasma reduction, (iii) masking pattern and deposition of the gold current collector, (iv) oxidative etching in an oxygen plasma, (v) drop-casting of the H_2SO_4/PVA gel electrolyte and (vi) solidification of the gel electrolyte. Reproduced with permission from [96]. Copyright 2013 Springer Nature. (c) Schematic representation of experimental setup for gold IDEs fabrication through femtosecond laser ablation. Reproduced with permission from [62]. Copyright 2015 Elsevier. (d) Inkjet printing fabrication of interdigitated electrodes showing (i) Ag ink deposition, (ii) ink solvent evaporation at 110 °C for 1 h, (iii) SU-8 ink deposition as a sensitive layer, and (iv) curing parameters of SU-8 at 110 °C for 15 min for the first and second layer, followed by room temperature cure for 16 h and a final step of 110 °C for 15 min. Reproduced with permission from [97]. Copyright 2020 MDPI.

and gold combined with metal oxides [101]. The gas mixture used for plasma generation can be chosen according to the characteristics of the composing materials, as illustrated in figure 6.5(b). For instance, reducing gases such as methane are utilized to treat graphene oxide films to reduce oxygen species and increase electrical conductivity [96]. Laser etching is another dry etching technique to fabricate micropatterned electrodes onto a film [62, 102–105]. The system represented in figure 6.5(c) consists of an XY-controlled laser spot to etch the material following a route designed via software, which is simpler and less expensive than using shadow masks. These features are attractive to design IDEs, including wearable electronic devices [106].

As for wet etching techniques, in microfluidic etching, the interdigit pattern is obtained with a PDMS stamp pressed onto the film and forming microfluidic channels between the flat film and the PDMS stamp [87, 107, 108]. IDEs are created by circulating aqueous etching solutions through the channels to etch the material away. The exposure time to the etching solution requires fine control and immediate washing should be performed to prevent etchant diffusion to unexposed regions. The microfluidic etching is a cost-effective method for IDE patterning since PDMS stamps are cheap and robust with good stability over a wide range of etchants [75, 107].

6.3.2.3 Direct printing

Direct printing is a simple, low-cost method to fabricate IDEs with varied designs [72, 109, 110]. It is based on the transfer of the electrode material, as ink suspension, to the substrate according to the chosen design. The technique has evolved to encompass inkjet printing [64], screen printing [111], gravure printing [112], and extrusion printing [113]. These printing techniques require inks with suitable viscosity and adequate dispersion of the electrode material, which remains challenging. In inkjet printing, the IDE design is provided by a digital file and transferred via software to the printer head controlled by an XY-motor [110, 114]. The digital format of the design allows easy and fast adjustments or changes to the layout. The ink with specific characteristics, stability, and particle size is continuously ejected from the head nozzle generating droplets that are propelled out of a nozzle by heat or piezo effect. To reach adequate viscosity with good stability of the particles in suspension, inks are prepared with surfactants or polymeric additives as particles dispersant as well as with a mixture of solvents to accomplish the expected surface tension [113, 115]. Inkjet printing has been used to print not only the IDEs but also the sensitive layer [97], as shown in figure 6.5(d). This strategy allows one to produce fully-printed devices on a large scale, opening new opportunities for IDE fabrication and application in the next generation of electronic devices.

The screen-printing method is simpler than inkjet printing and works based on the ink transfer through a planar pre-patterned screen mesh by a squeegee leaving the pattern imprinted on the substrate [85, 87, 111, 116]. Screen printable inks are prepared by dispersing the active material with a polymer binder in a proper solvent to reach the slurry feature. Such composition was optimized over time, aiming at continuous printed patterns rather than separated dots. One of the main advantages of the screen-printing method is the variety of substrates that can be employed, such as paper [117, 118], glass [119], polymers [120], ceramic [120], textiles [121, 122], and nonwoven fabrics [123]. The main limitation is related to the poor resolution compared to other printing techniques, in addition to the waste of materials, e.g., screens, inks, and cleaning solutions. Another configuration consists of roll-to-roll screens, known as gravure printing, where the squeegee is replaced by a roller with the pattern notched [112, 123]. The ink goes inside the roller and the fabrication process is continuous, having a higher speed than planar screen-printing. However, gravure printing is more expensive and it is more difficult to clean the equipment, causing waste of ink [123]. Extrusion printing [109, 114, 124] requires a viscous ink to be extruded through a nozzle with an XYZ-positioning stage to print the IDE pattern controlled via software. It is scalable with a high loading of active materials, and similarly to inkjet printing, the pattern design in extrusion printing can be easily adjusted by editing the project. For extrusion printing, high viscosity inks are required to prevent the ink from spreading on the substrate and a constant line can be formed along the direction of the nozzle as the ink is extruded. A similar apparatus can be employed for 3D-printing of IDEs through layer-by-layer ink stacking with a designed model [109, 124, 125]. 3D-printing technology, compared to other printing methods, has a unique advantage for fabricating 3D electrodes. Other remarkable advantages of 3D-printing methods are the high-resolution the electrode patterns can achieve and the electrode increased surface area [114].

Overall, a variety of techniques are useful to fabricate IDEs from many materials, which is relevant to optimize the performance of e-tongues. Table 6.1 displays examples of IDEs fabrication techniques used for impedimetric e-tongues and modified with diverse sensing materials.

6.4 Materials employed to modify IDEs

6.4.1 Sensing materials used in impedimetric e-tongues

The choice of materials is key in determining e-tongue performance. Though e-tongues do not require specific interactions with the analytes, the sensing layer should be sensitive to small electrical changes in the liquid media [17, 126]. In addition, the sensing material should be stable and versatile, allowing a fine-tuning

Table 6.1. Examples of sensing materials, modification techniques, and data analysis employed in impedimetric e-tongues, and their applications for detecting varied analytes.

Sensing materials	Modification technique	IDE fabrication technique	Data analysis	Application	References
CNT, PCL	Drop-casting	Vacuum evaporation	Raw data analysis	Basic tastes analysis	[144]
PEDOT:PSS, PPy, AuNPs, rGO	Drop-casting	Photolithography	PCA	Organophosphate pesticides detection	[37]
MoS_2, GO	Drop-casting	Photolithography	PCA	Antibiotics detection	[30]
GO, ZrO_2, CeO_2, PEDOT: PSS	Dip coating	Chemical etching	PCA	Detection of diabetes biomarker	[132]
PEDOT:PSS, AuNPs, $LuPc_2$	Spin coating		PCA and PLS	Wine discrimination analysis	[31]
TiO2, TiO_2-Ag, GaN	Spin coating and sputtering	Photolithography	PCA	Determination of water content in ethanol	[141]
Au, Pt, Ni, Fe, Al_2O_3	Electron beam deposition		IDMAP, PCA, PC and Sammon's Mapping	Classification of basic tastes and of compounds of the oil industry	[42]
Phenyl boronic acid substituted hydrogels	Polymerization	Direct printing	PCA and IDMAP	Different types of sugars distinction and apple juice analysis	[179]
SA, 16-mer, PPy	LB	Photolithography	PCA	Inorganic contaminants detection, beverages and basic tastes analysis	[130, 133, 180]
PPy, PS-PMMA, lignin, perylene derivative, chitosan, PSS	LB	Photolithography	PCA	Analysis of water samples	[55]
PEI, PSS, PAH, GO	LbL	Photolithography	PCA	Detection of triclosan	[142]
POEA, chitosan, PMMA	LbL		PCA	Copper ions detection in aqueous solutions	[134]
PAH, AgNPs, cardiolipin, DPPG	LbL	Photolithography	PCA	Methylene blue detection	[126]

(*Continued*)

Table 6.1. (*Continued*)

Sensing materials	Modification technique	IDE fabrication technique	Data analysis	Application	References
AgCl-PANI, PSS	LbL	Photolithography	PCA	Basic tastes analysis	[153]
PAH, NiTsPc, PEDOT:PSS, PPy	LbL	Photolithography	PCA	Basic tastes analysis	[56]
NiTsPc, CuTsPc, PAH, AuNPs	LbL	Photolithography	PCA	Analysis of fat content in milk samples	[138]
PANI, PAH, PSS, PPy	LbL	Photolithography	PCA	Detection of geosmin and 2-methylisoborneol	[131]
PAH, PEDOT:PSS, PPy, CuTsPc	LbL	Photolithography	IDMAP and PCA	Detection of gliadin in gluten-containing foodstuff	[181]
PDDA, montmorillonite, CuTsPc, PEDOT:PSS	LbL	3D printing	PCA	Analysis of sandy and clayey soils enriched with macronutrients	[29]
PDDA, montmorillonite, CuTsPc, PEDOT:PSS	LbL	Photolithography	IDMAP, PCA and Sammon's Mapping	Detection of plant macronutrients in soil samples	[182]
CuTsPc, PANI, rGO, PAH, AgNPs	LbL	Photolithography	PCA	Commercial salts and flavor enhancers similarity analysis	[60]
PEI, PSS, PEDOT:PSS, PPy, PANI, chitosan, NiTsPc	LbL	Photolithography	IDMAP, PCA and Sammon's Mapping	Taste masking evaluation	[183]
PANI, PSS	LbL	Photolithography	PCA	Taste masking evaluation	[184]
16-mer, PPy, Ru-Pic, SAP	LB and LbL	Photolithography	PCA	Inorganic contaminants detection and basic tastes distinction	[32]
PPV derivative, azo compound, FeTsPc, CuTsPc	LB and LbL	Photolithography	PCA	Phenolic compounds detection in water	[185]

Chitosan, PSS, PANI, PPy, SA	LB and LbL	Photolithography	PCA and ANN	Red wines classification	[186]
PANI, POEA, PPy, SA, Rupy, lignin	LB and LbL	Photolithography	PCA	Wine and basic tastes analysis	[187]
Bis(benzimidazo)perylene	LB and PVD	Photolithography	PCA and Force Scheme	Detection of Cu(II), methylene blue and saccharose	[157]
PA6, PANI	Electrospinning	Photolithography	PCA	Detection of tetracycline residue in milk samples	[38]
PA6, chitosan, cellulose nanowhiskers, rGO, AgNPs, AuNPs	Electrospinning	Photolithography	PCA	Detection of isoborneol in water	[145]
PA6, cellulose nanowhiskers, AgNPs	Electrospinning	Photolithography	PCA, PLS, IDMAP and PC	Heavy metal ions solutions discrimination and lead aqueous solutions analysis	[161]
PA6, PPy, POEA	Electrospinning and LbL	Photolithography	PCA	Detection of paraoxon pesticide in corn crop	[160]
PA6, PPy, PANI	Electrospinning and drop-casting	Photolithography	PCA	Detection of geosmin and 2-methylisoborneol	[159]
PLA, MWCNT	SBS	Direct printing	PCA	Analysis of water contaminated with metals or pesticides	[18]
Gold, carbon, ITO, doped silicon			PCA	Honey discrimination	[188]

CNT—multiwalled carbon nanotube; PCL—polycaprolactone; PEDOT:PSS—poly(3,4-ethylenedioxythiophene)-poly(styrenesulfonate); PPy—polypirrole; AuNPs—gold nanoparticles; rGO—reduced graphene oxide; $LuPc_2$—lutetium bisphthalocyanine; SA—stearic acid; 16-mer—polyaniline oligomer; PS-PMMA—poly(styrene-co-methyl methacrylate) ionomer; PSS—poly(styrenesulfonate); PAH—poly(allylamine hydrochloride); POEA—poly(o-ethoxyaniline); PMMA—poly(methacrylic acid); AgNPs—silver nanoparticles; DPPG—dipalmitoyl phosphatidyl glycerol; PANI—polyaniline; NiTsPc—nickel(II) tetrasulfonated phthalocyanine; CuTsPc—copper(II) tetrasulfonated phthalocyanine; PDDA—poly(diallyldimethylammonium chloride); PEI—poly(ethyleneimine); Ru-Pic—complex mer-[$RuCl_3$(dppb)(4-Mepy)]; SAP—sulfonatedazobenzene polymer; PPV—(p-phenylenevinylene); FeTsPc—iron(II) tetrasulfonated phthalocyanine; Rupy—(mer-[$RuCl_3$(dppb)(py)] (dppb = PPh2(CH2)4PPh2; py = pyridine)); PVD—physical vapor deposition; PA6—polyamide 6; PLA—poly(lactic acid); MWCNT—multiwalled carbon nanotubes; ITO - indium tin oxide; LbL—layer-by-layer; LB - Langmuir-Blodgett; SBS - solution blow spinning; PCA - principal components analysis; PLS - partial least squares; IDMAP - interactive document mapping; ANN - artificial neural network; PC - parallel coordinates.

of its sensitivity by chemical modification or functionalization [61]. To accomplish these requirements, a variety of materials have been used in e-tongues, including conducting polymers, polyelectrolytes, carbon-based nanomaterials, inorganic nanoparticles, and metal complexes, as discussed below.

Conducting polymers (CPs) are characterized by a polymer backbone consisting of alternating single and double bonds [127]. Their overlapping *p*-orbitals create delocalized π-electrons, leading to useful electronic properties for impedimetric sensors [128, 129]. Other advantages of CPs are controllable physical/chemical properties, low cost, reversible doping/de-doping process, and adjustable electrical conductivities from 10^{-11} to 10^{5} S cm^{-1} upon doping [16, 129]. Polypyrrole (PPy) [32, 130], polyaniline (PANI) [130, 131], poly(3,4-ethylenedioxythiophene) (PEDOT) [29, 132], and their derivatives [133, 134] are the most representative examples of CPs in e-tongues. Using CPs nanostructured films is efficient to obtain sensitive e-tongues [17, 29, 130, 133]. For instance, Silva *et al* functionalized 3D-printed graphene IDEs with a film of poly(diallyldimethylammonium chloride)/poly(3,4-ethylenedioxythiophene)-poly(styrenesulfonate) (PDDA/PEDOT:PSS) and increased the sensitivity of an e-tongue up to 60% compared to the bare electrode, due to the electrical nature of PEDOT:PSS [29]. Polyelectrolytes, i.e., polymers with ionizable functional groups, may increase charge transfer in the sensing layer, thus improving the film response [135, 136]. Magro *et al* developed a sensor array with the polyelectrolytes polyethyleneimine (PEI), poly (sodium 4-styrenesulfonate) (PSS) and poly(allylamine hydrochloride) (PAH) to analyze triclosan in water samples [137]. The sensing films were deposited onto commercial gold IDEs patterned onto glass slides comprising 250 fingers with 5 μm width and distance between fingers of 5 μm. The films based on strong polyelectrolytes, e.g., PEI and PSS, led to a higher performance in quantifying triclosan in alkaline aqueous matrices, such as wastewater.

Metal complexes, such as metalloporphyrins and metallophtalocyanines, are suitable for chemical sensors, including e-tongues [31, 32, 61, 135, 138], owing to their semiconducting properties associated with their conjugated system delocalized within the entire molecule. Their sensing properties can be modulated by changing the metal ion or by introducing peripheral substituents in the macrocycles [139, 140]. Inorganic materials are employed in e-tongues [138, 141, 142] either in the form of nanoparticles or as thin films. The usefulness for e-tongues is mainly related to the increased surface area which provides further active sites for the analytes [143]. In addition, the film conductivity may be modulated by adding gold nanoparticles, which increase the sensitivity of an e-tongue [138]. Using metallic nanoparticles can also be convenient when the liquids to be analyzed may dissolve the sensing layer. Indeed, an e-tongue made with gold IDEs coated with inorganic TiO_2, TiO_2-Ag, and GaN films could discriminate ethanol/water mixtures, while polymer films could not be employed for continuous monitoring as they are dissolved by ethanol [141].

Carbon-based nanomaterials such as carbon nanotubes [18, 116, 144] and graphene-based materials [29, 30, 60, 132, 145], and 2D transition metal dichalcogenide nanomaterials (e.g., molybdenum disulfide (MoS_2)) [30, 146–149], are also promising for electrodes and sensing layers due to their high surface area, chemical stability, and electronic properties. Hybrid nanomaterials may have their electronic properties tuned [15, 150, 151] to enhance the sensing performance [152]. These hybrids can be obtained from the materials mentioned above. For instance, the functionalization of conducting polymers with inorganic nanoparticles and carbon nanomaterials can enhance the sensitivity, stability, and reproducibility of the electrode, improving the overall sensor performance [151]. The incorporation of AgCl nanoparticles into a PANI matrix, for example, prevented the de-doping of PANI in basic media and increased the reusability of the sensing units [153]. High sensitivity was obtained with an e-tongue produced with reduced graphene oxide (rGO) nanocomposites with conducting polymers (PEDOT:PSS and PPy) and gold nanoparticles (Au NPs) [37]. The functionalization of graphene oxide (GO) with metal oxide nanoparticles has also been applied to increase the sensitivity of an e-tongue for ammonia [132].

6.4.2 Techniques used to modify IDEs

IDEs can be modified with micro- and nanodeposition techniques, including layer-by-layer assembly [137, 154–156], Langmuir–Blodgett (LB) technique [130, 157], spin-coating [102, 158], drop-casting [30, 37], sputtering [132, 142], physical vapor deposition [28] and nanofiber coating techniques [18, 145, 159–161]. The most used is the LbL technique due to its ease of fabrication and ability to produce films with customizable thickness, composition, and structure [16, 162, 163]. The technique is based on the alternate adsorption of chemical species onto a solid substrate driven by electrostatic forces and secondary interactions such as hydrogen bonding and coordination bonding [162, 164]. Its simplicity, low cost, and versatility make the LbL assembly attractive for sensing platforms [155, 163], particularly since the film properties may be tuned by controlling the experimental conditions such as concentration and ionic strength [155, 165]. Of special relevance are the film thickness and roughness, which are critical for the sensitivity of the devices. In addition to the original dipping procedure, the variant of the LbL technique with spraying [156] may speed up the process up to 500 times, which is relevant to fabricate sensing units for commercial applications [154]. The other method used to deposit multilayers in sensing units is the LB technique [16, 155], with which supramolecular structures can be obtained with precise control over film thickness and architecture [166]. The LB technique was originally designed to fabricate films from amphiphilic molecules, such as fatty acids and phospholipids, but it has been extended to various materials, including conducting polymers [32, 130] and metal complexes [167, 168].

To produce sensing units with large specific surface areas and the capability of mass production, various methods have been used to modify the IDEs. For example, electrospinning (ES) is a method through which nanofibers with high surface area and high porosity can be obtained for sensor applications [160, 169]. In this direction, an e-tongue where IDEs were modified with electrospun nanofibers presented better performance than its counterpart where IDEs were modified by drop-cast films [159]. The nanofiber films can be produced from various conducting materials, including

CPs [38, 129, 160, 170], inorganic nanoparticles [145, 161, 171, 172] and carbon-based nanomaterials [145, 173–175], and may be functionalized with LbL films [160]. An alternative to ES is solution blow spinning (SBS) with which nanofibers are formed from a polymer solution through the use of pressurized gas [176–178]. As an illustration, IDEs coated with SBS films of poly(lactic acid)/multiwalled carbon nanotubes (PLA/MWCNT) were able to discriminate neat water samples from those contaminated with traces of pesticides and heavy metals [18].

Examples of sensing materials used to modify IDEs in impedimetric e-tongues are shown in table 6.1, which also brings the techniques to perform such modification and to fabricate the IDEs.

6.5 Examples of impedimetric e-tongues to evaluate complex matrices

The global selectivity and cross-sensitivity inherent in impedimetric e-tongues combined with their versatility afforded their use in complex liquid systems [13, 17, 189], as exemplified in table 6.1 and described in this section.

6.5.1 Analyses of basic tastes

The main principle of detection for e-tongues involved electrochemical measurements [17] before the first reports of impedimetric e-tongues [32, 130]. For both types of e-tongues, it is essential that their cross-sensitivity and global selectivity are sufficient to mimic the human gustatory system, whose first test is to discriminate the five basic tastes (sweet, salty, sour, bitter, and umami) [17, 32, 55, 130]. This discrimination was made with several impedimetric e-tongues, in some cases with reports of taste suppression [32, 130, 133, 180]. For instance, an e-tongue made with LbL and LB films of conducting polymers, sulfonated azobenzene polymer (SAP), a Ru complex, and stearic acid (SA) could determine the suppression of the bitter taste by sweet [32]. Suppression of the sour taste by adding sucrose is indicated in figure 6.6(a) [133], while a sweet taste was suppressed by adding sour compounds or salt [180]. These results are consistent with the perception by the human gustatory system, in which taste sensation in a mixture of compounds with different tastes is perceived with less intensity than the expected sensation of these compounds individually or summed together [33, 190]. The versatility of impedimetric e-tongues in identifying the basic tastes was shown by Borato *et al* [187], in which two sets of sensor arrays based on nanostructured films were equally able to detect the basic tastes using the PCA to treat data. In further studies, the umami basic taste was incorporated in the analyses [144], the influence of the pH variation was evaluated and minimized [153], and a microfluid impedimetric e-tongue was proposed to allow for faster measurements and reduced volume for sampling [56].

6.5.2 Taste masking evaluation for pharmaceuticals

The enhanced performance of impedimetric e-tongues in recognizing flavors provides the opportunity to evaluate taste-masking of unpleasant tasting and bitter

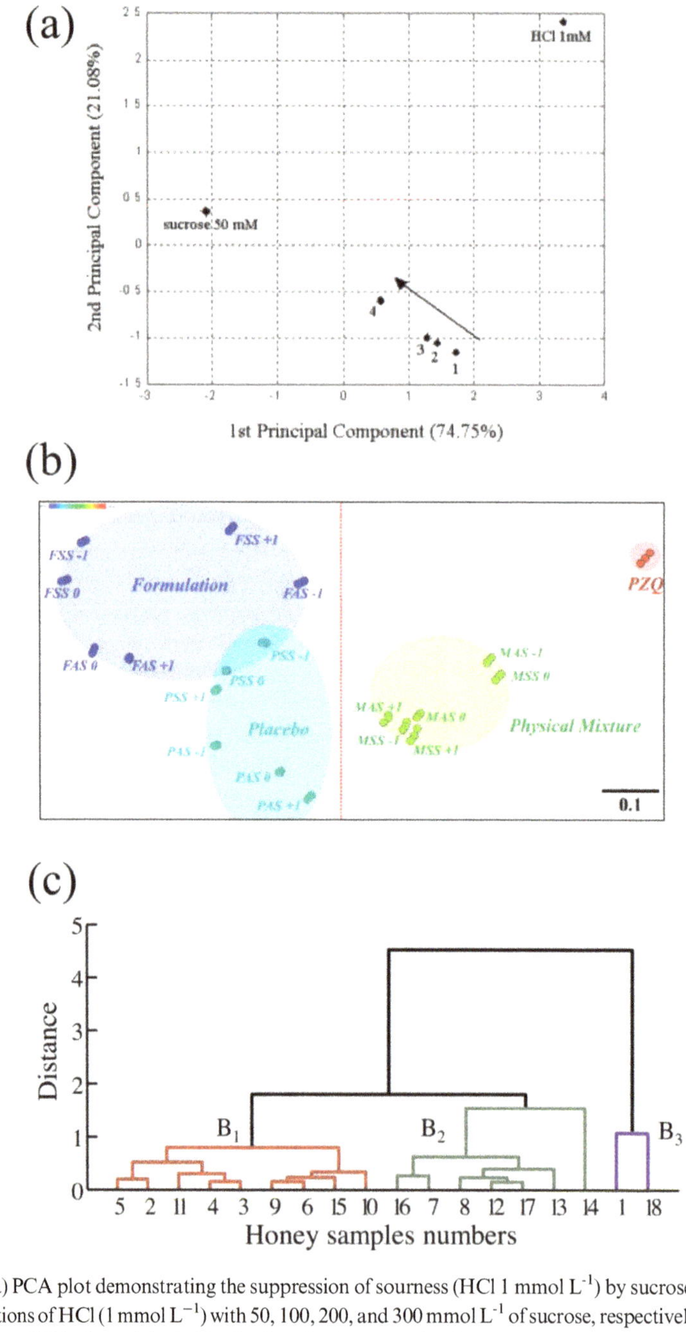

Figure 6.6. (a) PCA plot demonstrating the suppression of sourness (HCl 1 mmol L^{-1}) by sucrose addition. 1, 2, 3, and 4 are solutions of HCl (1 mmol L^{-1}) with 50, 100, 200, and 300 mmol L^{-1} of sucrose, respectively. Reprinted with permission from [133]. Copyright 2003 Elsevier. (b) IDMAP plot for encapsulated praziquantel (formulation), placebo, physical mixtures, and PZQ alone. Reprinted with permission from [183]. Copyright 2018 The Chemical Society of Japan. (c) Dendrogram of *Bupleurum* honey. The three subgroups were labeled as B_1, B_2, and B_3. Adapted and reproduced with permission from [188]. Copyright 2019 Elsevier.

drugs, which is relevant to increase children's adherence to treatments where the drug tastes unpleasant. An e-tongue made of gold IDEs coated with LbL films of PSS, PEI, PPy, PEDOT:PSS, PANI, chitosan (Chit), and tetrasulphonated nickel (II) phthalocyanine (NiTsPc) was used to evaluate the masking effect in the bitter-tasting drug praziquantel (PZQ) encapsulated with polymer microcapsules and microspheres [183]. Using the interactive document mapping (IDMAP) technique to analyze the e-tongue spectra, the encapsulate PZQ could be distinguished from placebo formulations, PZQ alone, and their physical mixtures [183]. As shown in figure 6.6(b), samples in which the bitter taste was not expected to be pronounced (formulation and placebo samples) are located on the left side of the IDMAP plot, while those samples with intense bitter taste are on the right side of the plot. Furthermore, samples of the physical mixtures appear between the microparticles samples and PZQ alone, demonstrating suppression of the bitter taste.

Using an e-tongue comprising five sensing units with PANI/PSS LbL films, the taste-masking ability by encapsulating Saquinavir, an antiretroviral drug used in HIV treatment, with the polymers Eudragit RS100® and Pullulan was verified [184]. In a PCA plot, the data for encapsulated Saquinavir could be distinguished from the data for Saquinavir alone. The taste masking of efavirenz (EFV), also used in HIV treatment, was achieved by adding flavorizers, and this was detected with an e-tongue made with five sensing units, including a bare gold interdigitated electrode and four other such electrodes coated with 1, 3, 5, and 7 bilayers of PANI/PSS LbL films [191]. According to the PCA plot, the formulations with vanilla and lemon flavorizers had higher efficiency in taste masking. Furthermore, the e-tongue was also able to discriminate low modifications in the drug formulation.

6.5.3 Analysis of beverages and foodstuff

The high sensitivity of impedimetric e-tongues, which are capable of detecting basic tastes well below the human threshold for taste detection [13, 17, 32], has been exploited to analyze a variety of beverages and food. The list is extensive and includes coconut water [32], isotonic drinks with fruit flavors [133], wines from different vintages, brands, and types of grapes [32, 186, 187]. For instance, the data from an e-tongue was used to predict the chemical characteristics of five red wines using regression techniques [31]. A detailed classification of honey from two botanic origins was made possible from data with an impedimetric e-tongue, as indicated in the dendrogram of figure 6.6(c) for the honey with *Bupleurum* pollen prevalence [188]. Interestingly, all subgroups presented almost identical water and sugar content. Furthermore, the small changes in impedance caused by changes in the electrolyte solution could be related to the pollen composition. This is a demonstration of the e-tongue ability in discriminating honey of different floral origins, which can also be used to detect honey adulteration.

Tetracycline was detected in fat and skimmed milk samples at concentrations from 1 to 300 ppb using an e-tongue comprising seven sensing units of IDEs modified with PA6 and PANI nanofibers [38]. The detection of trace amounts of gliadin, a protein of gluten, was performed with a microfluidic impedimetric e-tongue [181]. The e-tongue was capable of distinguishing gliadin in ethanol solutions from 0.005 mg kg^{-1} up to 1310 mg kg^{-1}, with a practical application demonstrated in tests with samples of gliadin and gluten-free foodstuff deliberately contaminated. Different types of sugar and apple juices were studied with an e-tongue made of carbon interdigitated electrodes printed on paper, with the juices being distinguished based on their sugar content [179].

6.5.4 Quality control

Impedimetric e-tongues have been used in quality control analysis, i.e. to check the conditions or parameters of a product or process to reach the required level of quality. For instance, to determine the water content in ethanol an e-tongue with ultrathin films of gallium nitrate (GaN) and titanium dioxide (TiO_2) was used [141]. The PCA plot in figure 6.7(a) shows the discrimination of the various contents of

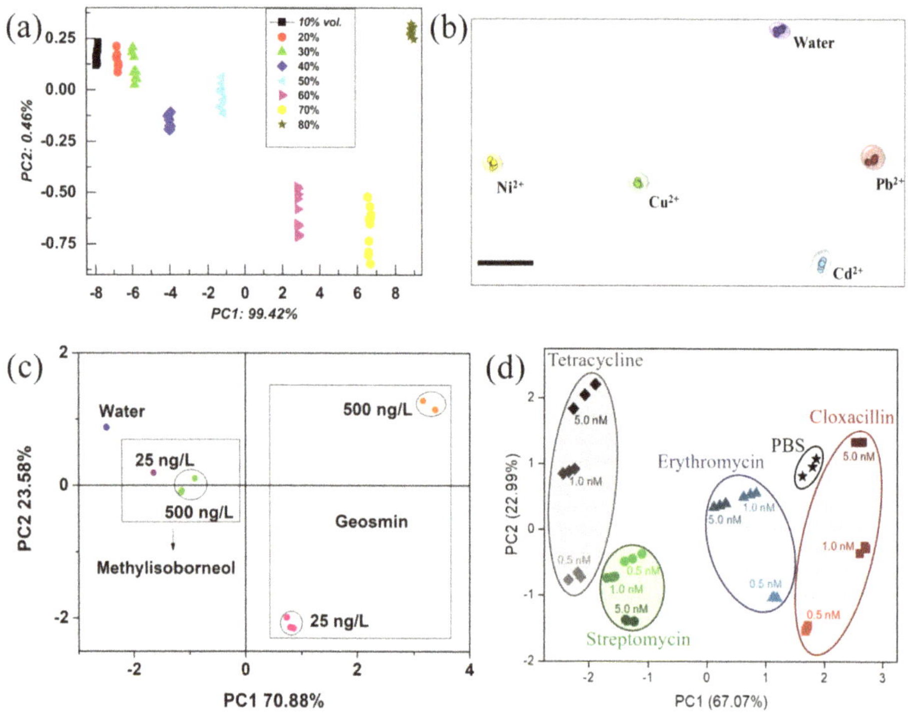

Figure 6.7. (a) PCA plot for the electrical capacitance data for ethanol/water mixtures showing the correlation between PC1 and the amount of water in the sample. Reprinted with permission from [141]. Copyright 2016 Elsevier. (b) PLS plot for data with heavy metal ions (Ni^{2+}, Cu^{2+}, Cd^{2+}, Pb^{2+}) at a concentration of 1 mmol L^{-1} and water. Reprinted with permission from [161]. Copyright 2019 Elsevier. (c) PCA plot from the electrical resistance data at 1 kHz using the e-tongue of electrospun materials for geosmin and 2-methylisoborneol. Reprinted with permission from [159]. Copyright 2020 Wiley. (d) PCA plot for PBS and four antibiotic solutions with different concentrations using the capacitance data at 1 kHz. Reprinted with permission from [30]. Copyright 2020 Elsevier.

water, and a 'calibration curve' could be established, as the e-tongue was useful for identifying fuel adulteration. The quality of soil in terms of plant macronutrients (N, P, K, Ca, Mg and S) was evaluated with an e-tongue containing four sensing units of LbL films [182]. It is worth mentioning that a flow system was chosen to avoid decantation of the soil in the solutions analyzed. The impedance measurements were performed with soil samples enriched with one of the macronutrients, in addition to a control sample (soil without fertilizer), and distilled water. These different samples could be distinguished when the impedance spectra were treated with PCA and the non-linear statistical techniques (IDAMAP and Sammon's mapping), with IDMAP providing the most efficient discrimination with feature selection (by choosing the frequencies at which distinction was highest).

6.5.5 Environmental monitoring

A great advantage of e-tongues is the possibility of analyzing non-edible media, such as toxic and pollutant compounds [189]. This is relevant due to the increasing concern about water quality, which motivated the use of impedimetric e-tongues to detect hazardous components and contaminants in water, such as pharmaceutics [30], dyes [126, 157], heavy metals [133, 134, 161], and pesticides [18, 37, 160]. For instance, nylon nanofibers modified with conducting polymers were used in an e-tongue to detect different concentrations of the paraoxon pesticide in water, with a clear distinction from non-contaminated water samples [160]. Organophosphate pesticides were detected using an e-tongue based on graphene hybrid nanocomposites, which represents an alternative to the biosensors normally employed to detect such pesticides [37]. Impedimetric e-tongues are also useful for heavy metals, as demonstrated with an array of sensing units made with electrospun nanofibers, cellulose nanowhiskers (CNW), and silver nanoparticles to discriminate heavy metal ions (Pb, Cu, Ni, Cd) in water [161]. Figure 6.7(b) shows the discrimination in the partial least square (PLS) plot after feature selection where optimization was reached by choosing the most efficient sensing units and frequency values.

The detection of phenolic compounds is important since they may be toxic even at very low concentrations. The detection of such compounds could be performed with an e-tongue made with ten sensing units modified with LB and LbL films of a tetrasulfonated phthalocyanine, a PPV (p-phenylenevinylene) derivative, and an azo compound [185]. The e-tongue could detect similar phenolic compounds in water at nanomolar levels, including the discrimination between isomers, without the need for enzymatic reactions or solvent extractions. Other water contaminants as geosmin (GSM) and 2-methylisoborneol (MIB) were detected using two e-tongues that differed by the method employed to deposit the films on IDEs [159]. The e-tongue with sensing units modified with electrospun nanofibers was more efficient than that obtained with dropcasting films. Figure 6.7(c) shows the PCA plot obtained with the electrical resistance data in which 94.46% (PC1 + PC2 values) of the total variance were accounted for. The more efficient e-tongue was able to detect GSM and MIB in river water at concentrations down to ng L^{-1}.

Four antibiotics with similar structures were detected with an e-tongue made with five sensing units of drop-cast films of varying proportions of graphene oxide (GO) and molybdenum disulfide (MoS_2) [30]. The antibiotics could be distinguished at various concentrations in aqueous solutions. Figure 6.7(d) shows the PCA plot with clear separation, including from a PBS sample without antibiotics. Successful analyses were also obtained using river water and urine-contaminated samples.

The examples mentioned before illustrate that impedimetric e-tongues are versatile for the analysis of a plethora of analytes.

6.6 Final remarks

The development of fast, sensitive, and reliable analytical techniques for evaluating water and food quality is of paramount importance. Chemical sensors, including impedimetric e-tongues, may fulfill the stringent requirements for monitoring water and the environment, as discussed in this book chapter. It is significant that the performance of e-tongues in discriminating samples, relying on principles of cross-sensitivity and global selectivity, is sufficient for effective monitoring. Furthermore, there is ample literature demonstrating that impedimetric e-tongues can discriminate the basic tastes and classify food and beverages, in many cases with superior ability to traditional analytical techniques. This high performance is achieved by optimizing parameters such as those involved in fabricating IDEs, the materials and techniques to functionalize the electrodes. Throughout the chapter, examples were given of such materials as conducting polymers, polyelectrolytes, nanoparticles, carbon nanotubes, graphene, etc. In the last two decades, isolated efforts in the commercialization of e-tongues have been made, but the very few systems available do not rely on impedimetric measurements. On the other hand, the miniaturization of electronics towards micro- and nanoscale associated with microfluidics systems and more precise film-deposition techniques are expected to boost these devices with portability for real-time analysis and industrial applications.

Acknowledgments

The authors are grateful for the financial support from Fundação de Amparo à Pesquisa do Estado de São Paulo (FAPESP) (grant numbers: 2017/10582-8, 2017/12174-4, 2018/22214-6, 2016/23793-4) Conselho Nacional de Desenvolvimento Científico e Tecnológico (CNPq), CAPES and MCTI-SisNano (CNPq/402.287/2013-4), and Rede Agronano-EMBRAPA from Brazil.

References

[1] Méndez E, González-Fuentes M A, Rebollar-Perez G, Méndez-Albores A and Torres E 2017 *J. Environ. Sci. Heal. Part* A **52** 235–53

[2] Koyuncu I, Arikan O A, Wiesner M R and Rice C 2008 *J. Memb. Sci.* **309** 94–101

[3] Shimizu F M, Braunger M L and Riul A 2019 *Chemosensors* **7** 1–19

[4] Sidhu G K, Singh S, Kumar V, Dhanjal D S, Datta S and Singh J 2019 *Crit. Rev. Environ. Sci. Technol.* **49** 1135–87

[5] Lari S Z, Khan N A, Gandhi K N, Meshram T S and Thacker N P 2014 *J. Environ. Heal. Sci. Eng.* **12** 11

[6] Islam M S, Ahmed M K, Raknuzzaman M, Habibullah -Al- Mamun M and Islam M K 2015 *Ecol. Indic.* **48** 282–91

[7] Sharma S and Bhattacharya A 2017 *Appl. Water Sci.* **7** 1043–67

[8] Hussain M A 2016 *Foods* **5** 21

[9] Nerín C, Aznar M and Carrizo D 2016 *Trends Food Sci. Technol.* **48** 63–8

[10] Andre R S, Mercante L A, Facure M H M, Pavinatto A and Correa D S 2021 *Woodhead Publishing Series in Composites Science and Engineering* ed Y Dong, A Baji, S B T-E P and C Ramakrishna (Cambridge: Woodhead Publishing) pp 261–86

[11] Law J W-F, Ab Mutalib N-S, Chan K-G and Lee L-H 2015 *Front. Microbiol.* **5** 770

[12] Vlasov Y G, Legin A V and Rudnitskaya A M 2008 *Russ. J. Gen. Chem.* **78** 2532–44

[13] Jiang H, Zhang M, Bhandari B and Adhikari B 2018 *Food Rev. Int.* **34** 746–69

[14] Mercante L A, Scagion V P, Migliorini F L, Mattoso L H C C and Correa D S 2017 *Trends Anal. Chem.* **91** 91–103

[15] Dewangan K, Shrivas K and Kurrey R 2020 *Hybrid Nanomaterials as Chemical Sensors* (Amsterdam: Elsevier)

[16] Correa D S, Medeiros E S, Oliveira J E, Paterno L G and Mattoso L H C 2014 *J. Nanosci. Nanotechnol.* **14** 1–19

[17] Riul A, Dantas C A R A R, Miyazaki C M M, Oliveira O N N, Riul-Jr A, Dantas C A R A R, Miyazaki C M M and Oliveira-Jr O N 2010 *Analyst* **135** 2481–95

[18] Oliveira J E, Grassi V, Scagion V P, Mattoso L H C, Glenn G M and Medeiros E S 2013 *IEEE Sens. J.* **13** 759–66

[19] Campos I, Alcañiz M, Aguado D, Barat R, Ferrer J, Gil L, Marrakchi M, Martínez-Mañez R, Soto J and Vivancos J-L 2012 *Water Res* **46** 2605–14

[20] Winquist F 2008 *Microchim. Acta* **163** 3–10

[21] Winquist F, Wide P and Lundström I 1997 *Anal. Chim. Acta* **357** 21–31

[22] Scampicchio M, Ballabio D, Arecchi A, Cosio S M and Mannino S 2008 *Microchim. Acta* **163** 11–21

[23] Scampicchio M, Benedetti S, Brunetti B and Mannino S 2006 *Electroanalysis* **18** 1643–8

[24] Buratti S, Benedetti S, Scampicchio M and Pangerod E C 2004 *Anal. Chim. Acta* **525** 133–9

[25] Gallardo J, Alegret S and del Valle M 2005 *Talanta* **66** 1303–9

[26] Gutiérrez M, Alegret S, Cáceres R, Casadesús J, Marfà O and del Valle M 2007 *Comput. Electron. Agric.* **57** 12–22

[27] Ciosek P, Mamińska R, Dybko A and Wróblewski W 2007 *Sens. Actuators B Chem.* **127** 8–14

[28] Mimendia A, Gutiérrez J M, Leija L, Hernández P R, Favari L, Muñoz R and del Valle M 2010 *Environ. Model. Softw.* **25** 1023–30

[29] Americo da Silva B, Neris Coutinho R D A, Rodrigues R J, da Silva T A, Braunger M L, Coutinho M A N, do Amaral L R, Rodrigues V and Riul A 2019 *Chemosensors* **7** 50

[30] Facure M H M, Schneider R, dos Santos D M and Correa D S 2020 *Talanta* **217** 121039

[31] Garcia-Hernandez C, Salvo Comino C, Martín-Pedrosa F, Rodriguez-Mendez M L and Garcia-Cabezon C 2018 *Sens. Actuators B Chem.* **277** 365–72

[32] Riul A, Dos Santos D S S, Wohnrath K, Di Tommazo R, Carvalho A C P L F C P L F, Fonseca F J J, Oliveira O N N, Taylor D M M and Mattoso L H C H C 2002 *Langmuir* **18** 239–45

[33] Breslin P A S 1996 *Trends Food Sci. Technol.* **7** 390–9
[34] Sapsanis C, Omran H, Chernikova V, Shekhah O, Belmabkhout Y, Buttner U, Eddaoudi M and Salama K 2015 *Sensors* **15** 18153–66
[35] Wiziack N K L, Paterno L G, Fonseca F J and Mattoso L H C 2007 *Sens. Actuators, B Chem.* **122** 484–92
[36] Varshney M and Li Y 2009 *Biosens. Bioelectron.* **24** 2951–60
[37] Facure M H M, Mercante L A, Mattoso L H C and Correa D S 2017 *Talanta* **167** 59–66
[38] Scagion V P, Mercante L A, Sakamoto K Y, Oliveira J E, Fonseca F J, Mattoso L H C, Ferreira D and Correa D S 2016 *RSC Adv.* **6** 103740–6
[39] Breijo E G, Pinatti C O, Peris R M, Fillol M A, Martínez-Máñez R and Camino J S 2013 *Sensors Actuators A Phys.* **192** 1–8
[40] Moreno-Barón L, Cartas R, Merkoçi A, Alegret S, del Valle M, Leija L, Hernandez P R and Muñoz R 2006 *Sens. Actuators B Chem.* **113** 487–99
[41] Moreno-Barón L, Cartas R, Merkoçi A, Alegret S, Gutiérrez J M, Leija L, Hernandez P R, Muñoz R and del Valle M 2005 *Anal. Lett.* **38** 2189–206
[42] Shimizu F M, Todão F R, Gobbi A L, Oliveira O N, Garcia C D and Lima R S 2017 *ACS Sensors* **2** 1027–34
[43] Kao K C 2004 *Dielectric Phenomena in Solids: with Emphasis on Physical Concepts of Electronic Processes* (San Diego, CA: Elsevier)
[44] Barsoukov E and Macdonald J R 2005 *Impedance Spectroscopy Theory, Experiment, and Applications* (Hoboken, NJ: Wiley)
[45] Macedo P B, Moynihan C T and Bose R 1972 *Phys. Chem. Glas.* **13** 171–9
[46] Alexander C L, Tribollet B and Orazem M E 2016 *Electrochim. Acta* **201** 374–9
[47] Kerner Z and Pajkossy T 1998 *J. Electroanal. Chem.* **448** 139–42
[48] Cole K S and Cole R H 1941 *J. Chem. Phys.* **9** 341–51
[49] Davidson D W and Cole R H 1951 *J. Chem. Phys.* **19** 1484–90
[50] Davidson D W 1961 *Can. J. Chem.* **39** 571–94
[51] Warburg E 1899 *Ann. der Phys. Chemie* **303** 493–9
[52] Tolouei N E, Ghamari S and Shavezipur M 2020 *J. Electroanal. Chem.* **878** 114598
[53] Cho S 2011 *BioChip J.* **5** 327–32
[54] Gimsa J and Wachner D 1998 *Biophys. J.* **75** 1107–16
[55] Cabral F P A, Bergamo B B, Dantas C A R, Riul A and Giacometti J A 2009 *Rev. Sci. Instrum.* **80** 2–5
[56] Daikuzono C M, Dantas C A R, Volpati D, Constantino C J L, Piazzetta M H O, Gobbi A L, Taylor D M, Oliveira O N and Riul A 2015 *Sens. Actuators B Chem.* **207** 1129–35
[57] Brosel-Oliu S, Abramova N, Uria N and Bratov A 2019 *Anal. Chim. Acta* **1088** 1–19
[58] Mazlan N S, Ramli M M, Abdullah M M A B, Halin D S C, Isa S S M, Talip L F A, Danial N S and Murad S A Z 2017 *AIP Conf. Proc.* 1885 020276
[59] Podrazka M, Báczyńska E, Kundys M, Jeleń P S and Nery E W 2017 *Biosensors* **8** 1–24
[60] Fukushima K L, Scagion V P, Facure M H M, Pinheiro A C M, Correa D S, Nunes C A and Oliveira J E 2020 *IEEE Sens. J.* **1**
[61] Lvova L, Pudi R, Galloni P, Lippolis V, Di Natale C, Lundström I and Paolesse R 2015 *Sens. Actuators B Chem.* **207** 1076–86
[62] Manzoli A, De Almeida G F B, Filho J A, Mattoso L H C, Riul A, Mendonca C R and Correa D S 2015 *Opt. Laser Technol.* **69** 148–53
[63] Alexander F, Price D T and Bhansali S 2010 *J. Phys. Conf. Ser.* **224** 012134

[64] Rivadeneyra A, Fernández-Salmerón J, Banqueri J, López-Villanueva J A, Capitan-Vallvey L F and Palma A J 2014 *Sens. Actuators B Chem.* **204** 552–60
[65] Sharma N K, Nain A, Singh K, Rani N and Singal A 2019 *Int. J. Innov. Technol. Explor. Eng.* **8** 4015–25
[66] Dizon A and Orazem M E 2019 *Electrochim. Acta* **327** 135000
[67] Rana S, Page R H and McNeil C J 2011 *Electrochim. Acta* **56** 8559–63
[68] Yan X-F, Wang M-H and An D 2011 *Chinese J. Anal. Chem.* **39** 1601–10
[69] Yang L and Guiseppi-Elie A 2008 *Encyclopedia of Microfluidics and Nanofluidics* (Boston, MA: Springer US) 811–23
[70] Zhang J, Zhang G, Zhou T and Sun S 2020 *Adv. Funct. Mater.* **30** 1910000
[71] Li D, Batchelor-McAuley C, Chen L and Compton R G 2019 *ACS Sensors* **4** 2250–66
[72] del Barrio J and Sánchez-Somolinos C 2019 *Adv. Opt. Mater.* **7** 1900598
[73] Nag A and Mukhopadhyay S C 2018 *Sens. Actuators A Phys.* **269** 53–61
[74] Liang J, Mondal A K, Wang D and Iacopi F 2019 *Adv. Mater. Technol.* **4** 1800200
[75] Araromi O A, Rosset S and Shea H R 2015 *ACS Appl. Mater. Interfaces* **7** 18046–53
[76] Nam I, Kim G-P, Park S, Park J, Kim N D and Yi J 2012 *Nanoscale* **4** 7350
[77] Suh Y, Shin D and Chun Y T 2019 *RSC Adv.* **9** 38085–104
[78] Lee D, Lee S, Rho J, Jang W, Han S H and Chung T D 2018 *Biosens. Bioelectron.* **101** 317–21
[79] Kamath R R and Madou M J 2014 *Anal. Chem.* **86** 2963–71
[80] Voitechovič E, Bratov A, Abramova N, Razumienė J, Kirsanov D, Legin A, Lakshmi D, Piletsky S, Whitcombe M and Ivanova-Mitseva P K 2015 *Electrochim. Acta* **173** 59–66
[81] Guimerà A, Gabriel G, Prats-Alfonso E, Abramova N, Bratov A and Villa R 2015 *Sens. Actuators B Chem.* **207** 1010–8
[82] Kechagias J, Iakovakis V, Katsanos M and Maropoulos S 2008 *J. Mater. Sci.* **43** 2522–35
[83] Ludwig B, Zheng Z, Shou W, Wang Y and Pan H 2016 *Sci. Rep.* **6** 23150
[84] Gong Y, Li C, Huang X, Luo Y, Li D, Meng Q and Iversen B B 2013 *ACS Appl. Mater. Interfaces* **5** 795–800
[85] Zhao C, Liu Y, Beirne S, Razal J and Chen J 2018 *Adv. Mater. Technol.* **3** 1800028
[86] Wu J *et al* 2015 *Langmuir* **31** 1210–7
[87] Liu N and Gao Y 2017 *Small* **13** 1701989
[88] Cook M M, Rahman M D and Lu P-H 1998 Advances in resist technology and processing XV *Proc. SPIE* **3333** 1180
[89] Liu J, Zeng X and Ai J 2016 *Chinese Sci. Bull* **61** 610–5
[90] Roy D, Basu P K, Raghunathan P and Eswaran S V 2003 *Magn. Reson. Chem.* **41** 84–90
[91] Vlnieska V, Zakharova M, Mikhaylov A and Kunka D 2020 *Polymers (Basel)* **12** 2359
[92] Hemanth S, Anhøj T A, Caviglia C and Keller S S 2017 *Microelectron. Eng.* **176** 40–4
[93] Muaz A K M, Hashim U, Liu W, Ibrahim F, Thong K L and Mohktar M S 2014 *2014 IEEE Conf. on Biomedical Engineering and Sciences (IECBES)* (IEEE) 146–50
[94] Yang L, Wei J, Ma Z, Song P, Ma J, Zhao Y, Huang Z, Zhang M, Yang F and Wang X 2019 *Nanomaterials* **9** 1789
[95] Mentzel T S, Wanger D D, Ray N, Walker B J, Strasfeld D, Bawendi M G and Kastner M A 2012 *Nano Lett.* **12** 4404–8
[96] Wu Z, Parvez K, Feng X and Müllen K 2013 *Nat. Commun.* **4** 2487
[97] Mojena-Medina D, Hubl M, Bäuscher M, Jorcano J L, Ngo H and Acedo P 2020 *Sensors* **20** 5711
[98] Sassa F, Biswas G C and Suzuki H 2020 *Lab Chip* **20** 1358–89

[99] Wei L, Nitta N and Yushin G 2013 *ACS Nano* **7** 6498–506
[100] Zhang C, Pei X, Xu Z, Wang C, Zhang G, Guo X, Fu H and Wu N 2020 *Mater. Today Commun.* **22** 100723
[101] Shi X, Zeng Z, Liao C, Tao S, Guo E, Long X, Wang X, Deng D and Dai Y 2018 *J. Alloys Compd.* **739** 979–86
[102] Paula K T, Gaál G, Almeida G F B, Andrade M B, Facure M H M, Correa D S, Riul A, Rodrigues V and Mendonça C R 2018 *Opt. Laser Technol.* **101** 74–9
[103] Huang G, Li N, Du Y, Feng Q, Xiao H, Wu X and Fu S-Y 2018 *ACS Appl. Mater. Interfaces* **10** 723–32
[104] Kumar R, Joanni E, Savu R, Pereira M S, Singh R K, Constantino C J L, Kubota L T, Matsuda A and Moshkalev S A 2019 *Energy* **179** 676–84
[105] Wang C-P, Chou C-P, Chang T-L and Chou C-Y 2018 *Microelectron. Eng.* **189** 69–73
[106] Yun X, Xiong Z, Tu L, Bai L and Wang X 2017 *Carbon N. Y.* **125** 308–17
[107] Wang C-K, Liao W-H, Wu H-M, Lo Y-H, Lin T-R and Tung Y-C 2017 *J. Micromech. Microeng.* **27** 115003
[108] Anbari A, Chien H-T, Datta S S, Deng W, Weitz D A and Fan J 2018 *Small* **14** 1703575
[109] Cheng M, Deivanayagam R and Shahbazian-Yassar R 2020 *Batter. Supercaps* **3** 130–46
[110] Salim A and Lim S 2017 *Sensors* **17** 2593
[111] Wang Y, Shi Y, Zhao C X, Wong J I, Sun X W and Yang H Y 2014 *Nanotechnology* **25** 094010
[112] Bariya M *et al* 2018 *ACS Nano* **12** 6978–87
[113] Sang Tran T, Dutta N and Roy Choudhury N 2019 *Materials (Basel)* 12 978
[114] Li Q, Zhang J, Li Q, Li G, Tian X, Luo Z, Qiao F, Wu X and Zhang J 2019 *Front. Mater.* **5** 1–14
[115] Hartwig M, Zichner R and Joseph Y 2018 *Chemosensors* **6** 66
[116] Migliorini F L, Santos D M, dos, Soares A C, Mattoso L H C, Oliveira O N and Correa D S 2020 *Chemosensors* **8** 87
[117] Cinti S, Mazzaracchio V, Cacciotti I, Moscone D and Arduini F 2017 *Sensors* **17** 2267
[118] Yáñez-Sedeño P, Campuzano S and Pingarrón J M 2020 *Biosensors* **10** 76
[119] Ghosale A, Shrivas K, Deb M K, Ganesan V, Karbhal I, Bajpai P K and Shankar R 2018 *Anal. Methods* **10** 3248–55
[120] Tehrani Z, Korochkina T, Govindarajan S, Thomas D J, O'Mahony J, Kettle J, Claypole T C and Gethin D T 2015 *Org. Electron.* **26** 386–94
[121] Paul G, Torah R, Beeby S and Tudor J 2014 *Sens. Actuators A Phys.* **206** 35–41
[122] Achilli A, Bonfiglio A and Pani D 2018 *IEEE Sens. J.* **18** 4097–107
[123] Rivadeneyra A and López-Villanueva J A 2020 *Micromachines* **11** 367
[124] Zhang M, Mei H, Chang P and Cheng L 2020 *J. Mater. Chem.* A **8** 10670–94
[125] Gaál G, da Silva T A, Gaál V, Hensel R C, Amaral L R, Rodrigues V and Riul A 2018 *Front. Chem.* **6** 1–8
[126] Aoki P H B, Alessio P, Riul A, De Saja Saez J A and Constantino C J L 2010 *Anal. Chem.* **82** 3537–46
[127] Guo X and Facchetti A 2020 *Nat. Mater.* **19** 922–8
[128] Le T H, Kim Y and Yoon H 2017 *Polymers (Basel)* 9
[129] Wang X X, Yu G F, Zhang J, Yu M, Ramakrishna S and Long Y Z 2021 *Prog. Mater. Sci.* **115** 100704
[130] Riul A, Soto A M G, Mello S V, Bone S, Taylor D M and Mattoso L H C 2003 *Synth. Met.* **132** 109–16

[131] Braga G S, Paterno L G and Fonseca F J 2012 *Sens. Actuators B Chem.* **171–172** 181–9
[132] Cavallari M R, Braga G S, Da Silva M F P, Izquierdo J E E, Paterno L G, Dirani E A T, Kymissis I and Fonseca F J 2017 *IEEE Sens. J.* **17** 1971–80
[133] Riul A, Malmegrim R R, Fonseca F J and Mattoso L H C 2003 *Biosens. Bioelectron.* **18** 1365–9
[134] Borato C E, Leite F L, Mattoso L H C, Goy R C, Campana Filho S P, De Vasconcelos C L, Da Trindade Neto C G, Pereira M R, Fonseca J L C and Oliveira O N 2006 *IEEE Trans. Dielectr. Electr. Insul.* **13** 1101–9
[135] Rončević I Š, Krivić D, Buljac M, Vladislavić N and Buzuk M 2020 *Sensors* **20** 1–67
[136] Andre R S, Pereira J C, Mercante L A, Locilento D, Mattoso L H C and Correa D S 2018 *J. Alloys Compd.* **767** 1022–9
[137] Magro C, Zagalo P, Pereira-da-Silva J, Pires Mateus E, Branco Ribeiro A, Ribeiro P and Raposo M 2020 *Nanomaterials* **10** 640
[138] Mercante L A, Scagion V P, Pavinatto A, Sanfelice R C, Mattoso L H C and Correa D S 2015 *J. Nanomater.* **2015** 1–7
[139] Paolesse R, Natale C D, Burgio M, Martinelli E, Mazzone E, Palleschi G and D'Amico A 2003 *Sens. Actuators B Chem.* **95** 400–5
[140] Paolesse R, Lvova L, Nardis S, Di Natale C, D'Amico A and Lo Castro F 2008 *Microchim. Acta* **163** 103–12
[141] De Queiroz D P, Florentino A D O, Bruno J C, Da Silva J H D, Riul A and Giacometti J A 2016 *Sens. Actuators, B Chem.* **230** 566–70
[142] Magro C, Mateus E P, Paz-Garcia J M, Sério S, Raposo M and Ribeiro A B 2019 *Sensors (Switzerland)* **19**
[143] Heuer-Jungemann A *et al* 2019 *Chem. Rev.* **119** 4819–80
[144] Pioggia G, Di Francesco F, Ferro M, Sorrentino F, Salvo P and Ahluwalia A 2008 *Microchim. Acta* **163** 57–62
[145] Migliorini F L, Teodoro K B R, Scagion V P, dos Santos D M, Fonseca F J, Mattoso L H C and Correa D S 2019 *Surfaces* **2** 432–43
[146] Eftekhari A 2017 *J. Mater. Chem.* A **5** 18299–325
[147] Hu Y, Huang Y, Tan C, Zhang X, Lu Q, Sindoro M, Huang X, Huang W, Wang L and Zhang H 2017 *Mater. Chem. Front.* **1** 24–36
[148] Samadi M, Sarikhani N, Zirak M, Zhang H, Zhang H L and Moshfegh A Z 2018 *Nanoscale Horizons* **3** 90–204
[149] Loo A H, Bonanni A, Sofer Z and Pumera M 2015 *Electrochem. Commun.* **50** 39–42
[150] Correa D S, Pavinatto A, Mercante L A, Mattoso L H C, Oliveira J E and Riul A 2017 *Nanobiosensors: Nanotechnology in the Agri-Food Industry* ed A M B T-N Grumezescu (New York: Academic) pp 205–44
[151] Dakshayini B S, Reddy K R, Mishra A, Shetti N P, Malode S J, Basu S, Naveen S and Raghu A V 2019 *Microchem. J.* **147** 7–24
[152] Li C, Bai H and Shi G 2009 *Chem. Soc. Rev.* **38** 2397–409
[153] Manzoli A, Shimizu F M M, Mercante L A, Paris E C, Oliveira O N, Correa D S and Mattoso L H C 2014 *Phys. Chem. Chem. Phys.* **16** 24275–81
[154] Aoki P H B, Alessio P, Volpati D, Paulovich F V, Riul A, Oliveira O N and Constantino C J L 2014 *Mater. Sci. Eng.* C **41** 363–71
[155] De Oliveira R F, De Barros A and Ferreira M 2017 *Nanostructures* (Amsterdam: Elsevier) pp 105–23

[156] Aoki P H B, Volpati D, Cabrera F C, Trombini V L, Riul A and Constantino C J L 2012 *Mater. Sci. Eng.* C **32** 862–71
[157] Volpati D, Aoki P H B, Dantas C A R, Paulovich F V, De Oliveira M C F, Oliveira O N, Riul A, Aroca R F and Constantino C J L 2012 *Langmuir* **28** 1029–40
[158] Castiello F R, Porter J, Modarres P and Tabrizian M 2019 *Phys. Chem. Chem. Phys.* **21** 15787–97
[159] Migliorini F L, Teodoro K B R, dos Santos D M, Fonseca F J, Mattoso L H C and Correa D S 2020 *Polym. Adv. Technol.* **31** 2075–82
[160] Oliveira J E, Scagion V P, Grassi V, Correa D S and Mattoso L H C 2012 *Sens. Actuators, B Chem.* **171–172** 249–55
[161] Teodoro K B R, Shimizu F M, Scagion V P and Correa D S 2019 *Sens. Actuators, B Chem.* **290** 387–95
[162] Yuan W, Weng G M, Lipton J, Li C M, Van Tassel P R and Taylor A D 2020 *Adv. Colloid Interface Sci.* **282**
[163] Rivero P J, Goicoechea J and Arregui F J 2019 *Sensors (Switzerland)* **19**
[164] Lipton J, Weng G M, R�hr J A, Wang H and Taylor A D 2020 *Matter* **2** 1148–65
[165] Gittleson F S, Kohn D J, Li X and Taylor A D 2012 *ACS Nano* **6** 3703–11
[166] da Rocha Rodrigues R, da Silva R L C G, Caseli L and Péres L O 2020 *Adv. Colloid Interface Sci.* **408** 102277
[167] Rubira R J G, Aoki P H B, Constantino C J L and Alessio P 2017 *Appl. Surf. Sci.* **416** 482–91
[168] Ferreira M, Constantino C J L, Riul A, Wohnrath K, Aroca R F, Giacometti J A, Oliveira O N and Mattoso L H C 2003 *Polymer (Guildf)* **44** 4205–11
[169] Yang T, Zhan L and Huang C Z 2020 *TrAC Trends Anal. Chem.* **124** 115813
[170] Mirzaei A, Kumar V, Bonyani M, Majhi S M, Bang J H, Kim J-Y, Kim H W, Kim S S and Kim K-H 2020 *Asian J. Atmos. Environ.* **14** 85–104
[171] Homocianu M and Pascariu P 2020 *Polym. Rev.* **60** 493–541
[172] Zhang C L and Yu S H 2014 *Chem. Soc. Rev.* **43** 4423–48
[173] Al-Dhahebi A M, Gopinath S C B and Saheed M S M 2020 *Nano Converg.* **7**
[174] Agyemang F O, Tomboc G M, Kwofie S and Kim H 2018 *Electrochim. Acta* **259** 1110–9
[175] Lee J K Y, Chen N, Peng S, Li L, Tian L, Thakor N and Ramakrishna S 2018 *Prog. Polym. Sci.* **86** 40–84
[176] Medeiros E S, Glenn G M, Klamczynski A P, Orts W J and Mattoso L H C 2009 *J. Appl. Polym. Sci.* **113** 2322–30
[177] dos Santos D M, Corrêa D S, Medeiros E S, Oliveira J and Mattoso L H C 2020 *ACS Appl. Mater. Interfaces* **12** 45673–701
[178] Daristotle J L, Behrens A M, Sandler A D and Kofinas P 2016 *ACS Appl. Mater. Interfaces* **8** 34951–63
[179] Daikuzono C M, Delaney C, Morrin A, Diamond D, Florea L and Oliveira O N 2019 *Analyst* **144** 2827–32
[180] Riul A, Malmegrim R R, Fonseca F J and Mattoso L H C 2003 *Artif. Organs* **27** 469–72
[181] Daikuzono C M, Shimizu F M, Manzoli A, Riul A, Piazzetta M H O, Gobbi A L, Correa D S, Paulovich F V and Oliveira O N 2017 *ACS Appl. Mater. Interfaces* **9** 19646–52
[182] Braunger M L, Shimizu F M, Jimenez M J M, Amaral L R, Piazzetta M H D O, Gobbi Â L, Magalhães P S G, Rodrigues V, Oliveira O N and Riul A 2017 *Chemosensors* **5** 1–10
[183] Machado J C, Shimizu F M, Ortiz M, Pinhatti M S, Carr O, Guterres S S, Oliveira O N and Volpato N M 2018 *Bull. Chem. Soc. Jpn.* **91** 865–74

[184] Krieser K, Emanuelli J, Daudt R M, Bilatto S, Willig J B, Guterres S S, Pohlmann A R, Buffon A, Correa D S and Külkamp-Guerreiro I C 2020 *Mater. Sci. Eng.* C **117** 111315
[185] Olivati C A, Riul A, Balogh D T, Oliveira O N and Ferreira M 2009 *Bioprocess Biosyst. Eng.* **32** 41–6
[186] Riul A, de Sousa H C, Malmegrim R R, dos Santos D S, Carvalho A C P L F, Fonseca F J, Oliveira O N and Mattoso L H C 2004 *Sens. Actuators B Chem.* **98** 77–82
[187] Borato C E, Riul A Jr, Ferreira M, Oliveira O N Jr and Mattoso L H C 2004 *Instrum. Sci. Technol.* **32** 21–30
[188] Elamine Y, Inácio P M C, Lyoussi B, Anjos O, Estevinho L M, da Graça Miguel M and Gomes H L 2019 *Sens. Actuators B Chem.* **285** 24–33
[189] Kirsanov D *et al* 2019 *Sensors* **19** 1–20
[190] Eric W D 1996 *Trends Food Sci. Technol.* **7** 399–403
[191] Hoffmeister C R D *et al* 2017 *Eur. J. Pharm. Sci.* **99** 310–7

Chapter 7

Recent developments on devices applied to impedimetric electronic tongues

Flavio M Shimizu, Gabriel Gaál, Maria L Braunger and Antonio Riul Jr

Since the first publication of an impedimetric e-tongue in 2002, efforts have been driven towards establishing this method as a standard analytical tool for the discrimination and quality control of foodstuffs, drinks, environmental and pharmaceutical samples. Impedimetric e-tongues are comprised of an array of sensing units, usually interdigitated electrodes, that are further covered with different organic and inorganic materials to generate an electrical fingerprint of a liquid sample under analysis. Then, a high amount of capacitance or resistance data is collected during the analysis and advanced computational techniques are successfully applied to facilitate data visualization. Despite being in its infancy, many works are reporting innovative ideas, clarifying concepts, and encouraging future research because there are still limitations and open questions to be addressed. In this book chapter, we will explore emergent technologies paving the way for innovation in the development of impedimetric e-tongues using microfluidics, 3D-printing, and paper-based devices.

7.1 Introduction

The first electronic tongues (e-tongues) were developed based on the potentiometric response of sensing units immersed in liquid samples [1, 2]. Sequentially, other devices were developed based on voltammetric and impedimetric responses [3, 4]. Since the publication of the first impedimetric e-tongue in 2002 [4], efforts have been made towards establishing this method as a standard analytical tool for the discrimination and quality control of liquids or diluted samples. Despite being in its infancy, innovative ideas and clarifying concepts encourage future research although there are still limitations and open questions to be addressed.

As presented in the previous chapter, standard devices comprising an impedimetric e-tongue are based on metallic interdigitated electrodes (IDEs) fabricated by

doi:10.1088/978-0-7503-3687-1ch7

photolithography [5], that are modified by materials having distinct electric features, and then immersed in the analyte for evaluation. Then, a high amount of data is collected during the analysis and computational techniques are successfully applied to facilitate the data visualization. Therefore, several parameters in the design of the device and data processing can be modified to optimize discrimination, reusability, lifetime, scalability, reducing the cost of fabrication, aiming for commercial implementation.

Considering the design of the device, the first step is usually the fabrication and test of alternative electrodes. In this context, the recent literature presents developments with inkjet and 3D-printed IDEs based on carbon materials, and laser ablation for patterning electrodes of different materials. The second step is looking for alternatives for the procedure of dipping the sensing unit in the analyte, which is time-consuming and uses high amounts of the analyte, especially considering independent measurements for statistical validation. That can be replaced by the integration of microfluidic channels in an e-tongue platform for possible improvements in reproducibility, reduction in measurement time and consumption of reagents and samples, and drifts in the sensor response [6].

Each has its pros and cons, and some studies suggest options for the development of microfluidic devices using low-cost fabrication techniques to optimize the design aiming to reduce manual operations. Real-life applications require simple procedures and fast analysis not reliant on trained personnel. Therefore, reducing manual operations and automating procedures are important to achieve commercial platforms for impedimetric e-tongues.

In this book chapter, we will explore emergent technologies paving the way for innovation in the development of impedimetric e-tongue platforms, including alternatives for electrode fabrication, microfluidic devices, 3D-printing approaches, and paper-based devices.

7.2 Alternative approaches for electrode fabrication

Usually, e-tongues transducers are based on gold interdigitated electrodes (IDEs) fabricated via photolithography processes, which are multistep, time-consuming, and costly. In these processes, a photoresistor layer is coated on a glass surface and the IDEs pattern is engraved on it. The system is then placed for metallization, where an adhesive thin layer (generally Cr or Ti) and gold are evaporated onto the entire surface. Finally, the substrate is immersed in a specific solution that reacts with the photoresist and removes the unpatterned places, leaving the gold IDEs on top of the glass surface. Each step requires precise control over several parameters, and specific equipment with highly trained users. Also, to ensure the quality of the IDEs these steps must be performed within a highly controlled environment, most of the time needing expensive cleanroom facilities.

Efforts have been made to reduce the complexity involved on the IDE patterning process. Among them, femtosecond laser ablation of different materials was proposed to fabricate high-quality IDEs for e-tongue application. In this technique, femtosecond pulses are focused on a narrow spot, which leads to localized high peak intensity and direct vaporization of the material used for the IDE. By automating the dislocation of the ablated material, one can create a different electrode pattern without the need of new photomasks, as required on photolithography methods, which facilitates the prototyping process. Also, the resolution of the IDE depends only on the laser and optical system employed, eliminating the need for cleanroom facilities. Figure 7.1(a) illustrates the fs-pulsed laser setup used by Manzoli *et al* to pattern gold IDEs on top of glass substrates [7]. They were able to obtain uniform electrode distributions, with a finger width of 7.4 μm and spacing of 3.2 μm, with high reproducibility, in less than 25 min, reducing considerably the fabrication time. With such IDEs, they obtained similar AC responses compared to electrodes fabricates via traditional photolithography.

Paula *et al* also used femtosecond laser ablation to fabricate IDEs, however, they used a composite material based on an insulating polylactic acid (PLA) doped with graphene fibers [8], illustrated in figure 7.1(b) and (c). The PLA is a biocompatible and biodegradable polymer, which allows application in several biosensing devices. Although this material is a commercial filament for 3D printers, they proposed a coating material to cover a glass substrate, obtaining a thinner film when compared to printed tracks. So, they dissolved the filament in chloroform, and covered the glass substrate by spin coating. This substrate was then patterned via femtosecond

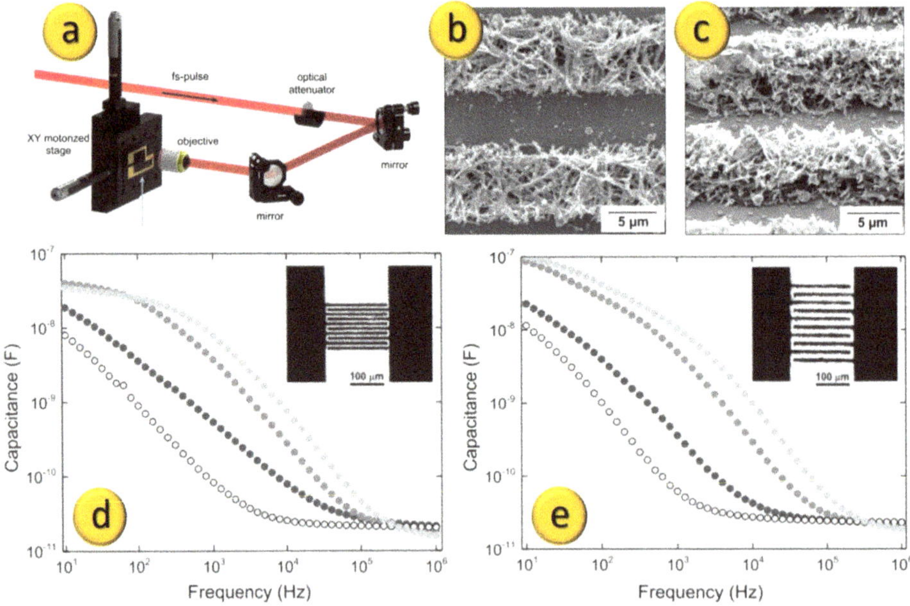

Figure 7.1. (a) Illustration of the optical apparatus used to micromachine the IDEs with the fs-pulsed laser. (b) Top view of scanning electron microscopy image of the PLA-dopped micromachined tracks with 5 μm gap, and (c) at 45° tilt angle. Impedance spectroscopy of two different IDE geometries (d) 10 digits with 5 μm gap, and (e) 10 digits with 10 μm gap. In both graphs: (open circles) Sucrose; (black circles) H_2O; (gray circles) NaCl; (light-gray circles) HCl. The insets illustrate an optical microscopic image of the respective PLA-dopped IDEs. Reproduced with permission from [7, 8]. Copyright 2015 and 2018 Elsevier.

laser micromachining in several IDEs geometries. Finally, they used four distinct geometries as sensing units and were able to distinguish basic tastes (sour, bitter, and salty) with concentrations below the human threshold. Figure 7.1(d) and (e) illustrates several spectra obtained for two distinct IDE geometries.

Despite the advantages of pulsed laser micromachining for IDE fabrication, some issues preclude its large application in electrode patterning. The main drawback is the cost of the optical system, which includes a laser oscillator to ensure a high-frequency laser pulse. The alignment process can also be time-consuming along with the expertise needed to maintain such a system working properly. Moreover, this technique does not fix some adhesion issues between the conductive material and the substrate, which can lead to a reduced lifetime of the devices.

Fused deposition modeling (FDM) 3D printing is a largely exploited alternative method to produce complex structures in a single-step process, potentiating several fields with a rapid prototyping process. This technology uses thermoplastics to fabricate different constructs. In the past few years, new types of thermoplastic composites have been developed to introduce new functionalities to the printed structures. Among them, conductive fillers allow the fabrication of conductive tracks fully integrated with the structural pieces of the construct, reducing interface problems such as low adhesion between the electrode and the substrate. Moreover, FDM technology is based on open source projects, which allows open discussion among the 3D printing community, leading to constant improvements on different printer setups [9].

Overall, FDM 3D printers use one or more hot nozzles to extrude thermoplastic filaments. The first layer of molten filament is deposited onto a heated bed that ensures good adhesion to the bed, enabling the construct to stand still during the printing process. The printing head and the heated bed are moved in perpendicular directions (usually XY Cartesian plane), creating a printing plane. After printing the specific plane, the hot extruder is moved upwards initiating a new printing plane, thus creating a layer-by-layer structure.

Gaál *et al* took advantage of the flexibility promoted by the FDM technology to fabricate a new type of 3D printer based on the H-frame type XY-positioning system [10], illustrated in figure 7.2(a). This system allows fixing the construct during the printing process of a specific layer, which may improve the overall resolution. They also used a dual extruder (two-nozzle printer) system to fabricate devices having different materials in a single printing process. Macro IDEs were fabricated by printing an insulating, transparent PLA layer used as the substrate, and then a conductive graphene-PLA filament was used to print the IDEs, as shown in figure 7.2(b). The fabrication time for each IDE is only 8 min, consolidating a rapid prototyping process. Moreover, a trial and error process can be employed in the prototyping procedure as the materials employed are relatively cheap and biodegradable. They compared the AC response of several printed IDEs with the AC response of traditional gold IDEs fabricated in the same geometry, with identical electrical responses. Different materials are deposited onto the IDEs using the layer-by-layer (LbL) self-assembly technique to obtain distinct sensing units for e-tongue application in soil analysis. Seven different soil samples were simply diluted

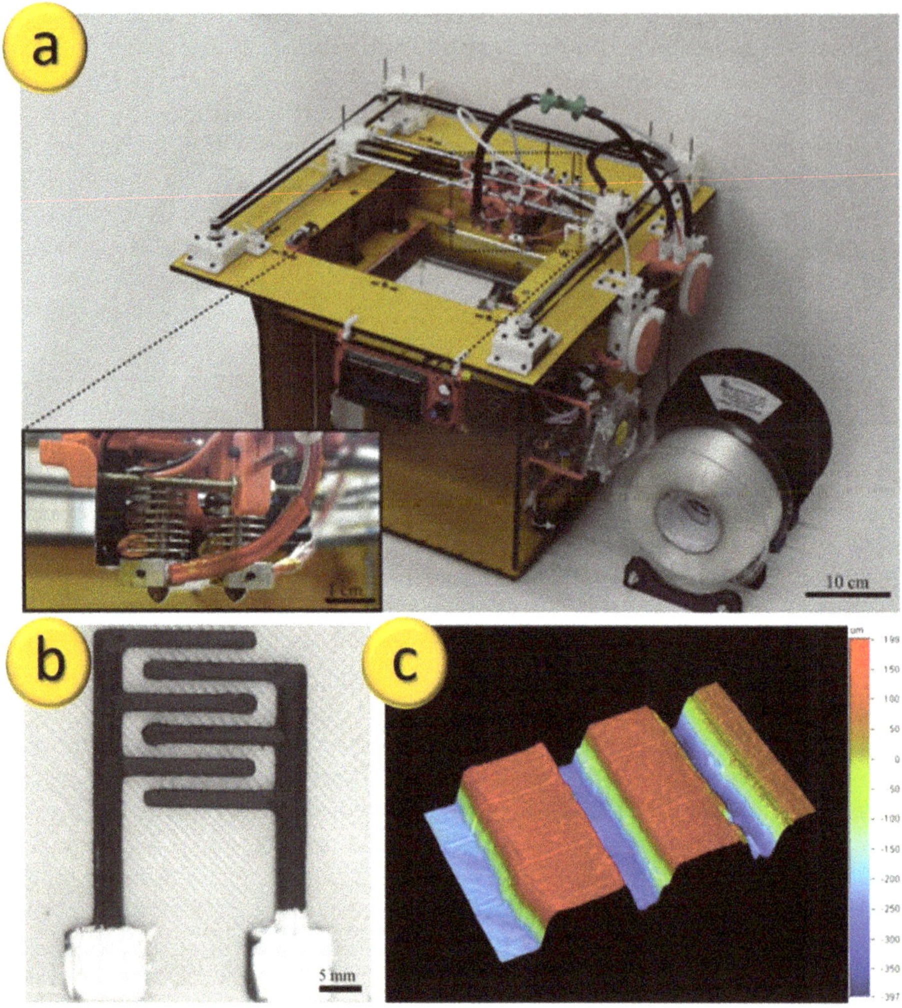

Figure 7.2. (a) Photography of the two-nozzle H-framed FDM 3D-printer used to fabricate the printed IDEs. The inset details the two-nozzle system. (b) Optical image of a printed IDE having 6 digits, with 2 mm thick and spaced 1.4 mm. (c) 3D profilometry mapping of a printed IDE. Reproduced with permission from [10]. Copyright 2018 Gaál, da Silva, Gaál, Hensel, Amaral, Rodrigues and Riul (CC BY 4.0).

in water and further analyzed with the 3D-printed e-tongue. PCA analysis was employed with good distinction of all samples, comparable to results previously obtained with gold IDEs fabricated by photolithography [11].

Using a similar sensor array, Americo da Silva *et al* studied sandy and clayey soil samples enriched with different concentrations of N, P, and K macronutrients [12]. Briefly, figure 7.3(a) shows one of the sensing units (PDDA/CuTsPc) immersed in a soil sample dispersed in water. Authors used a straightforward sample code for the soil samples (S-NPK) based on their fertility, in which 0 represents the lowest concentration of the added macronutrient in the NPK sequence (natural fertility), 1

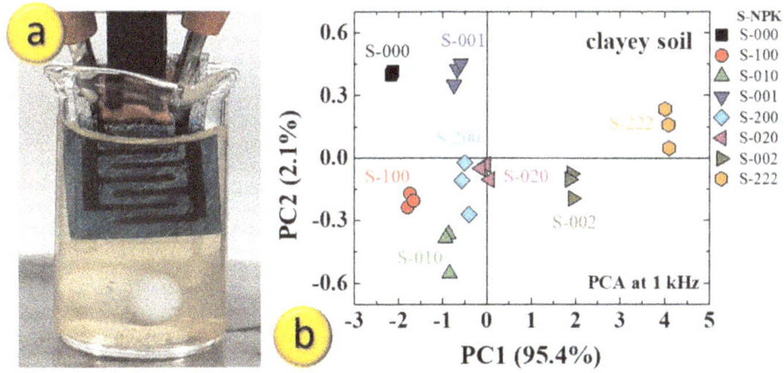

Figure 7.3. 3D-printed graphene electrodes applied in an impedimetric e-tongue for soil discrimination. (a) PDDA/CuTsPc sensing unit dipped in a soil sample dispersed in water to perform the impedance measurements. (b) PCA for the capacitance data of the e-tongue device applied to soil samples enriched with different amounts of N, P, and K. Reproduced with permission from [12]. Copyright 2019 MDPI.

represents medium concentration, and 2 is related to higher concentrations added. To illustrate, S-000 is the control soil sample, while S-222 is that in which higher amounts of all macronutrients (N, P, and K) are added. Figure 7.3(b) shows the PCA plot of the capacitance dataset for clayey soils, with clear distinction of all samples in eight distinct fertilities. The control sample (S-000) is located in the left border, while the S-222 is on the right corner, indicating that PC1 can be related to soil fertility. In this case, the sensor array seems to be especially sensitive to K, compared to N and P, considering that S-002 is closer to S-222 in PC1, while S-200 and S-020 are located near the samples with medium fertility (S-010 and S-001). With this study the authors showed that beyond the discriminant analysis of soil samples using e-tongues [10, 11], it is possible to correlate the impedance data with soils enriched with macronutrients added at distinct concentrations using a low-cost e-tongue setup.

7.3 Microfluidic devices fabricated by photolithography

Flow injection analysis was used for potentiometric and voltammetric e-tongues early in the 2000s [6]. The first microfluidic impedimetric e-tongue was reported in 2015 by Daikuzono *et al* [13], whose setup is composed of a sensor array having a microchannel stamped in a poly(dimethylsiloxane) (PDMS) matrix and sealed onto gold IDEs. The IDEs were then coated with different materials using the dynamic LbL technique by sequentially injecting polyelectrolytes solution inside the microchannel. Three LbL films were fabricated (PAH/NiTsPc, PAH/PPy, and PAH/PEDOT:PSS) and one device was left uncoated, forming an e-tongue with four sensing units. Details on this impedimetric microfluidic e-tongue can be found in [14], and a schematic diagram of the experimental apparatus is shown in figure 7.4(a–b). The sensor array was successfully applied to distinguish basic tastes relevant to human gustative perception (sweetness, saltiness, sourness, bitterness, and umami) at 1 mM using PCA of the capacitance data, figure 7.4(c). The authors

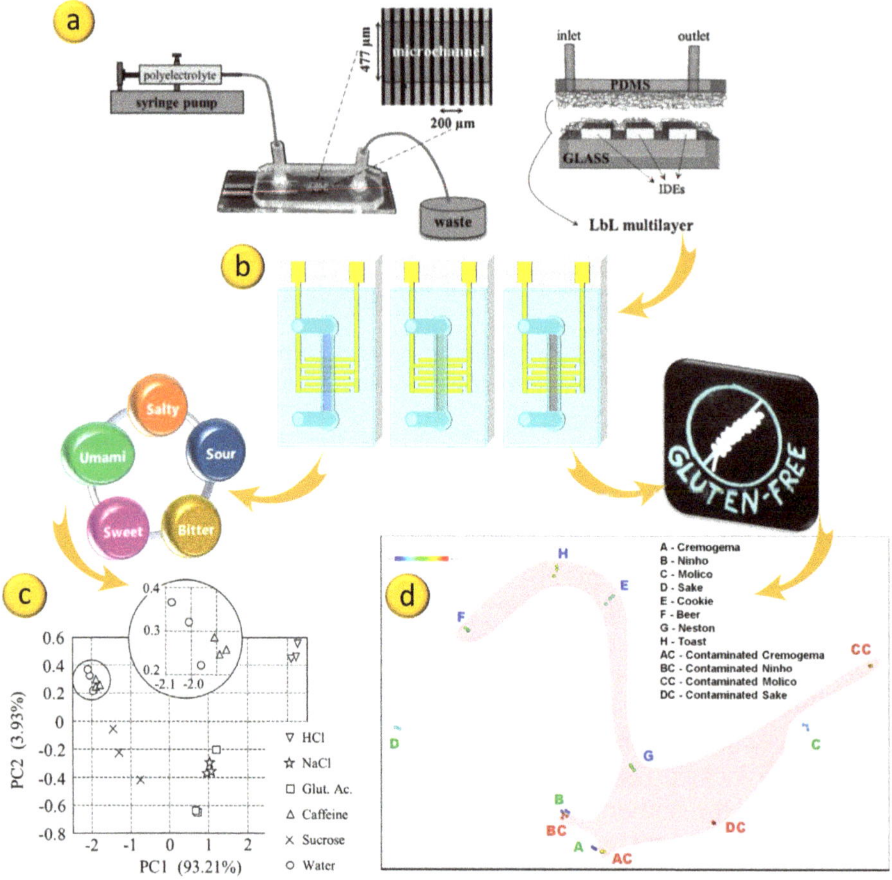

Figure 7.4. Microfluidic impedimetric e-tongue. (a) Experimental apparatus, including microfluidic device and syringe pump. The inset is a zoom in the intersection of the IDE covered by the LbL film, thus forming one of the four sensing units. In the bottom right corner is a schematic side view of the IDE covered with the polyelectrolytes and then sealed with the PDMS microchannel. (b) Schematic representation of the sensing units modified with LbL films. (c) Microfluidic e-tongue differentiating basic tastes under flow (1000 $\mu L\ h^{-1}$) conditions inside a PDMS microchannel. 1 mM aqueous solutions of (∇) HCl, (☆) NaCl, (□) l-glutamic acid monosodium salt hydrate, (Δ) caffeine, (×) sucrose, (○) Milli-Q water. Reproduced with permission from [13]. Copyright 2014 Elsevier. (d) IDMAP plot for gluten-free, gluten-containing, and spiked gliadin-free foods with frequency selection for the normalized capacitance data obtained with PAH/PPy and PAH/PEDOT:PSS electrodes. Reproduced with permission from [18]. Copyright 2017 American Chemical Society.

were also able to visualize the suppression effect of sweetening sour, as shown in figure 7.4(b).

PCA is a common method to facilitate the visualization of the data information obtained by sensors. Briefly, the original data points are projected onto a new coordinate system called principal components, therefore reducing dimensions and preserving the variation of the dataset. Other methods can be applied for data visualization, such as interactive document map (IDMAP) and Sammon's mapping. One of the benefits of using information visualization methods is the possibility to optimize outcome performance [15]. This can be made by choosing the most appropriate multidimensional projection method or applying feature selection, in which only part of the data is used for discrimination. Besides, the quality of data discrimination can be measured by the silhouette coefficient (SC) [16], a quality index that reflects the clustering strength. The SC method is based on the average silhouette value, which measures how close one point is to a cluster when compared to all others, considering all points in the dataset [16, 17]. Briefly, the following interpretation has been reported for positive SC values: strong (0.71–1.0); reasonable (0.51–0.70); weak (0.26–0.50); and no substantial structure has been found ($\leqslant$ 0.25) [17].

In another study, Daikuzono *et al* used a similar microfluidic array to evaluate gliadin in gluten-free and gluten-containing foodstuff [18]. Through IDMAP, the authors observed the e-tongue ability to distinguish different concentrations of gliadin in an ethanol solution, gluten-free (cremogema, ninho, molico, and sake) and gluten-containing (cookie, beer, neston, and toast) foodstuffs, figure 7.4(d). The IDMAP method was able to group samples in ethanol solution with SC = 0.97, and discriminate the gluten-free food from samples deliberately contaminated with the best result (SC = 0.86) obtained with data processed with only two sensing units (PAH/PPy and PAH/PEDOT:PSS), rather than the four original ones. The authors also used UV–vis absorption spectroscopy to ensure the gliadin presence in the analyzed samples (absorption band at 280 nm), even considering intentionally contaminated samples.

Another crucial problem nowadays is the need to increase food production without expanding agricultural areas [19]. In this direction, precision agriculture explores the development of sensors to locally monitor the level of plant nutrients available in the soil, directing on-site on-demand application of fertilizers according to the real demand of macronutrients [20, 21]. Traditional chemical soil analysis is often time-consuming and costly [22], leading to an evaluation of samples below the optimum amount to make adequate the mapping of the overall spatial variability of soil fertility [23], thus motivating the development of alternative sensing in agricultural practices [24]. In this context, Braunger *et al* proposed the first impedimetric e-tongue to discriminate nutrients in soil samples, figure 7.5(a) [11]. The microfluidic e-tongue was based on gold IDEs and PDMS microchannels design as those previously described in [13]. However, a distinct sensor array is used (bare IDE, PDDA/CuTsPc, PDDA/MMt-K, and PDDA/PEDOT:PSS), figure 7.5(b). The soil samples were enriched with different macronutrients (N, P, K, Ca, Mg, and S) and then simply diluted in water for the electrical analysis, figure 7.5(c). The sensor array was thoroughly rinsed with water after measurements in each soil sample, and sequentially measured in distilled water to check possible cross-contamination. The cleaning procedure indicated no contamination of the sensing units, making possible the reuse of the device for almost 1 year. Linear (PCA) and non-linear (IDMAP and Sammon's mapping) statistical techniques were employed to find out the best frequency to analyze the electrical response of the soil set tested. Firstly, the authors processed the full impedance spectra to visualize the data, and the techniques were able to successfully distinguish the soil samples. Therefore, to better compare the quality of data discrimination achieved by the techniques, the authors also used

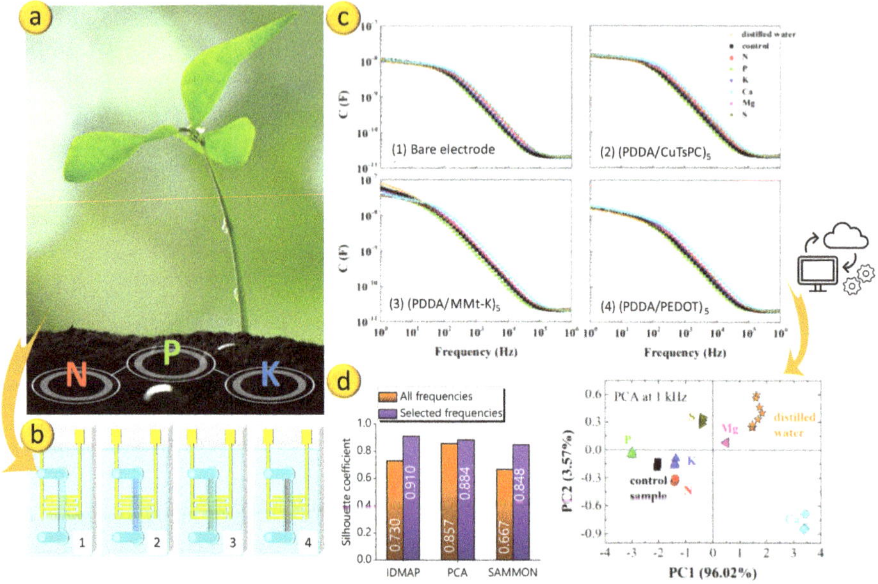

Figure 7.5. Microfluidic e-tongue based on impedance spectroscopy for soil discrimination. Schematic representation of (a) the soil nutrients assessed in this study, and (b) of the sensing units (1) non modified, and ones modified with (2) $(PDDA/CuTsPc)_3$, (3) $(PDDA/MMt\text{-}K)_3$, and (4) $(PDDA/PEDOT:PSS)_3$ LbL films. (c) Capacitance versus frequency spectra obtained for seven soil samples (control, N, P, K, Ca, Mg, and S) in aqueous solutions using bare electrode; $(PDDA/CuTsPc)_3$; $(PDDA/MMt\text{-}K)_3$, and $(PDDA/PEDOT:PSS)_3$ sensing units. (c) SC values for the three multidimensional projection data considering the whole frequency spectra, the analysis with the selected feature, and the improvement in the discrimination performance. (d) PCA of the capacitance data at 1 kHz for the soil samples dispersed in water at 5 ml/h flow rate. Reproduced with permission from [11]. Copyright 2017 MDPI.

SC values [16, 17]. Considering the whole capacitance spectra, the higher SC value was obtained with PCA (0.857), indicating strong discrimination among all soil samples. Sequentially, the authors selected capacitance values in the frequency range 79 Hz - 25 kHz, as indicated by the parallel coordinates technique, processing the original data with the information visualization methods once again. There was an enhancement in the SC values with the feature selection for all three methods applied, as shown in figure 7.5(d). Non-linear techniques (IDMAP and Sammon) showed an increase from 25 and 27%, respectively, while PCA improved only 3%, with its SC value already high for the whole spectra. Considering the kHz region enhanced the discrimination of the soil samples on this e-tongue, the authors used the well-established PCA plot at 1 kHz [4] to evaluate the dataset, as shown in figure 7.5(d). The distribution of clusters on PCA at the selected frequency is similar to those discussed above, with good discrimination of all samples. This outcome indicates that it is viable to develop a rapid and low-cost device for on-site evaluation of soil samples considering only one frequency, instead of the whole impedance spectra, enabling faster analysis.

7.4 Microfluidic devices fabricated by alternative methods

Photolithography is the standard method to fabricate devices on an industrial scale since it yields microstructures with high resolution with high reproducibility [25]. However, not all research groups have access to this type of laboratory facilities, and alternative methods that do not require clean-room laboratory facilities have been proposed to overcome these drawbacks and develop devices for rapid prototyping. In this trend, Shimizu *et al* used a simple technique to produce microchannels by inserting iron steel microwires transversally to the microchannel (scaffolds) during the PDMS curing process, referred to as polymerization and scaffold removal (PSR) [26], figure 7.6(a). The stainless steel microwires were coated with gold (Au), platinum (Pt), nickel (Ni), iron (Fe), and aluminum oxide (Al_2O_3) films to promote chemical diversification and thus ensuring diversified electrical signals to the e-tongue system. The microwires were further short-circuited to promote a single response of the five resistors in parallel, with each resistor having a particular resistance and interaction with the analyte, creating distinct sensing units, figure 7.6(b). The key feature of this e-tongue setup is a single measurement of the real admittance, which already accounts for all individual admittances of the microwires inside the microchannel. Another interesting feature is that the active surface of the microwires can be replaced at any time, with no need for a new calibration procedure by just displacing the microwire in the microchannel. As a proof-of-concept, the basic tastes salt, bitter, sweet, and sour at concentrations below the human threshold, were successfully distinguished with PCA analysis. The system was also able to discriminate the electrical response, figure 7.6(c), of diverse chemical inputs used in the oil industry, as depicted in the PCA plot in figure 7.6(d). It is worth mentioning that this microfluidic platform can operate even in turbulence flow rate condition [27].

Aiming to increase the sensitivity of this e-tongue, a more sensitive configuration was obtained by arranging the electrodes in a parallel association of capacitors [28], figure 7.6(e-1). To fabricate this new system four sets of stainless steel microwires are placed on top and in the bottom of the microchannel, creating four parallel plate capacitors, figure 7.6(e-2). Two different arrays of sensing units were evaluated in this work, one comprised of Au, Pt, Ni, Fe, and Al_2O_3 films and another set of stainless steel microwires coated with silicon (SiO_2), nickel (NiO_2), aluminum (Al_2O_3), and iron (Fe_2O_3) oxides films, further short-circuited to promote a single electrical response and ensure diversified signal response. As a proof of concept, the platform was applied for monitoring of modified nanomaterials. In figure 7.6(f) is shown the capacitance spectra of pristine silica nanoparticles (SiO_2NP) and ones functionalized with 1-[3-(trimethoxysilyl)propyl]urea (TMSPU) and tetraethyl orthosilicate (TEOS) at molar ratios of molar ratios of 0.5, 1.0, 1.5, and 2.0. To facilitate the analysis, capacitance data were processed by multivariate techniques; figure 7.6(g) shows that an IDMAP plot can discriminate the different functionalization level of TMSPU. Multiwalled carbon nanotubes (MWCNTs) with different oxidation extents and a multiplex platform were also evaluated, demonstrating potential for industrial application. Moreover, in pursuit of novel materials, the

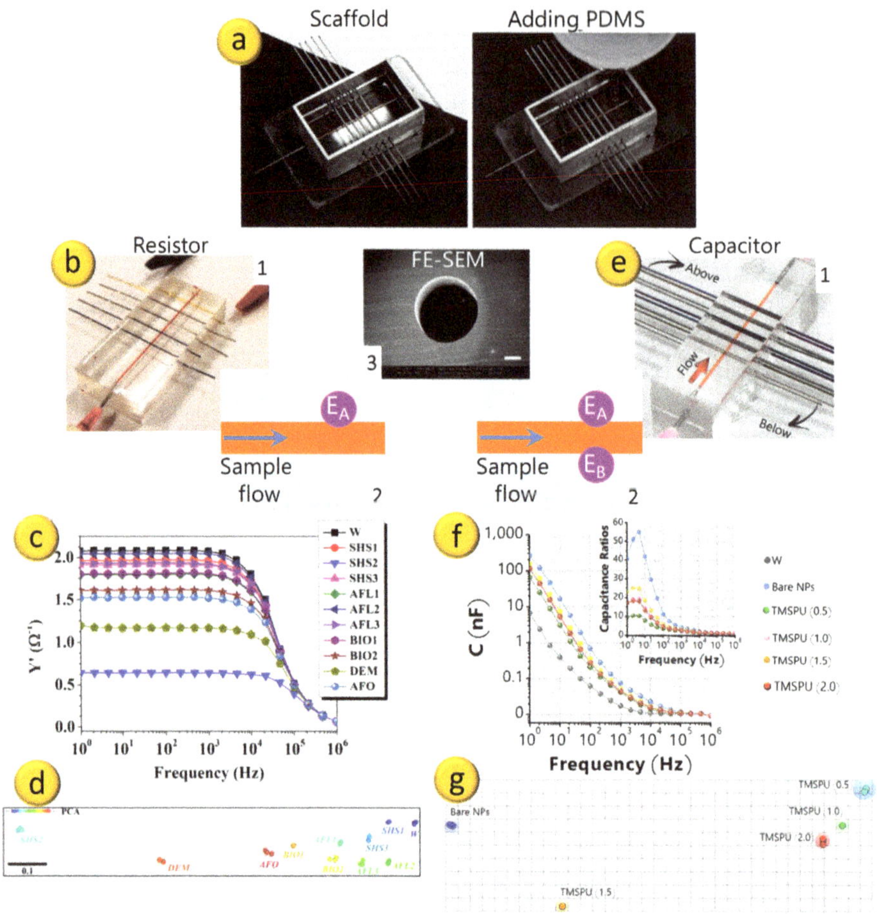

Figure 7.6. (a) Scaffold and addition of PDMS to construct the chip. (b-1) Final device filled with red dye with integrated electrodes after the polymer cure and scaffold removal. (b-2) Illustration of the lateral view of microfluidic channel with the cross section of the channel in which electrodes are inserted above (E_A) the sample. (b-3) FE-SEM depicts the cross section of the sample channel used in the analyses for only one sample. (c) Real admittance spectra, and (d) PCA plot for pure petrochemical inputs and water (W) using the e-tongue with all the microwires in short circuit. The signals (c) and dots (d) for each compound are related to replicated measurements. The inputs are utilized as antifouling (AFL1, AFL2, and AFL3), scavengers of hydrogen sulfide (SHS1, SHS2, and SHS3), biocides (BIO1 and BIO2), demulsifier (DEM), and antifoam (AFO) in the oil industry. (e-1) Final chip with electrodes insulated by different oxide films placed below and above the sample channel that was filled with red dye. (e-2) Illustration of the lateral view of this channel with the cross section of the two channels (filled with red dye) where electrodes are inserted above (EA) and below (EB) the sample. (f) Spectra of real capacitance and (g) IDMAP plot using the e-tongue of capacitors with different metals and SiO_2 as dielectric. Reproduced with permission from [26, 28]. Copyright 2017, 2018 American Chemical Society.

same platform was built with commercial pencil graphite cores with different specific values of surface roughness and graphite to clay (G/C) ratio to create ready-to-use sensing units for the e-tongue analysis, which was capable of distinguishing different brands of mineral waters, distilled, tap, filtered tap, and lake waters [29].

3D printers are also used to prototype microfluidic devices. Several works exploited the strengths offered by this additive manufacturing process to fabricate different microfluidic devices. Gaál *et al* proposed an interesting alternative to integrate different materials into printed microchannels maintaining them sealed at the end of the process [30]. As described above, the FDM 3D printing technique builds the construct stacking consecutive printing planes, following an LbL method. So, they stopped the printing process at the desired layer to insert the materials and then resumed the printing process. Among the several materials inserted into the microchannel, they proposed the integration of gold IDEs evaporated onto transparency sheets. These electrodes were further coated with LbL films (PDDA/GPSS, PDDA/CuTsPc, PDDA/MMT-K, and a bare IDE) to assemble four sensing units. This microfluidic e-tongue was applied to distinguish basic tastants (salt, bitter, sweet, and sour) at concentrations below the human threshold. Figure 7.7 illustrates the printed microfluidic e-tongue setup.

Despite the undeniable advantages of using microfluidic devices combined with e-tongues based on IDEs, the experimental procedure using this type of device is usually time-consuming. There are several operational steps considering the microfluidic e-tongue, once the user must handle a syringe pump for changing analytes, cleaning the microchannels, and manually change cables to connect each sensing unit to the electrical testing equipment for data acquisition. Considering four sensing units to evaluate only one analyte, the user must handle the setup at least four times, and eight times more considering the cleaning procedure. In this context, Braunger *et al* designed an e-tongue device with collinear IDEs comprised of a single microchannel (figure 7.8) aiming to decrease the number of manual operations during data acquisition [31]. The authors used FDM 3D printing technique to fabricate a mold for casting the PDMS microchannel instead of conventional photolithography, with the IDEs fabricated in printed circuit boards. They developed a multiplexing circuit to automate the signal's routing from the sensor array to the impedance analyzer, and all instruments, including the syringe pump, were controlled by software.

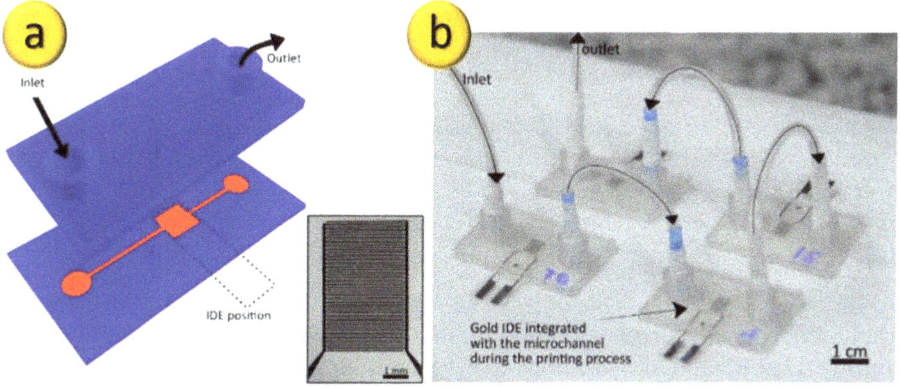

Figure 7.7. (a) Illustration of the printed e-tongue system, with the microchannel highlighted in red. The IDEs were manually inserted at the outlined site. (b) Printed e-tongue system. Reproduced with permission from [30]. Copyright 2017 Elsevier.

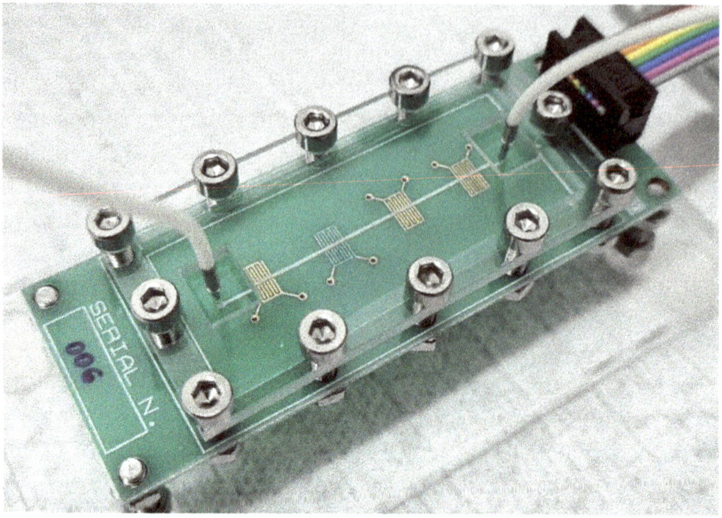

Figure 7.8. Impedimetric e-tongue in a single microchannel. The sensor array is formed by four IDEs, one bare and the others covered with LbL films of PDDA/CuTsPc, PDDA/MMt-K, and PDDA/PEDOT:PSS. The white-colored analyte was used to highlight the single microchannel and the analyte flowing over the four sensing units.

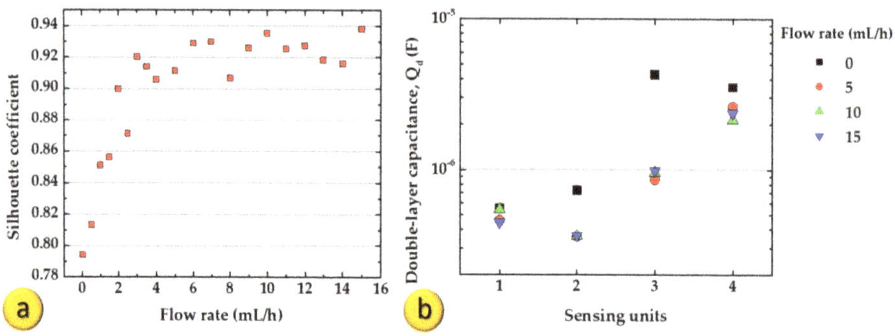

Figure 7.9. Microfluidic e-tongue in a single channel. (a) Silhouette coefficient (SC) for the PCA analysis of the impedance data for sucralose-based sweeteners obtained with a microfluidic e-tongue in a single channel for distinct flow rates. (b) Double-layer capacitance obtained by circuit equivalent fitting of each sensing unit for one brand of sucralose-based sweetener in four distinct flow rates. Reproduced with permission from [31].

The e-tongue was tested to distinguish four brands of sucralose-based sweeteners and the setup was handled only four times to alternate analytes. That is a considerable improvement considering the previous microfluidic e-tongue setup requiring sixteen manual operations to evaluate four samples [14]. The experimental procedure was carried out in 20 distinct flow rates, with the results compared by the SC value [16], as shown in figure 7.9(a). There is an uptrend and then stabilization (at ~ 3 mL h^{-1}) of SC values with the flow rate, which possibly mitigates the formation of the double-layer leading to faster stabilization and precision response in the e-tongue device. The authors used an equivalent electric circuit model to evaluate the impedance data, thus confirming a decrease in the double-layer capacitance at higher flow rates, as shown in figure 7.9(b). Therefore, it is possible to achieve higher discrimination between samples using a microfluidic e-tongue than with static impedance measurements. Moreover, by recognizing the best flow rate for the experimental procedure it is possible to optimize the sample volume for analysis and discharge, which is highly desirable for the evaluation of expensive and/or biohazard analytes. The value for the optimized flow rate obtained by the authors is valid for their specific microfluidic e-tongue and sweeteners. However, we believe this is an expected trend for analysis involving e-tongues with equivalent sensing units and analyte samples having similar viscosity, considering a small margin of error. Therefore, more studies on the flow rate using different devices and analyte samples are required, especially considering distinct electrolytes and higher viscosities.

Although various advantages are mentioned for microfluidic platforms, there are no studies evaluating the electrical double-layer formation at the electrode surface, one of the most important parameter in impedimetric experiments. Some articles report drift [6] or long waiting time (up to dozens of minutes) to stabilize the electrical response [32]. Those problems are mitigated in flow condition, decreasing the thickness of the electrical double-layer and leading to rapid steady-state in the electrical response. Figure 7.10(a) schematically depicts the formation of the electrical double layer at the surface of gold electrodes under flow (5 ml h^{-1}) and static conditions. Riul *et al* reported such a time study using the same microfluidic impedimetric e-tongue configuration within a single channel to evaluate the impedance at 1 kHz in 10 mM of sucrose solution using a bare microfluidic IDE. As shown in figure 7.10(b), the impedance value is constant, $|Z| = 3$ MΩ, under the flow condition since the beginning of the experiment. Under static condition the

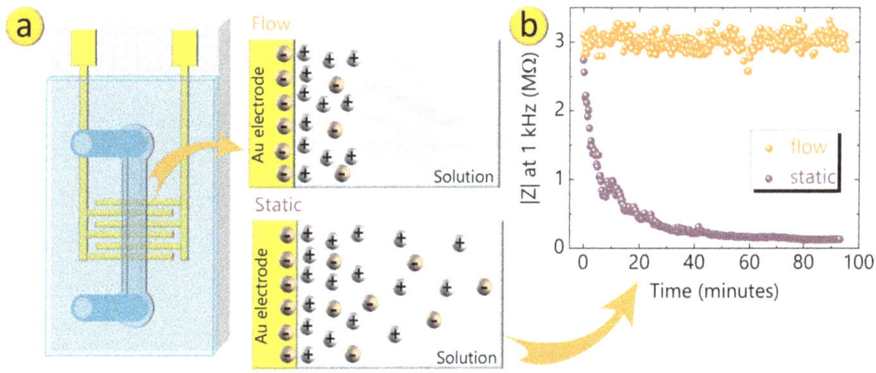

Figure 7.10. a) Scheme of the formation of the electrical double at the surface of the electrode under flow and static conditions. b) Impedance magnitude at 1 kHz under flow (5 mL h^{-1}) and static conditions in sucrose solution.

impedance value decreased exponentially, saturating after only 32 min, indicating that the analysis should be performed only after this time value.

As already mentioned, one of the main goals in the development of new platforms for impedimetric e-tongue is the reduction of manual operations to optimize the overall time of analysis. One approach is the integration of passive mixers inside microchannels to speed up the evaluation of suppression/enhancement effects of flavors applied in pharmaceuticals, foodstuff, and beverages. In this context, Braunger *et al* used the stereolithography technique to fabricate microfluidic devices using 3D-printed molds for PDMS, with a passive mixer geometry to be integrated into an e-tongue platform [33]. The device is based on a Y-shaped microchannel with one electrode at each branch, thus allowing the evaluation of the injected analytes just after the inlets and their mixture before the outlet, as shown in figure 7.11(a). In this sensor array, each sensing unit is formed by one Y-shaped microchannel with three electrodes, therefore, we have a total of 12 electrodes and four Y-shaped sensing units in this platform. One Y-shaped sensing unit is used as received (bare IDE), with the three others functionalized with materials having distinct electric features (PDDA/CuTsPc, PDDA/MMt-K, and PDDA/PEDOT:PSS). In the same platform, there is a multiplexing system to route the signal of each electrode to a pair of coaxial connectors, and then the electrodes are selected using computer software also developed by the authors. As a proof-of-concept, this microfluidic mixer e-tongue was successfully tested to distinguish basic tastes relevant to human gustative perception (sweetness, saltiness, sourness, bitterness, and umami) at 1 mM. To evaluate the mixing outcome of the analytes, sucrose flow in one arm of the Y-shape microchannels and NaCl, HCl, or caffeine in the other arm, as the sugar addition commonly suppresses salty, sour, and bitter tastes. The authors also tested the mixing of the two strong electrolytes, NaCl, and HCl, and the PCA plot shows that the mixtures are halfway from their respective pure solutions (figure 7.11(b)), indicating a proper mixing of the analytes inside the microchannels.

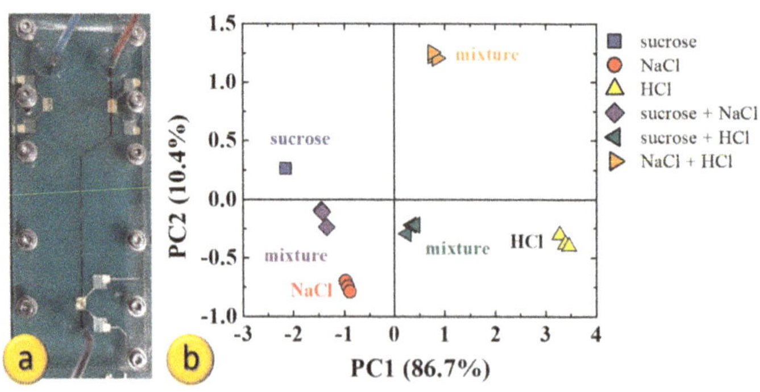

Figure 7.11. Microfluidic e-tongue with passive mixers. (a) Picture of one Y-shaped microchannel with passive mixer flowing through electrodes at the device inlets (before the mixer: blue/red dyes) and outlet (after the mixer: purple mixture). (b) PCA score plots of the capacitance data to evaluate sucrose, NaCl, HCl, and their mixtures. Reproduced with permission from [33]. Copyright 2020 MDPI.

7.5 Paper-based impedimetric electronic tongue

Another interesting approach for sensors is paper-based devices. Some advantages of this method include low-cost fabrication, scalability with screen printing processes, resulting in disposable and often biodegradable devices for point-of-care analysis. In this context, Mondal *et al* developed an impedimetric e-tongue using different conducting polymers deposited onto filter paper [34]. The polymers used as sensing units were poly(aniline) and poly(pyrrole), both doped and de-doped, and two Ag electrodes provided the electrical contact. The e-tongue was able to distinguish the basic tastes and two commercial brands of orange juice (Tropicana and Real) as well as freshly squeezed juice from the fruit.

In this trend, Daikuzono *et al* developed a paper-based sensor for the detection of monosaccharides based on electrical impedance spectroscopy [35]. Hydrogels are hydrophilic polymeric materials that can swell and retain large amounts of water, without dissolving. Chemical (pH, solvent composition, molecular species, etc) and physical stimuli (temperature, electric, magnetic, pressure and light) can influence the swelling of hydrogels, and the research, tests, and application of hydrogels are of great relevance in the development of new sensor architectures. Carbon interdigitated electrodes were screen printed on paper (figure 7.12(a)) and then modified with a layer of hydrogel containing acrylamide copolymerized with 20 mol% 3-(acrylamido) phenylboronic acid (PBA). The hydrogel layer swells in the presence of glucose and fructose and changes in its dielectric properties are related to distinct concentrations using impedance spectroscopy (figure 7.12(b)). The relative capacitance plots of the

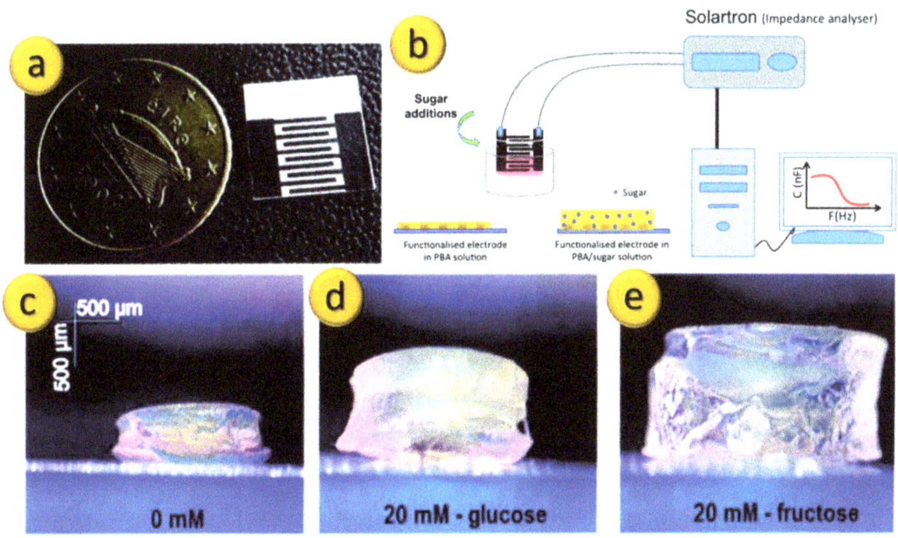

Figure 7.12. Paper-based device for monosaccharides detection. (a) Screen-printed carbon interdigitated electrodes with PBA-containing hydrogel layer. (b) Schematic of the experimental setup for impedance spectroscopy measurements. Microscope images of the device when immersed in (c) PBS buffer, (d) glucose in PBS and (e) fructose in PBS. Reproduced with permission from [35]. Copyright 2017 The Royal Society of Chemistry.

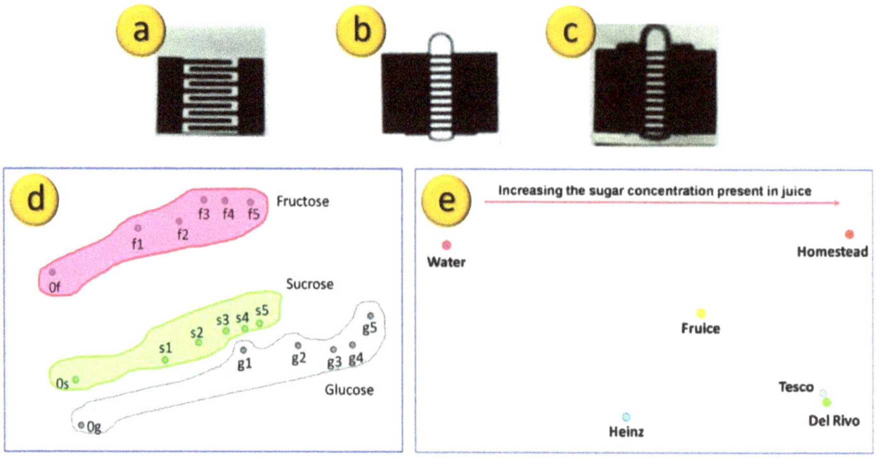

Figure 7.13. Paper-based microfluidic e-tongue. Images of (a) the screen-printed carbon interdigitated electrode on paper, (b) the wax-printed channel onto the electrode and (c) the wax-printed channel on electrode after melting. (d) IDMAP plot of the capacitance data for the e-tongue in different concentrations of glucose, sucrose, and fructose. (e) IDMAP plot of the capacitance data for the e-tongue applied to apple juices and water samples. Reproduced with permission from [36]. Copyright 2019 The Royal Society of Chemistry.

hydrogel in the presence of fructose and glucose exhibit greater swelling in fructose (figure 7.12(e)) than glucose (figure 7.12(d)) at the same concentration of the monosaccharides, which can be used for selective detection after calibration.

In a new approach, Daikuzono *et al* developed a paper-based microfluidic e-tongue based on the same carbon interdigitated electrodes (figure 7.13(a)) [36]. The channels of the devices were made along the center of the interdigitated electrodes (figure 7.13(b)) using a wax printer, and then heated for wax melting and also to improve adhesion onto the paper (figure 7.13(c)). The four sensing units were formed by hydrogels containing different types and concentrations of PBA, able to bind the saccharide molecules. The e-tongue was able to distinguish glucose, fructose, and sucrose in six different concentrations (from 0 to 0.056 g mL^{-1}) based on the capacitance data, as shown in the IDMAP plot of figure 7.13(d). Using the same approach, the authors were also able to differentiate five brands of commercial apple juice, as shown in figure 7.13(e). The IDMAP explained differences among brands, correlating them to the overall sugar content at each brand, increasing from the left to the right in figure 7.13(e).

7.6 Future trends

Compared to potentiometric and voltammetric electronic tongues, the impedimetric technology is still in its infancy, but even in a short period (~ 20 years) it has been successfully used to discriminate foodstuff, beverages, pharmaceuticals, and pollutants, and more recently samples of clinical interest such as in early cancer detection.

In this chapter, we have focused on the insights of novel designs and devices aiming for sensitive sensors and ways to automate the exhaustive laboratory routine.

From our point of view, efforts should be driven in four main directions: (i) the development of simple electrodes operating under flow condition, (ii) the search of adequate sensing nanomaterials for specific analytes, (iii) development of mathematical and computational tools, such as machine learning to facilitate the information visualization of impedimetric data, and (iv) miniaturization for a portable, commercial impedimetric device. All these points represent a niche in the impedimetric technology, among them, the development of machine learning combined with the Internet of Things (IoT) represent a hot topic that is being exhaustively exploited in all research areas and is well presented in chapter 10. The solution of mentioned challenges will help on the consolidation of the electronic tongue technology in the market, because up to now, there are only two commercial potentiometric e-tongues from Alpha MOS and Insent presented in chapters 2 and 3.

Acknowledgments

The authors are grateful for the financial support of the Brazilian agencies São Paulo Research Foundation (FAPESP, grant numbers 2015/14836-9 and 2017/19862-3), Coordenação de Aperfeiçoamento de Pessoal de Nível Superior (CAPES, grant number 88887.583380/2020-00), and National Council for Scientific and Technological Development (CNPq, grant number 308570/2018-9).

References

[1] Hayashi K, Yamanaka M, Toko K and Yamafuji K 1990 *Sens. Actuators* B **2** 205–13

[2] Vlasov Y, Legin A, Rudnitskaya A, Di Natale C and D'Amico A 2005 *Pure Appl. Chem.* **77** 1965–83

[3] Winquist F, Wide P and Lundström I 1997 *Anal. Chim. Acta.* **357** 21–31

[4] Riul A Jr., dos Santos D S Jr., Wohnrath K, Di Tommazo R, Carvalho A C P L F and Fonseca F J *et al* 2002 *Langmuir* **18** 239–45

[5] Raymundo-Pereira P A, Shimizu F M, Coelho D, Piazzeta M H O, Gobbi A L and Machado S A S *et al* 2016 *Biosens. Bioelectron.* **86** 369–76

[6] Gutés A, Céspedes F and del Valle M 2007 *Anal. Chim. Acta* **600** 90–6

[7] Manzoli A, de Almeida G F B, Filho J A, Mattoso L H C, Riul A and Mendonca C R *et al* 2015 *Opt. Laser Technol.* **69** 148–53

[8] Paula K T, Gaál G, Almeida G F B, Andrade M B, Facure M H M and Correa D S *et al* 2018 *Opt. Laser Technol.* **101** 74–9

[9] Gaal G, Gaal V, Braunger M L, Riul A and Rodrigues V 2020 *3D Printing in Biomedical Engineering [Internet]* ed S Singh, C Prakash and R Singh (Singapore: Springer) 127–45 Available from: http://link.springer.com/10.1007/978–981-15–5424-7_6

[10] Gaál G, da Silva T A, Gaál V, Hensel R C, Amaral L R and Rodrigues V *et al* 2018 *Front. Chem.* **6** 1–8

[11] Braunger M, Shimizu F, Jimenez M, Amaral L, Piazzetta M and Gobbi Â *et al* 2017 *Chemosensors* **5** 14

[12] da Silva T A, Braunger M L, Coutinho M A N, do Amaral L R, Rodrigues V and Riul A Jr 2019 *Chemosensors* **7** 50

[13] Daikuzono C M, Dantas C A R, Volpati D, Constantino C J L, Piazzetta M H O and Gobbi A L *et al* 2015 *Sens. Actuators* B **207** 1129–35

[14] Braunger M L, Daikuzono C M and Riul A Jr. 2019 *Biomimetic Sensing: Methods and Protocols [Internet]* ed J E F H Fenniri (Berlin: Springer) pp 141–50
[15] Paulovich F V, Maki R M, De Oliveira M C F, Colhone M C, Santos F R and Migliaccio V *et al* 2011 *Anal. Bioanal. Chem.* **400** 1153–9
[16] Rousseeuw P J J 1987 *Comput. Appl. Math.* **20** 53–65
[17] Struyf A, Hubert M and Rousseeuw P 1996 *J. Stat. Softw.* **1** 1–30
[18] Daikuzono C M, Shimizu F M, Manzoli A, Riul A, Piazzetta M H O and Gobbi A L *et al* 2017 *ACS Appl. Mater. Interfaces* **9** 19646–52
[19] Crist E, Mora C and Engelman R 2017 *Science* **356** 260–4
[20] Mishra P, Mapara S and Vyas P 2015 *Int J. Appl. Innov. Eng. Manag.* **4** 114–7
[21] Rogovska N, Laird D A, Chiou C-P and Bond L J 2019 *Precis Agric.* **20** 40–55
[22] http://www.iac.sp.gov.br/produtoseservicos/analisedosolo/docs/Boletim_Tecnico_106_rev_atual_2009.pdf
[23] Nanni M R, Povh F P, Demattê J A M, Oliveira R B, de Chicati M L and Cezar E 2011 *Sci Agric.* **68** 386–92
[24] Viscarra Rossel R A, McBratney A B and Minasny B *Proximal Soil Sensing* (Progress in Soil Science) (Dordrecht: Springer)
[25] Aono M and Ariga K 2016 *Adv Mater.* **28** 989–92
[26] Shimizu F M, Todão F R, Gobbi A L, Oliveira O N, Garcia C D and Lima R S 2017 *ACS Sens.* **2** 1027–34
[27] de Camargo C L, Shiroma L S, Giordano G F, Gobbi A L, Vieira L C S and Lima R S 2016 *Anal. Chim. Acta* **940** 73–83
[28] Shimizu F M, Pasqualeti A M, Todão F R, de Oliveira J F A, Vieira L C S and Gonçalves S P C *et al* 2018 *ACS Sens.* **3** 716–26
[29] Nicoliche C Y N, Costa G F, Gobbi A L, Shimizu F M and Lima R S 2019 *Chem. Commun.* **55** 4623–6
[30] Gaal G, Mendes M, de Almeida T P, Piazzetta M H O, Gobbi Â L and Riul A Jr *et al* 2017 *Sens. Actuators* B **242** 35–40
[31] Braunger M L, Fier I, Shimizu F M, de Barros A, Rodrigues V and Riul A 2020 *Sensors* **20** 6194
[32] De Queiroz D P, Florentino A D O, Bruno J C, Da Silva J H D, Riul A and Giacometti J A 2016 *Sens. Actuators* B **230** 566–70
[33] Braunger M L, Fier I, Rodrigues V, Arratia P E and Riul A 2020 *Chemosensors* **8** 13
[34] Mondal D, Paul D and Mukherji S 2018 2018 12th Int. Conf. on Sensing Technology (ICST) [Internet] (Piscataway, NJ: IEEE) 7–12 https://ieeexplore.ieee.org/document/8603573/
[35] Daikuzono C M, Delaney C, Tesfay H, Florea L, Oliveira O N and Morrin A *et al* 2017 *Analyst* **142** 1133–9
[36] Daikuzono C M, Delaney C, Morrin A, Diamond D, Florea L and Oliveira O N 2019 *Analyst* **144** 2827–32

IOP Publishing

Electronic Tongues
Fundamentals and recent advances
Flavio M Shimizu, Maria L Braunger and Antonio Riul Jr

Chapter 8

Materials for sensing in electronic tongues

Leonardo G Paterno

The sensoactive layer coating the sensing unit is the first element in the e-tongue system with which the analyte establishes contact. The right choice of materials that will compose such a sensoactive layer is, therefore, a crucial aspect in the e-tongue design and fabrication. This chapter provides an overview of different materials for sensing in e-tongues, including polymeric membranes functionalized with lipids and ionophores, chalcogenide glasses, conducting polymers, metal complexes, and nanoparticles (metal, metal oxides, lamellar hydroxides, dendrimers, carbon nanotubes, graphenes, transition metal dichalcogenides). The methods by which these materials are deposited onto bare electrodes are also described in detail.

8.1 Introduction

The historical development of e-tongues shows that, in the very beginning, potentiometric sensors were surface-modified with polymeric membranes containing lipids and ionophores to mimic the architecture of the biological gustatory system. The field has rapidly evolved from that point, moving towards different transduction principles, which consequently has opened the use of an even greater number of alternative sensing materials. From poly(vinyl chloride) (PVC) membranes to chalcogenide glasses and conventional metallic electrodes, the e-tongue sensing units are nowadays sensitized with several types of polymers, including insulating polyelectrolytes and dendrimers, as well as conducting polymers, semiconducting porphyrins and phthalocyanines, metal and metal oxide nanoparticles, carbon-based nanomaterials, and transition metal dichalcogenides. Elements displaying biochemical activity such as enzymes, proteins, and antigens have been employed as well, providing e-tongues with selectivity for detecting several diseases. Moreover, with the introduction of microfabrication processing and the availability of a plethora of multivariate data analysis and artificial intelligence methods, the field has been set to an unprecedent advancement level.

The next sections, therefore, will provide a detailed description on sensing materials for current e-tongues and emphasis will be given to structural features, synthesis procedures, and thin-film processing.

8.2 Polymeric membranes modified with lipids and ionophores

The first e-tongues were essentially based on arrays of potentiometric sensors. These sensors were made of different chalcogenide glasses or thus, conventional glass electrodes modified with polymeric membranes enriched with either lipids or ionophores, as schematically shown in figure 8.1(a). Despite that the e-tongue's working principle relies on an array of low-selectivity and high cross-sensitivity sensors, the main idea behind using lipids and ionophores was to, somehow, mimic the functioning of the biological gustatory system. In practice, however, surface modification of potentiometric electrodes with these materials has essentially led to several cross-sensitivity sensing units rather than enhanced selectivity.

Lipids are either hydrophobic or amphiphilic molecules exerting several biological functions, depending upon their molecular structure. For example, fats and oils store energy, whereas phospholipids and sterols are major

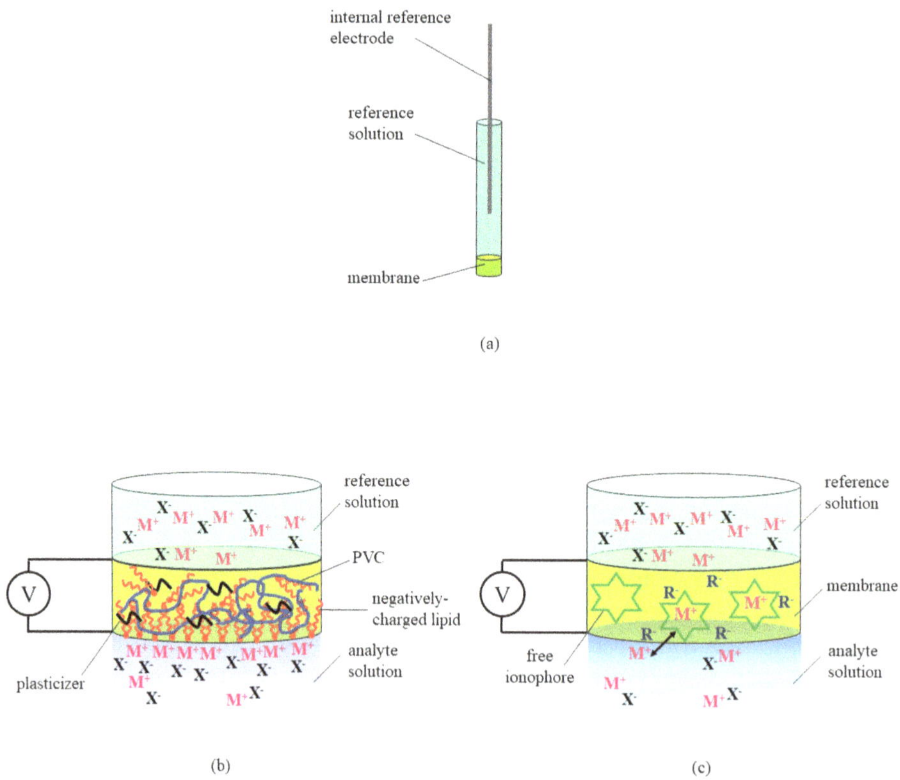

Figure 8.1. Schematic view of (a) a potentiometric electrode modified with membranes of PVC with (b) lipids and (c) ionophores.

components of biological membranes. Other lipids may play different roles, spanning from electron carriers, enzyme cofactors, light absorbing, emulsifying agents, to hormones and signaling agents [1]. Amphiphilic lipids are composed by a lengthy nonpolar tail and a hydrophilic head and although insoluble in water, they are capable of forming structured vesicles, multilamellar/unilamellar liposomes, or membranes in aqueous environment. In e-tongue applications, lipids are employed to mimic taste receptor membranes found in biological gustatory systems. In practical terms, a change on the membrane's double-layer electric potential will be elicited whenever it interacts with molecules of different substances showing specific hydrophilicity/hydrophobicity degrees [2]. The ionization degree and hydrophobicity of the chosen lipid will therefore determine the level of interaction with a specific molecule and the magnitude of the electric potential change. In the pioneer work carried out by Toko's group [3], different membranes were produced based upon PVC plasticized with dioctyl phenylphosphonate and enriched with a series of lipids (figure 8.1(b)). Whereas pristine PVC is a rigid solid at room temperature because of its glass transition temperature (T_g) of about 80 °C, plasticized PVC is a viscous liquid, exhibiting a T_g of −50 °C. Therefore, it is easily shaped into thin films (membranes), which are more permeable to small chemical species. Several lipids have been tested, including trioctylmethylammonium chloride, tetradodecylammonium bromide, oleic acid, and phosphoric acid di(2-ethylhexyl) ester (figure 8.2(a–d)). They were designed to discriminate the five basic taste patterns that are perceived by humans, namely sourness, bitterness, saltiness, sweetness, and umami. Sensitivity to a determined taste was achieved by a proper combination of lipids in each PVC membrane.

Ionophores are natural or synthetic compounds forming lipid-soluble complexes that act as vehicles for transporting ions across biological membranes [4]. They can be classified according to their specific function, namely carrier ionophore, which binds to a specific ion and transports it through the membrane, and channel ionophore, which provides the ion with a hydrophilic channel inside the membrane. In the uncomplexed form, they can also be classified as charged or electrically neutral [5]. For usage in e-tongues, they are dispersed within viscous films, such as plasticized PVC (figure 8.1(c)) [6]. The sensor as-formed responds to the activity of the target ion, according to a modified Nernst equation as follows:

$$E_M = E^0 + \frac{RT}{zF} \ln a_{I(aq)} \tag{8.1}$$

In (8.1), E_M is the measured potential, z is the valency and $a_{I(aq)}$ is the activity of the uncomplexed ion I, whereas R, T, and F are the universal gas constant, absolute temperature, and Faraday constant, respectively. E^0 is the potential due to the activity of the ion within the organic polymer matrix, which is assumed to be unaltered regardless of the ion concentration. Although these types of membranes are designed with a particular ionophore to sense a single type of ion, the interference from other ions is quite usual. It is, therefore, necessary to introduce a correction into equation (8.1)

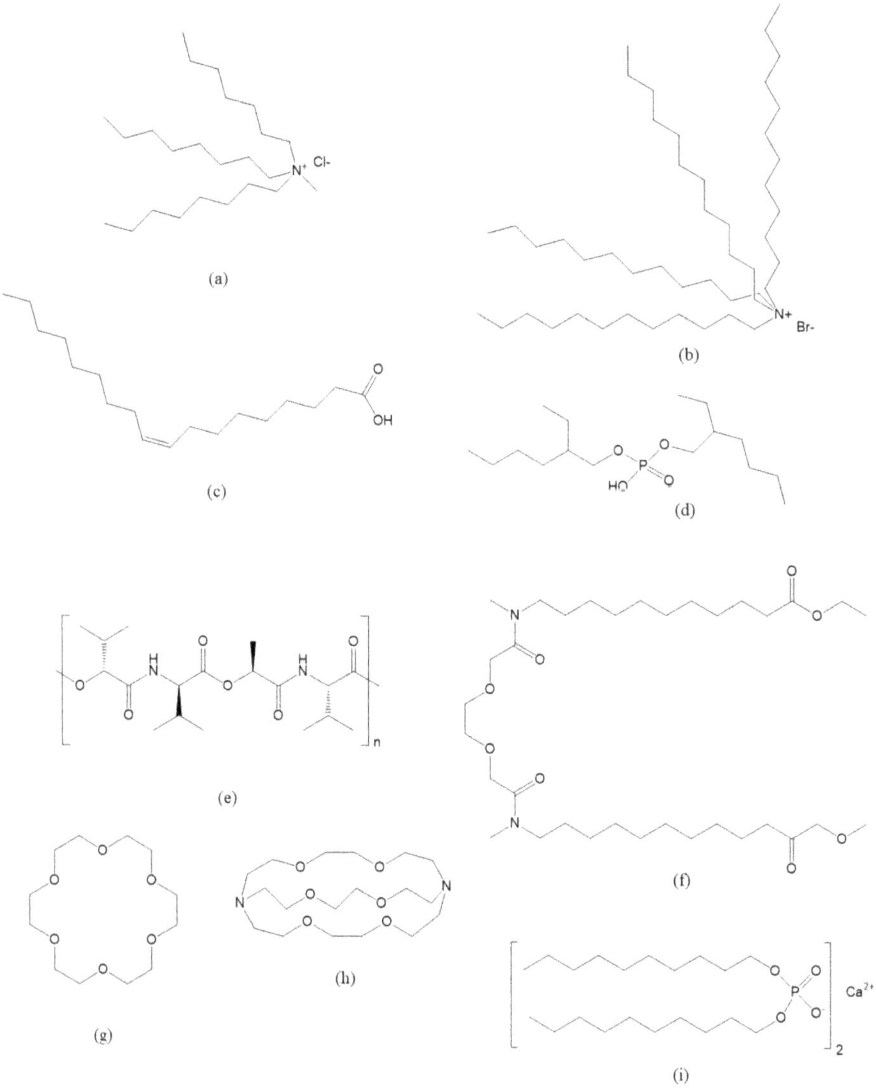

Figure 8.2. Molecular formulae of some lipids and ionophores: (a) trioctylmethylammonium chloride, (b) tetradodecylammonium bromide, (c) oleic acid, (d) phosphoric acid di(2-ethylhexyl) ester, (e) valinomycin, (f) Ca^{+2} selective ionophore, (g) 18-crown-6, (h) cryptand [2,2,2], and (i) calcium didecyl phosphate.

in order to account for the influence of interfering ions to E_M. This correction is provided in the Nicolsky–Eisenman equation as follows [7]:

$$E_M = E_I^0 + \frac{RT}{z_I F} \, ln \, [a_I(IJ) + K_{IJ} a_J \, (IJ) \, ^{z_I/z_J}] \tag{8.2}$$

The terms $a_I(IJ)$ and $a_J(IJ)$ are the activities of I and J in the mixed sample, z_I and z_J are valences of I and J, whereas K_{IJ} is the selectivity coefficient, sometimes called

Nicolsky coefficient. In practice, K_{IJ} accounts the ability of the membrane in measuring a single ion in the presence of others. For example, if an electrode has an equivalent response to two different ions, $K_{IJ} = 1.0$. Therefore, the lower K_{IJ} is, the lower is the interference of the interfering ion to E_M.

Natural and synthetic ionophores have been employed to build ion-selective electrodes, which are further assembled in e-tongues. As shown in figure 8.2(e–i), the molecular structures of these substances share specific sites capable of complexing the ions. Nonetheless, for a specific cation or anion, there exists a specific ionophore molecule.

8.3 Chalcogenide glasses

Chalcogenide glasses are non-crystalline materials comprised by one or more chalcogens (except oxygen), such as sulfur, selenium and tellurium, combined in binary, ternary and multicomponent compounds with boron, aluminum, gallium, silicon, germanium, phosphorous, arsenic, and antimony. They are usually referred to as covalently bonded networks, exhibiting only short-range order. Despite that, they are chemically much more stable than crystalline counterparts, thereby making them very suitable for building up ion-selective electrodes for operation in a wide range of pHs and media compositions. Besides that, incorporation of other metallic ions, specifically silver, copper, cadmium, lead and iron, can be properly performed to provide electrodes with selectivity to these ions [8].

Production of such glasses relies on thermal processing; pure elements in specific concentrations are alloyed in evacuated quartz ampoules (0.1–0.01 Pa) at a temperature range between 600 K and 1200 K. In order to ensure homogeneous mixing of all glass components, a few thermal holdings are performed at temperatures about 40 °C–50 °C above the melting points of each initial, intermediate and final component. Finally, the melt is rapidly quenched to room temperature in air, ice water, or liquid nitrogen. Residual mechanical stress is further removed by keeping the glass under temperatures below T_g for few hours. The glassy state is verified by x-ray and electron microscope analyses. [9].

The electrode architecture of ion-selective electrode based on chalcogenide glass is comprised by a glass tubing ended by the chalcogenide glass membrane. The selectivity of the membrane is imparted by its composition, containing the ion that is intended to be detected. Finally, the electrode is filled with an electrolyte solution saturated with the target ion plus a metallic wire to establish the electrical contact between solution and the external circuit. The detection mechanism relies on the ion-exchange process occurring at the glass membrane/solution interface. The partitioning of the ion of interest between the membrane and solution depends on the ion's activity, and the resulting charge separation results in a phase-boundary potential. If the ion's concentration at each side of the membrane is not equal, a potential difference across the membrane is elicited, which will correlate with the ion's activity through the modified Nernst equation (equation (8.1)). As in the case of ionophore-based electrodes, interference of other ions is recurrent. Thus, it is taken into account by the Nicolsky–Eisenman equation (equation (8.2)). One of the main advantages of chalcogenide glasses in ion-selective electrodes is the possibility of performing controlled doping of the glass with the ion of interest [10]. Moreover, a proper cross-sensitivity analysis can be made in order to enable the right choice of different electrode compositions for assembling efficient multisensor arrays for liquid analysis [11].

8.4 Organic semiconductors and coordination compounds

Organic semiconductors can be formed by either small or polymeric conjugated molecules. In this wide group of materials, one can include macrocyclic compounds and coordination compounds, in particular, transition metal porphyrins, phthalocyanines, polypyridyl complexes, and naphthalimides, as well as polymers, such as poly(acetylene), poly(aniline), poly(pyrrole), poly(thiophene), poly(vinyl carbazole), poly(p-phenylene vinylene), and others. Although transition metal complexes are not organic compounds, they are still grouped within organic semiconductors in the literature for the sake of simplicity. The molecular formulae of some representative organic semiconductors are given in figure 8.3.

The semiconducting behavior of organic semiconductors stems from their extended π-orbitals that give rise to delocalized electronic states. Most of them show a p-type semiconductor behavior. In conjugated polymers, semiconductor-conductor transition may also occur whenever charges are injected into or withdrawn from the polymer backbone by means of appropriate electrodes, doping or photoexcitation [12]. Lengthening of the conjugation is accompanied by narrowing of the band gap. In the limiting case, one should expect the band gap to disappear. However, single and double bonds have dissimilar lengths and because dimerization

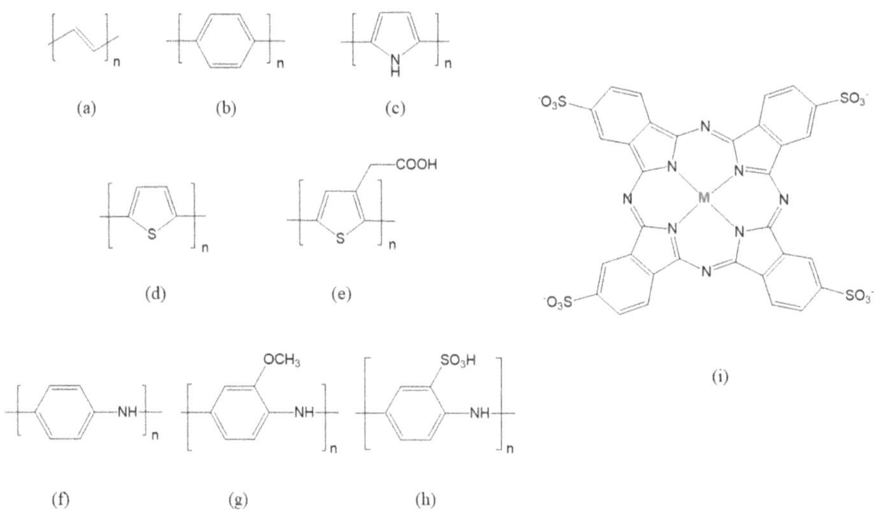

Figure 8.3. Molecular formulae of some organic semiconductors: (a) trans-poly(acetylene), (b) poly(p-phenylene), (c) poly(pyrrole), (d) poly(thiophene), (e) poly(3-acetic acid thiophene), (f) poly(aniline), (g) poly(o-methoxy aniline), (h) poly(o-sulfonic acid aniline), and (i) tetrasulfonated phthalocyanine. In (i), M stands for a divalent metal, including Fe^{2+}, Co^{2+}, Ni^{2+}, Cu^{2+}, and Zn^{2+}.

(Peierls distortion) is a common structural defect in these polymers, a band gap is still present (figures 8.4(a,b)) [13]. Further charge injection (or withdrawal) generates electronic states in the gap, which couple to vibrational modes. In non-aromatic polymers, such as poly(acetylene), these coupled states are called charged solitons (figure 8.4(c)), whereas in aromatic polymers, they are called polarons and bipolarons (figure 8.5). Since most solid-state conjugated polymers are amorphous, electrical conductivity is driven by hopping of these charge carriers among localized

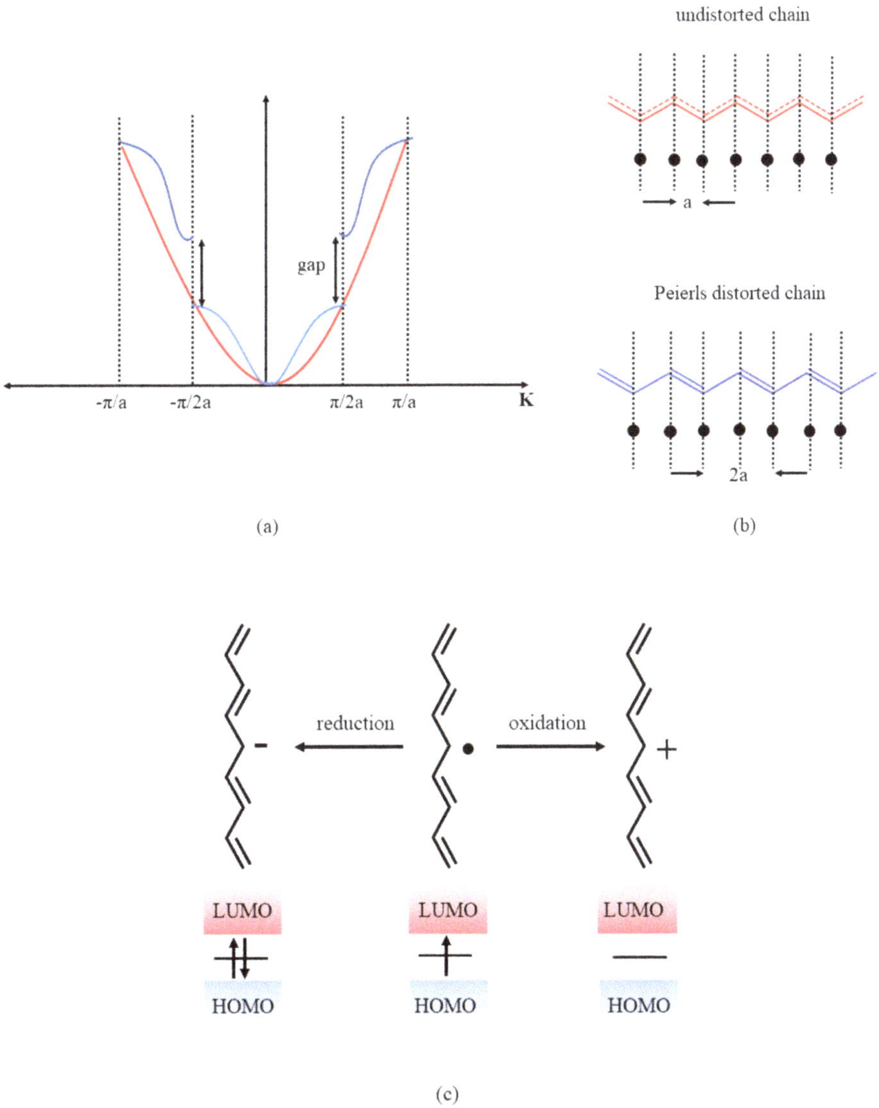

Figure 8.4. Electronic structure of poly(acetylene): (a) energy dispersion diagrams in the absence (in red) and presence (in blue) of a band gap, (b) corresponding molecular structures for undistorted (in red, top) and Peierls distorted (in blue, bottom) chain, and (c) neutral and charged-soliton electronic structures.

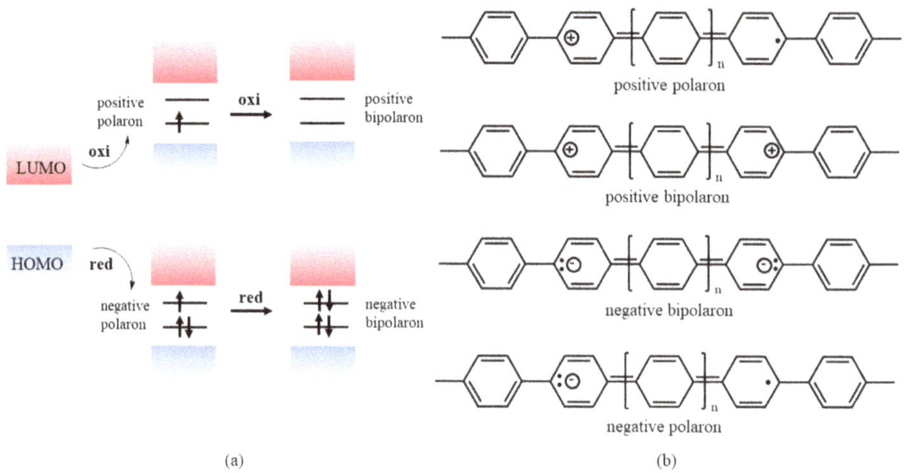

Figure 8.5. Electronic structure of an aromatic conjugated polymer: (a) energy diagrams and (b) molecular structures of positive/negative polaron and bipolaron states.

sites with statistically variable energies and is, therefore, significantly lower than that predicted for crystalline counterparts. In general, doping is performed by oxidation, thereby making positively-charged states as the prevalent charge carriers (figure 8.5). A main feature (and advantage) of semiconducting polymers is that doping is a reversible process and does not involve replacement of atoms from the original lattice as in the case of inorganic semiconductors.

Synthesis of macrocyclic compounds and conjugated polymers dates back to the 19th century, when they were extensively used in dyeing of textiles. Porphyrins and phthalocyanines are routinely produced by cyclotetramerization of pyrrole and phthalic acid derivatives, respectively [14, 15]. In a further step, metalation is conducted by mixing the macrocyclic base and a divalent transition metal (Lewis acid), mainly Fe, Co, Ni, Cu, and Zn. Al and rare earth ions can also be employed. Conjugated polymers are generally produced by radical polymerization of aromatic monomers such as aniline, pyrrole and thiophene [12]. The radical generation can be induced by chemical oxidants, such as ammonium persulfate or $FeCl_3$, or else by electrochemical processes, either in potentiostatic (fixed potential) or potentiodynamic (variable potential) conditions. Chemical synthesis allows for large-scale production, but the generated polymer needs to undergo further processing to attain thin films. Electrochemical synthesis is limited to small amounts of polymers, but they can be obtained directly on chosen electrodes. This latter aspect is very important since many conjugated polymers are poorly soluble in organic solvents. Nonetheless, the solubility of conjugated polymers is enhanced by the introduction of lateral groups, which preclude molecular packing and allow for solvent molecules to diffuse into the polymeric mass.

8.5 Carbon nanomaterials

Graphene (G), graphene oxide (GO), reduced graphene oxide (rGO), and carbon nanotubes of single and multiple walls (SWCNT and MWCNT), are the main representatives of carbon-based nanomaterials that have been employed in e-tongue sensors. Since they are composed by conjugated molecular networks, their electronic properties are greatly influenced by physicochemical properties of the surrounding environment, thereby making them very suitable for e-tongue purposes. Moreover, they can be obtained in either, two-, one-, and zero-dimensional nanostructures, therefore contributing for the enlargement of the electrode's surface area and, consequently, for enhancing the e-tongue's sensitivity. Nowadays, all these nanomaterials are commercially available in different sizes, shapes and purity degrees. This is very attractive, especially in the case of SWCNT and MWCNT, which are usually produced by costly and energy-demanding procedures. Conversely, production of GO and rGO under wet, chemical approaches, has been widely disseminated in research labs to reduce costs and enable molecular engineering of materials aiming at custom applications.

8.5.1 Graphene, graphene oxide and reduced graphene oxide

Graphene (G) is described by an infinite, one-atom thick sheet of sp^2 carbons arranged into a hexagonal honeycomb framework, in which the primitive rhombus unit cell is comprised by two atoms (figure 8.6(a)) [16]. The electronic and optical properties of G and derivatives stem from the extended π-orbitals over the graphitic sheet that contribute to a delocalized network of electrons. According to symmetry constraints, the first Brillouin zone in G has two inequivalent points K and K′ (called Dirac points) (figure 8.6(b)), where a band crossing occurs (figure 8.6(c)). G is, therefore, an exotic semiconductor, since it does not exhibit a band gap. Consequently, electrical conduction is due to electrons or holes, depending upon the sign of the applied voltage. Indeed, at room temperature and in the absence of structural defects, charge mobilities as high as 200 000 $cm^2\ V^{-1}\ s^{-1}$ have already been observed in high purity G samples [17]. In terms of optical properties, single-layer G thin-films are transparent (~ 97%) in the visible range, although transmittance gradually decreases with the number of layers.

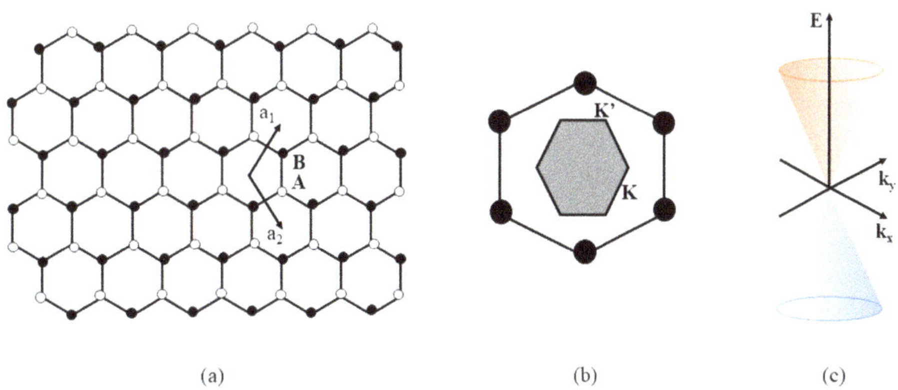

Figure 8.6. Some structural features of graphene: (a) representation of a graphene sheet showing two atoms (A and B) per unit cell and respective lattice vectors a_1 and a_2, (b) first Brillouin zone, and (c) energy dispersion spectrum.

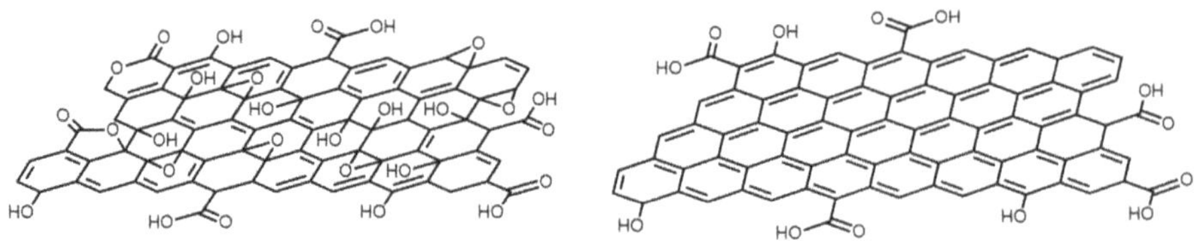

(a) (b)

Figure 8.7. Models for the molecular structure of (a) graphene oxide and (b) reduced graphene oxide.

When this material is oxidized, hydroxyl, epoxy, carbonyl and carboxyl groups are introduced at the sheet's basal plane and edges, resulting in a derivative called graphene oxide (GO) as shown in figure 8.7(a) [18]. Although this name is not rigorously correct, since it is very difficult to determine the extension of graphene oxidation and establish a molecular formula for GO, it has been formally accepted in literature. GO can be submitted to different reduction processes, which remove its oxygenated groups and reestablish part of sp^2 carbons that were lost after oxidation. The resulting material is called reduced graphene oxide (rGO), as shown in figure 8.7(b).

Unlike G, GO and rGO exhibit a distribution of mid-gap states centered at three different energies near the Dirac points, ranging from UV to near-infrared, owing to the presence of oxygenated groups [19]. The charge density and mobility are significantly decreased in GO, which can be in fact considered as an insulator. Because of the elimination of some oxygenated groups, rGO can exhibit measurable conductivity, whose magnitude will depend on the effectiveness of the reduction process. GO suspensions and thin films display a brown to yellow appearance, whose transmittance decreases with concentration. rGO suspensions and thin-films are black, because of partial restoration of the π-conjugated network.

Large-scale production of G is currently carried out by chemical vapor deposition (CVD) or liquid exfoliation of graphite. In CVD, a hydrocarbon source (e.g., acetylene, ethylene, benzene) fed under low pressure in a closed reactor is decomposed into small fragments that recombine onto epitaxially matched metal surfaces [20]. The method is very suitable for production of electronic devices even though it is not so useful for the attainment of large quantities of G in the powder form for further processing or other applications. Liquid exfoliation relies on cleaving graphite into individual graphene layers by intensive ultrasonic radiation within polar organic solvents (e.g. dimethylformamide or N-methyl pyrrolidone) [21]. Additional surfactant agents are usually necessary to preclude restacking of cleaved graphene layers. Nonetheless, it is worth mentioning that G itself is highly inert, thereby making its transfer to a particular support a quite challenging task.

In contrast, GO can be produced at a large scale by oxidative exfoliation of graphite [18]. This process is schematically shown in figure 8.8. Graphite oxidation was first investigated in 1850 by Brodie using a mixture of fuming nitric acid and potassium chlorate [22]. In 1898, Staudenmaier made Brodie's reaction safer and more

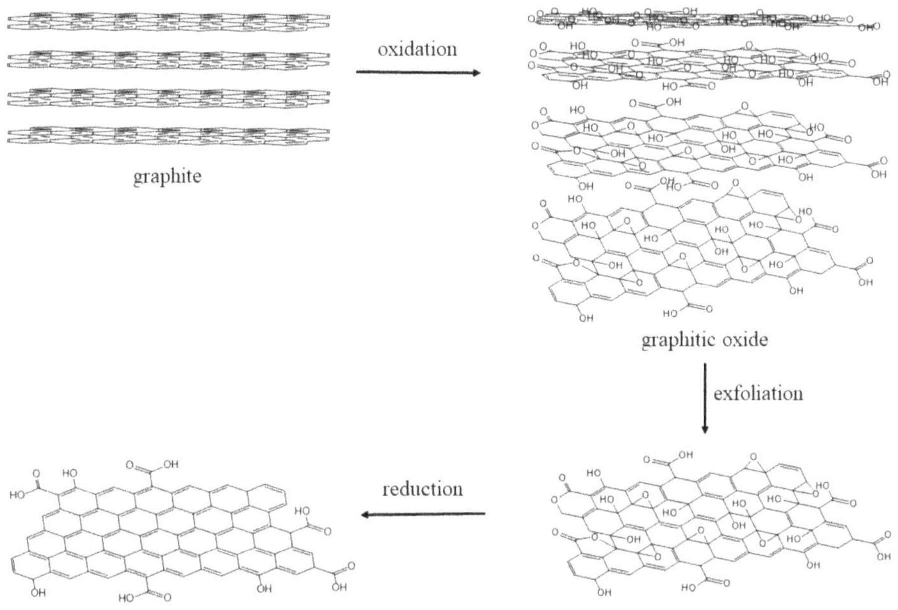

Figure 8.8. Sequential steps for production of graphene oxide and reduced graphene oxide.

effective by adding potassium chlorate in small portions [23]. Nitric acid is a strong oxidizing agent that reacts with aromatic compounds to produce carboxylic acid, lactone and ketone. The reaction is accompanied by the release of gaseous nitric/nitrous oxides, which help to exfoliate oxidized graphene sheets. The addition of potassium chlorate increases the medium oxidizing power even further. Almost 60 years later, Hummers and Offeman proposed an alternative method in which nitric acid and potassium chlorate were replaced by sulfuric acid and potassium permanganate [24]. In the presence of sulfuric acid, permanganate is converted to dimanganese heptoxide, which is a stronger oxidant [25]. After proper purification for elimination of acids, ions, and unoxidized graphite, graphitic oxide is obtained as a pasty product. Nonetheless, it has to be submitted to ultrasonic stirring in polar solvents, including water, to achieve full exfoliation into individual sheets. The as-produced suspensions are highly stable owing to the presence of carboxylate groups at the sheet edges (pKa ~ 4.3), which impose intermolecular electrostatic repulsion and prevent them from agglomeration [26]. Conversion of GO into rGO can be performed by a variety of methods, including chemical, thermal, electrochemical, and photo-thermal/photochemical [27]. Chemical reduction with hydrazine hydrate is among the most widespread methods, owing to its low cost and effectiveness [28]. However, hydrazine is quite toxic and explosive, it is selective to epoxy groups only.

8.5.2 Carbon nanotubes

A carbon nanotube can be viewed as a graphene sheet (monolayer of sp^2 carbons in a honeycomb network) being rolled-up into a tubule with nanometric diameter,

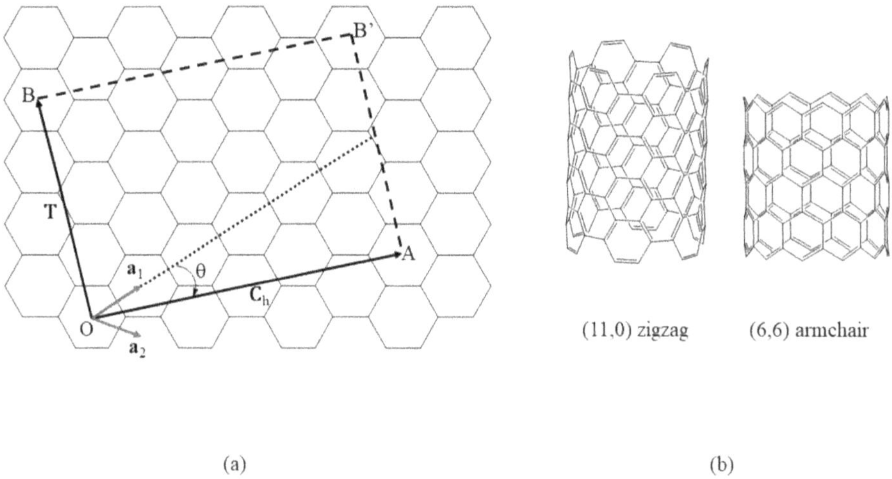

Figure 8.9. Schematic representation of a carbon nanotube: (a) graphene sheet described by chiral and translational vectors and (b) zigzag and armchair nanotubes.

ranging between 1 nm and 40 nm, and nano to micrometer lengths. As depicted in figure 8.9(a), the graphene sheet is described by two vectors; a chiral vector, $\boldsymbol{C}_h = n\mathbf{a}_1 + m\mathbf{a}_2$, which connects the two lattice points, O and A, and a primitive translational vector, $\boldsymbol{T} = t_1\mathbf{a}_1 + t_2\mathbf{a}_2$, by which the graphene sheet is rolled-up into a seamless tubule. The nanotube is thus specified by the pair of integer numbers n, m or, by its radius R ($R = \boldsymbol{C}_h/2\pi$) and chiral angle θ, which is formed between $\boldsymbol{C}_h$ and the nearest zigzag of C–C bonds. All possible tubes have chiral angles between zero and 30°. Based upon that, most nanotubes are found in three different geometries: achiral tubes (tubes with mirror symmetry), zigzag tubes (n, 0) (θ = 0°) and armchair tubes (n, n) (θ = 30°) (figure 8.9(b)). All other tubes are called chiral [29].

Mechanical, thermal, electrical, and optical properties of carbon nanotubes are by far much superior to those reported for most of engineering materials currently available. They are all related to the network of sp^2 carbon atoms and extended π-orbitals derived from it. In addition, they are also decisively dependent on the tubules' symmetry. Considering electrical and optical properties, which are more relevant for e-tongue applications than others, it is known that armchair SWCNT (when n = m) are metallic, whereas all others are semiconductors [30]. If n–m is a multiple of 3, the nanotube is a semiconductor displaying a narrow band gap, whereas all others are semiconductors with wider bandgap. Electrical conduction along the tube is ballistic, with electrons moving without resistance [31]. Nonetheless, inherent structural defects, such as vacancies and incomplete coordination, prevent unrestricted charge transportation. For example, armchair SWCNT shows electrical resistivity (~10^{-6} ohm m) two orders of magnitude greater than copper [32]. MWCNT and other SWCNT show even greater electrical resistivity. Optical properties derive from transitions within the one-dimensional density of states (DOS). They preferably occur between valence/conduction bands of the same symmetry, or, v1-c1, v2-c2, also known as first and second van Hove optical transitions, whereas crossed transitions such as v1-c2 or v2-c1 are very unlikely to occur [33]. Carbon nanotubes do not show fluorescence owing to the high electronic mobility that quenches optical excited states. The energy demanded for these optical transitions is highly dependent on n,m indexes and less dependent on the tube's diameter. Although the gap energy is inversely proportional to the tube's diameter, tubes with close diameters may display distinct behaviors. For instance, tubes of type (10,1) and (8,3) are, respectively, metallic and semiconductor [34].

The first successful attempt to obtain carbon nanotubes dates to 1991, when Iijima reported for the first time the production of MWCNT (4 to 30 nm in diameter and up to 1 μm in length) employing an arc-discharge evaporation method similar to that used for fullerene synthesis [35]. The reaction chamber was composed by a pair of graphite electrodes immersed within an argon-filled vessel (100 Torr). The electric-arc was then sustained by a dc power supply, which causes sublimation of graphite from the anode and deposition of as-formed nanotubes onto the cathode. The type of inert gas and chamber pressure was found to modulate the dimensions and quantity of nanotubes. The process was then modified by doping the anode with small amounts (< 5%) of transition metals, typically Fe, Co, Ni or a mixture of them [36]. This modification resulted in SWCNT, for which transition metals act as catalysts, as will be discussed later. Although the arc-discharge method was key for the attainment of the first carbon nanotubes, it was unable to produce them with fair structural quality at large scale. Moreover, carbon soot and amorphous carbon particles were inevitably produced along with nanotubes, thereby decreasing yield and purity of the samples. In order to circumvent these limitations, laser-ablation [37] and CVD [38] methods were proposed. In the first method, laser-ablation, a graphite rod is placed within a quartz tube filled with inert gas at low pressure and the whole set is inserted into a furnace at high temperature (~1200 °C). The graphite rod is ablated with a neodymium-doped yttrium aluminium garnet (Nd:YAG) pulsed laser, which scans the rod continuously to enable homogeneous evaporation of carbon. The inert gas phase thus drags the carbon species (nanotubes) generated at the hot zones of the reaction chamber towards a water-cooled copper collector set at one end of the quartz tube. Compared to arc-discharge, laser ablation improves purity and yield by mitigating formation of soot and amorphous particles and can produce MWCNT from pure graphite as well as SWCNT from transition metal doped graphite rods. As described earlier, the CVD method relies on the decomposition of a hydrocarbon precursor (e.g. ethylene, acetylene, benzene) fed-up at either liquid or gas phase, whose molecular fragments are later recombined to grow carbon nanotubes at the surface of a chosen support (substrate). In this sense, the method offers the opportunity to grow carbon nanotubes on patterned surfaces, therefore making it suitable for fabrication of electronic devices. As in previous methods, MWCNT and SWCNT can be produced using either pure hydrocarbons or a mixture with transition metals, respectively. Optionally, transition metal seeds can be assembled directly onto the support where the nanotubes will be deposited. The entire process is conducted within a vacuum chamber equipped with a heating setup. Conventional CVD demands relatively high temperatures (> 500 °C) to decompose the hydrocarbons, thereby limiting the growth process to more heat-resistant supports (e.g. silicon, metals, or metal oxides). Alternatively, radio-frequency plasma-aided CVD allows for hydrocarbon decomposition at much lower temperatures (~100 °C), thus enabling the use of even plastic substrates [39].

The growth mechanism of carbon nanotubes was extensively investigated in the 1990s and it is well-established today. It is a consensus that, regardless of the synthesis method, SWCNT does not grow in the absence of transition metal catalysts. Conversely, catalysts are not required for production of MWCNT, even though they can be produced with catalysts as well. Secondly, different synthesis methods provide nanotubes with very close morphologies. MWCNT grows both lengthening and thickening owing to an inherent mechanism that keeps the nanotubes' end opened, the so-called lip–lip interaction [40]. In this interaction, adjacent walls are bridged through metastable bonds that continuously break and re-form, thus facilitating the rapid absorption of carbon atoms by the nanotube edges. On the other hand, SWCNT growth is decisively determined by the catalysts. The most accepted model hypothesizes that a solubility limit, which is dependent on temperature gradients inside the reaction chamber, initially allows evaporated carbon to dissolve into the catalyst within the hot zone, but it rapidly segregates it in the form of fullerene-like nuclei as the colder zone is reached. Once nuclei are formed, growth occurs at this interface by incorporation of more carbon evaporated from graphite or the hydrocarbon source. The metal catalysts stabilize nascent dangling bonds and prevent the formation of carbon pentagons and premature nanotube closure [41]. If the catalyst is supported on a selected surface, growth can occur through root or tip modes. In the root growth mode, as the evaporated carbon is continuously fed, lengthening of the nanotube occurs above the catalyst particle. In the tip growth, lengthening is accompanied by upward extrusion of the metallic particle up to the tip of the nanotube [42].

Pristine carbon nanotubes are not amenable to solution processing owing to their chemical composition made almost exclusively of carbon atoms, imposing van der Waals interactions that produce bundles held strongly together. Although they can be dispersed in high surface tension solvents using ultrasonic stirring, the suspension is unstable, and precipitation occurs whenever stirring is interrupted [43]. Nonetheless, carbon nanotubes are known to interact with different classes of compounds via non-covalent adsorption and are also susceptible to chemical derivatization in order to promote covalent attachment of different functional groups [44]. These approaches allow for processing carbon nanotubes via solution, thereby making the fabrication of sensing units for e-tongue applications a more affordable process.

All the production methods described above invariably produce nanotubes with some extent of structural defects alongside a considerable quantity of impurities, including amorphous carbon particles, carbon soot and residual metals. Therefore, purification procedures need to be performed in order to isolate carbon nanotubes from unwanted subproducts. Usually, they are submitted to reflux in oxidizing acids, such as sulfuric and/or nitric acids, which not only remove amorphous carbon and residual metals but also introduce carbonyl and carboxylic acid groups at side walls and ends of the nanotubes [45]. On one hand, these functional groups worsen the end properties of nanotubes, once part of the original sp^2 carbons is converted to sp^3 ones. Nevertheless, they decrease the nanotube's surface tension, thus making feasible its dispersibility in common solvents. Besides that, they can be easily derivatized through common esterification and amidation reactions, which are intermediary steps toward functionalization with simple and more sophisticated moieties, such as polymers and biomolecules [44]. Nanotubes are also susceptible to conventional organic reactions including, but not limited to: halogenation, hydrogenation, cyclo, radical, electrophilic and nucleophilic additions, and ozonolysis [44]. Alternatively, non-covalent approaches may also be carried out to facilitate the processing of nanotubes via solution-based methods. The easiest way of achieving that is by wrapping the nanotube's walls with polymers [46], aromatic compounds [47], surfactants [48], and biomolecules [49]. These surface-modifying agents interact with the nanotube's walls mainly through π–π stacking, which is sufficiently strong to mitigate the van der Waals interaction between alike nanotubes but weak enough to avoid interference on the nanotube's electronic properties. Since no chemical reactions need to be performed, non-covalent adsorption is highly advantageous in terms of cost and execution. Some examples of covalent and non-covalent modified carbon nanotubes are given in figure 8.10.

8.6 Metal and metal oxide nanoparticles

Metal and metal oxide nanoparticles (MNP and MONP) are zero-dimensional nanomaterials displaying high surface-to-volume ratios. This morphological feature implies a strong quantum-confinement effect and results in a myriad of new chemical, optical, electrical and magnetic properties that are not present in their bulk counterparts. Depending upon the chemical composition, MNP and MONP may offer several useful attributes for chemical sensing, including electrocatalytic activity [50], plasmonic resonance [51], superparamagnetism [52], to name a few. Besides that, they show an enlarged surface area and plenty of surface atoms with incomplete coordination, which may be even more decisive for their sensing performance.

These inorganic nanoparticles have been mainly prepared by bottom-up methods performed in liquid solution, which is significantly cheaper than physical methods. The LaMer theory [53] is often evoked to describe the synthesis process under a microscopy point of view (figure 8.11(a)). Accordingly, molecular precursors are initially decomposed into monomers; monomers thus get associated into small nuclei, whose concentration reaches a supersaturation limit after some elapsed time. When the saturation limit is reached, the thermodynamic stability of the solution is severely affected, and the nuclei cannot stand longer; they must grow or dissociate back into monomers to decrease the positive Gibbs energy brought by the large surface area. The nucleation is a statistical process, and its rate (J) follows an Arrhenius-like law:

$$J(T, \Delta G_c) = A\exp^{\left(-\frac{\Delta G_c}{k_B T}\right)} \tag{8.3}$$

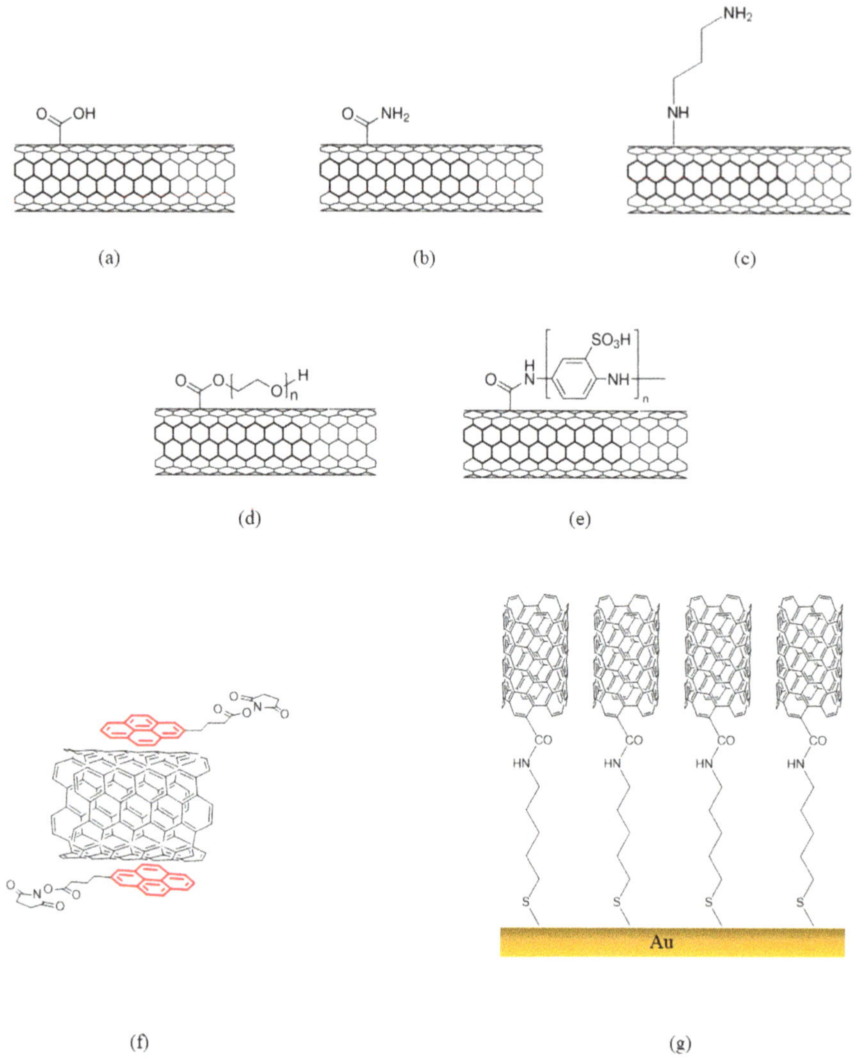

Figure 8.10. Some examples of carbon nanotubes modified by: (a) carboxylic acid, (b) amide, (c) ethylene diamine, (d) poly(ethylene glycol), (e) sulfonated poly(aniline), (f) π-stacking with a pyrene derivative, and (g) thiolated nanotube covalently attached to a gold surface.

In equation (8.3), ΔG_c stands for the Gibbs energy of the nanoparticle with a critical radius c as will be defined below, whereas A is the pre-exponential factor, k_B is the Boltzmann constant and T is the absolute temperature. Considering the nanoparticle as a spherical object for the sake of simplification, the variation on the Gibbs energy (ΔG) as a function of the nanoparticle's radius r is given by equation (8.4):

$$\Delta G = -\frac{4}{3}\pi r^3 \Delta G_V + 4\pi r^2 \gamma \tag{8.4}$$

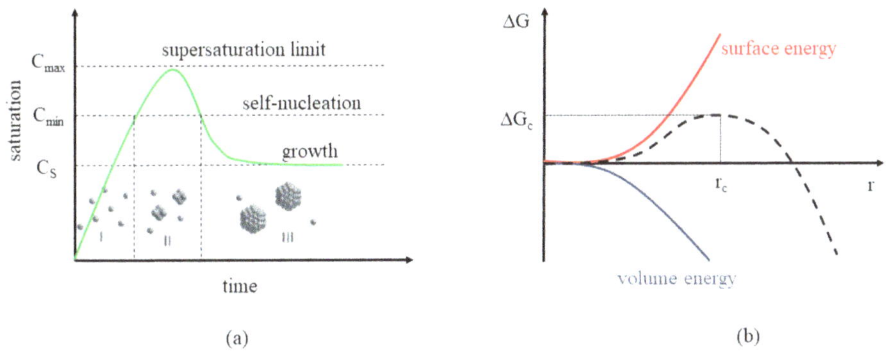

Figure 8.11. Kinetics and thermodynamics of nanoparticle formation in liquid medium: (a) LaMer diagram and (b) Gibbs energy of a spherical particle as a function of its radius.

In equation (8.4), ΔG_V stands for the volume Gibbs energy, whereas γ is the surface tension or, the surface Gibbs energy (ΔG_s) per unit area. As shown in figure 8.11(b), there is a critical radius, r_c, at which ΔG reaches a maximum (= ΔG_c). Therefore, ΔG_c and r_c can be estimated by calculating the derivative at this point and equating it to zero, dG/Dr =0, as follows:

$$\Delta G_c = \frac{16\pi\gamma^3}{\Delta G_V^2} \tag{8.5}$$

$$r_c = \frac{2\gamma}{\Delta G_V} \tag{8.6}$$

If the nanoparticle's radius grows above r_c, there is a gain of stability as ΔG decreases with the increase of the nanoparticle's volume (see that ΔG_V is negatively signed). In a first moment, growth is driven by adsorption of soluble monomers onto nascent nuclei. Later, growth is led by aggregation, which is significantly faster than adsorption of monomers. In this last step, smaller particles being unstable give place to bigger ones, since the increase of size is favored by a decrease of the Gibbs energy. This process is also known as Ostwald ripening [54].

The previous explanations make clear that preventing nanoparticles from agglomeration is a main concern; otherwise, the unusual physicochemical properties associated with nanoparticles are not fully developed. MNP and MONP are usually processed as colloidal suspensions, using either water or organic solvents. Since colloidal particles are subject to Brownian motion, the frequency of collisions is quite high at normal environmental conditions and they might lead to irreversible agglomeration. In fact, colloidal stability is kinetically controlled, and sedimentation of particles is a matter of time [55]. Thus, repulsive forces have to be introduced as a way to stabilize the nanoparticles. These repulsive forces can be of electrostatic, steric or electrosteric nature and are attained by modification of the nanoparticle's surface with specific functional groups [55]. Electrostatic stabilization is introduced by ionizable groups and is the main choice for water-based colloids (figure 8.12(a)).

(a) (b)

Figure 8.12. Models for colloidal stabilization via (a) electrostatic and (b) steric repulsive forces.

Steric hindrance is achieved by the adsorption of neutral polymers and surfactants and is largely employed in organic solvent-based colloids (figure 8.12(b)). Electrosteric stabilization can be achieved by polyelectrolytes, which simultaneously act as a steric barrier and electrostatic force against agglomeration. Besides stabilization, surface modification is also harnessed to add new functionalities to the nanoparticles and thereby widen their possible applications.

MNP are produced by chemical/electrochemical/photochemical reduction of metal salts in a straightforward manner [56]. For example, syntheses of spherical Ag and Au nanoparticles are easily carried out by boiling an aqueous mixture of the precursor salt, such as $AgNO_3$ or $HAuCl_3$, with an excess of sodium citrate, which acts simultaneously as reductant agent and stabilizer (ligand) [57, 58]. The reaction is completed in about 30 min, when an intense color develops because of plasmonic resonance bands in MNP. The resulting suspension is very stable and ready to use. Citrate adsorbed at the nanoparticles' surfaces makes them stable in physiological pH, but can be exchanged by other ligands, like amines, thiols, and so on. Alternatively, chemical reduction can be accomplished with $NaBH_4$ in the presence of alkanethiols, rendering the resulting particles dispersible in organic solvents [59]. Other ligands can be introduced during the reduction, like polymers [60], biomolecules [61] and other nanomaterials such as CNT [62], GO [63] or other inorganic nanoparticles. In general, these ligands are intentionally employed to control the nucleation rate, size, and shape of MNP.

MONP are mainly produced by alkaline hydrolysis of metal salts [64]. Although this method invariably produces polydisperse samples, it is of low-cost and easy execution. In aqueous medium, metal ions form aquo complexes that are rapidly converted to hydroxides as the pH is raised. The hydroxyl group in one complex can bridge to the metallic center in neighboring complexes. This process, called olation, can indeed continue and establish a polymer-like network through continuous hydroxo-bridges. Successive ionization and condensation reactions that take place in alkaline medium convert hydroxo- to oxo-bridges, a process called oxolation. The latter gives rise to the final metal oxide structure. At this stage, the MONP precipitates out of solution [65]. Since most MONP are amphoteric, they can be dispersed in both acidic and basic aqueous solutions, as surface groups can reversibly undergo protonation–deprotonation reactions under effect of pH changes, which thus leave their surface charged and promote electrostatic colloidal stabilization [66]. They may also be coated by organic surfactants, such as oleic acid, which make them stabilized by steric hindrance and thus dispersible in nonpolar organic solvents. MONP can also be produced by thermal decomposition of metal complexes [67]. Despite its higher cost, this method can produce monodisperse samples.

8.7 Biologically-active and molecularly-imprinted materials

Long ago, it was recognized that hybrid e-tongues comprising low and highly selective sensors may overperform those made of only low selective sensors when more complex liquids are analyzed [68]. Although e-tongues are defined as arrays of sensors showing negligible selectivity, the inclusion of selective sensors has been recently proposed. It has been demonstrated that biologically-active and molecularly-imprinted materials displaying high selectivity and specificity can widen the applicability of existing e-tongues to groups of substances that are generally not detected with arrays of conventional chemosensors [69]. Moreover, it is often observed that many specific sensors still respond to a group of substances rather than to a specific one. Therefore, they demand effective pattern recognition methods to provide reliable and accurate results, a strategy that is inherent to the e-tongues working principle.

Biologically-active materials for e-tongues include enzymes, nucleic acids and aptamers, antibodies, whole cells, tissues and phages. In common, these materials display unique selectivity for a sort of group of compounds. Provided that immobilization onto suitable electrodes does not impair their inherent selectivity, this property can be fully exploited in biosensors [70].

Enzymes are *per excellence* highly specific catalysts. The specificity stems from the ability that enzymes must bind the substrate molecule and stabilize subsequent transition states, thus determining which one of several potential chemical reactions takes place [71]. Antibodies, also called immunoglobulins, are proteins produced in the biological system in response to the presence of a foreign substance, called an antigen. They display specificity and affinity for the antigens that induced their syntheses, such as other proteins, polysaccharides, and nucleic acids [71]. Aptamers are oligonucleotides or peptides showing high affinity and specificity to target molecules. They are usually produced by *in vitro* selection or systematic evolution of ligands from a large random sequence pool. In contrast to antibodies, aptamers can be selected to bind essentially any target molecule. Moreover, the binding capacity of aptamers to specific targets is usually one magnitude greater than that exhibited by antibodies [72].

In the biological gustatory system, taste receptor cells (TRC) are responsible for tasting. They are clustered into taste buds in the tongue, which contain approximately 150 cells, including sensory neurons. They are further organized into small pegs of epithelium on the tongue called papillae. They sample many small molecules present in foodstuffs and report the sensation to the brainstem. Whereas the sense of smell permits us to sense thousands of odorants, the sense of taste is limited to discriminate between five primary tastes: bitter, sweet, sour, salty, and umami. Nonetheless, this apparent limitation is sufficient to classify compounds into potentially nutritive and safe (sweet, salty, umami) or potentially harmful or toxic (bitter and sour). For each primary taste, a particular group of receptors is evoked. For example, sweet, umami and bitter tastes are recognized by T1R and T2R receptors, which are types of G-proteins. On the other hand, sour and salty tastes are detected primarily by the passage of H^+ and sodium ions through channels expressed on the surface of cells in the tongue [71].

All the above-mentioned biologically-active materials have been immobilized within biosensor structures to be further assembled in e-tongues. As mentioned earlier, biosensors display high selectivity, but their responses are still subject to the interference of alike compounds. On the other hand, these apparent drawbacks should span the detection capabilities of the e-tongue, for example enabling detection of groups of similar substances, as well as the effect of inhibitory activities or even to correlate the e-tongue's response to the perception made by a human panel [73]. Moreover, conventional chemosensors can be added to the array to compensate any interference or matrix effect. In this regard, a few examples of hybrid e-tongues comprised by chemo- and biosensors are described below.

Molecularly-imprinted polymers (MIP) are generally based on cross-linked polymeric networks generated in the presence of the target analyte, which acts as a template. As the cross-linking process is completed, the analyte is extracted by a proper solvent (porogenic), living behind a cavity that shows affinity to the removed analyte (figure 8.13). Since the interactions between the polymeric matrix and template are relatively weak, such as electrostatic, H-bond, van der Waals and hydrophobic forces, the molecular-imprinting process is a way to produce synthetic receptors for a virtually unlimited number of target analytes [74]. Organic (mainly acrylate) and inorganic (mainly silicate) polymers can serve as matrixes for the construction of MIP. Nonetheless, these matrixes are inherently electrical insulators, thus hampering the possibility of using electrical/electrochemical transduction. This shortcoming can be overtaken using conducting polymers [75]. The main advantage of using MIP-based sensors in e-tongues is that several analytes can be selectively detected in complex media, whereas the conventional chemosensors composing the array compensate for limitation in cross-reactivity of MIP [76].

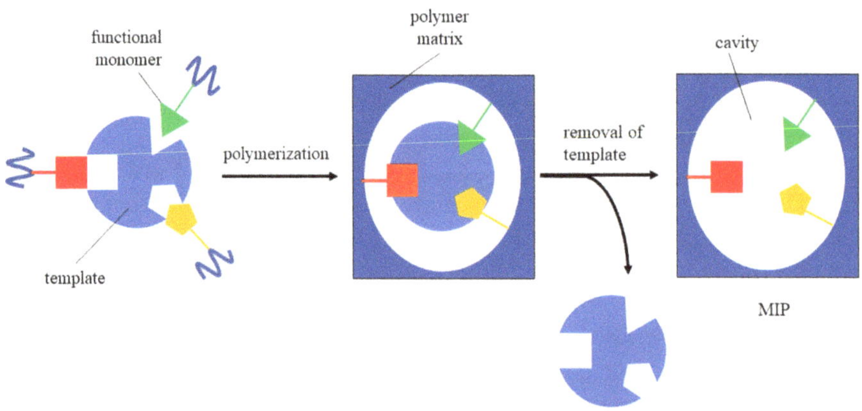

Figure 8.13. Schematic procedure of MIP production.

8.8 Thin-film deposition methods for integrating sensing materials in e-tongues

The previous subsections have provided an overview of different sensing materials that have been used for construction of the sensing units arrayed in e-tongues. Many of them are already available in the market, under different brands, prices and specificities. Nonetheless, implementation of these materials in e-tongues demand suitable processing methods. In this context, it is worth mentioning that chemo- and biosensors being recognized as low-cost analytical tools are envisaged as alternatives to more sophisticated analytical methods. Thus, as in the case of organic electronics, these sensors must be fabricated under low-cost procedures, whereby liquid solution based thin-film deposition methods are of outmost importance.

Thin-film deposition methods based on liquid solutions depend decisively on two important features: (i) materials displaying high solubility in common solvents, preferably in those showing negligible harm to humans and the environment, and (ii) materials provided with functional groups that enable their effective adsorption onto the sensor's electrode. As to feature (i), this can be a challenge for carbon nanotubes, transition metal dichalcogenides and some organic semiconductors. On the other hand, biologically-active materials are highly soluble in water. As to feature (ii), functionalization of pristine materials with pendant groups capable of attaching onto specific surfaces or blending/compositing with other materials can be very effective. Indeed, this strategy may also overcome the limited solubility. Of course, there are other options for circumventing these issues, for example, those methods in which synthesis and deposition occur simultaneously, such as electropolymerization. Among the most widespread thin-film deposition methods employed for the fabrication of e-tongue sensors can be highlighted casting, spin-coating, Langmuir–Blodgett (LB), layer-by-layer (LbL), electropolymerization and electrospinning. Next, will be given the main features of each of those methods.

Casting is the simplest and cheapest method for thin film deposition [77]. It consists of casting a diluted solution of the selected material onto a static substrate. A meniscus is thus formed, and the film is achieved after the solvent is gradually evaporated (figure 8.14(a)). Therefore, the solvent must have a sufficiently high vapor pressure otherwise evaporation will take too long. Despite its easy execution, casting cannot produce ultrathin films (thickness < 100 nm), which in the case of sensing applications is crucial to ensure higher sensitivity and fast signal recovery (reset).

Spin-coating is a subset of casting deposition, in which the substrate is rotated at high angular speed (~ 1000–5000 rpm) [77]. The rotation at high speeds submits the whole set to a high centrifugal force, which spreads the material solution toward the substrate edges, thus accelerating the solvent evaporation and attainment of ultra-thin films (figure 8.14(b)). The film formation takes place in four stages, as depicted in figure 8.14(c): (i) equilibrium stage, with the substrate stationary and wet with the solution; (ii) wave-formation stage, which occurs after a few revolutions (~ 2–3); (iii) corona stage, which occurs after tens of revolutions; and (iv) spiral stage, which occurs after thousands of revolutions. One of the main advantages of spin-coating is

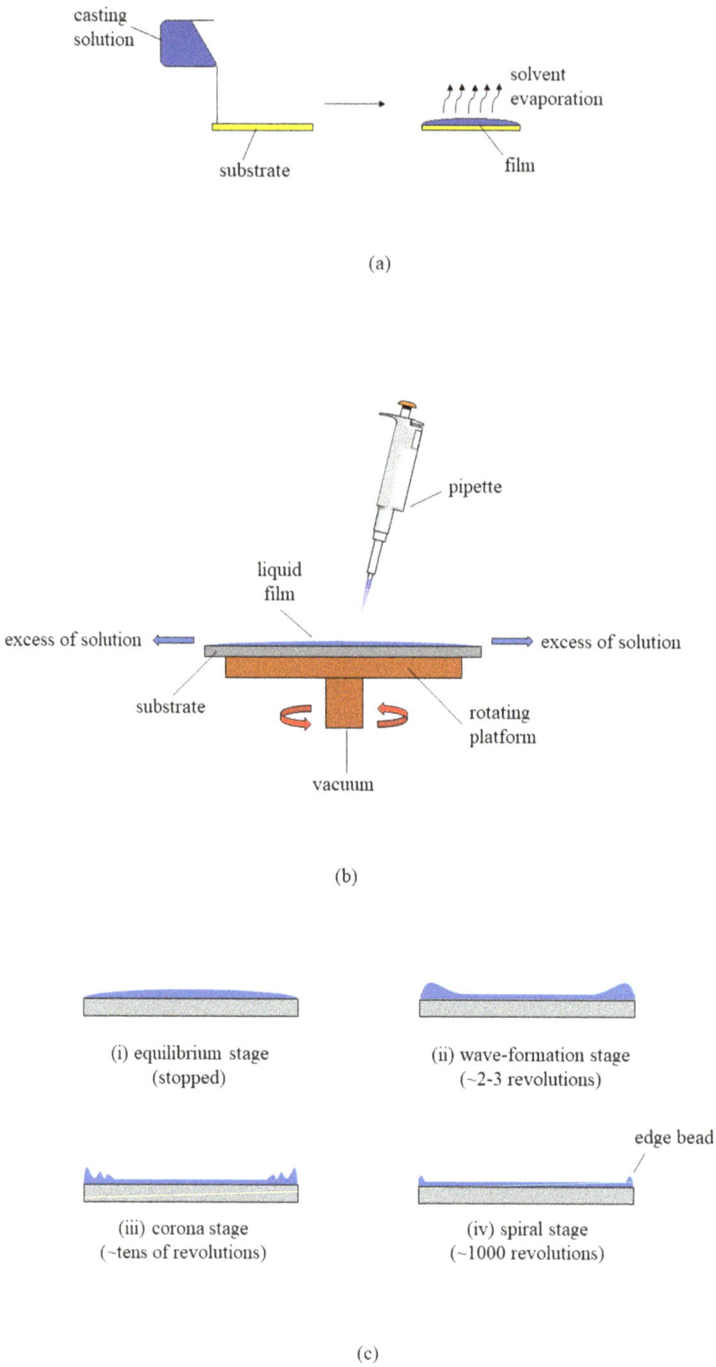

Figure 8.14. Some examples of thin-film deposition methods: (a) casting, (b) spin-coating, and (c) stages of film formation by spin-coating.

that the whole process is performed by automated and computer-driven equipment, the spin-coater. Moreover, the film thickness is externally controlled by setting-up the solution concentration, number of revolutions (expressed in rpm), time of spinning, and spinning ramp. As main drawbacks, the size of the substrate is limited by the spin-coater dimensions, a considerable amount of waste (liquid solution) is produced and because the solvent evaporation occurs at accelerated rates, there is no room for molecular orientation. The latter can be addressed by thermal annealing the film after deposition.

The LB technique [78] is entirely conducted with automated equipment called a Langmuir trough, which is schematically viewed in figure 8.15(a). The deposition solution containing the chosen material is spread over an aqueous subphase filled in the trough. Since the material is solely solubilized in organic solvents, the as-prepared solution does not mix with the subphase; instead, it forms a liquid

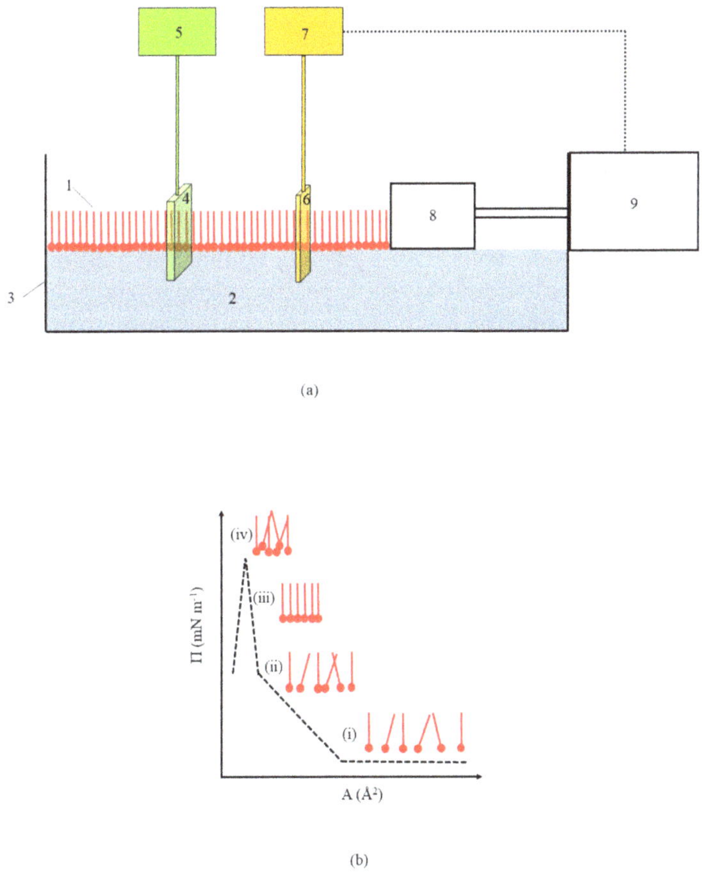

Figure 8.15. Schematic view of the LB deposition: (a) Langmuir trough (1—amphiphile monolayer, 2—liquid subphase, 3—Langmuir trough, 4—solid substrate, 5-dipping mechanism, 6—Wilhelmy plate, 7—electro-balance, 8—movable barrier, 9—barrier controller) and (b) surface pressure versus mean molecular area (Π x A) curve for each deposition step.

film over it. At this stage, molecules in the solution can be viewed as gas-phase particles. In a subsequent step, movable barriers horizontally compress the molecules until forming a condensed monolayer. The monolayers are finally transferred to the substrate that is mechanically immersed/emersed into/out of the subphase. The transfer can occur on the immersion as well as on the emersion of the substrate, depending upon the substrate and molecule structures. The Langmuir trough is usually equipped with surface pressure and surface potential measuring devices, which are used to monitor the formation and transfer of the monolayer. For example, surface pressure versus mean molecular area isotherms can monitor *in situ* the monolayer evolution, as shown in figure 8.15(b). The surface pressure Π is expressed as the difference between the surface tensions of pure water and water + monolayer. In step (i), molecules are disordered in a typical gaseous state that manifests as a plateau of low surface pressure and large molecular area. In step (ii), as the compressing barriers move toward molecules, they begin to organize, thereby occupying a smaller mean area and increasing the surface pressure. In step (iii), continuation of the barrier compression leads them to a condensed state, thus forming the monolayer. In step (iv), further compression leads the monolayer to collapse and the previous molecular organization is lost.

For a typical deposition involving amphiphilic molecules, multilayered films can be attained in three different types of supramolecular architectures, as shown in figure 8.16. Although the LB technique was originally devised for amphiphilic molecules [79], it was further adapted for semiconducting polymers [80] and small organic dyes [81], inorganic materials, G [82, 83], GO [84], hybrids of GO and MoS_2 [85]. The main advantage offered by the LB technique is its rigorous control of the film architecture at nanometric scale. Nonetheless, it is a time-consuming process because monolayer transfer has to be performed very slowly. In addition, Langmuir troughs are relatively expensive and usually require cleanroom facilities to operate.

In relative terms, the LbL technique is quite similar to the LB technique, since it is based on the sequential transfer of materials from solution to solid substrates [86]. However, it can be executed in common lab room conditions using much simpler and inexpensive glassware, it can be applied to a larger group of materials and has no restriction in terms of size, shape, and composition of the substrate, and, for many types of adsorbing materials, water can be used as the solvent. In practical terms, the LbL technique consists of the successive immersion of a chosen substrate into the solution/suspension of materials 'A' and 'B' in alternated fashion. Between each solution/suspension, the plated substrate is submitted to sequential rinsing and drying steps, which are performed to remove weakly adsorbed materials. At the end of a single deposition cycle, a (A/B) bilayered film is produced (figure 8.17(a)). In order to sustain a stepwise increase of the film thickness or attain multilayered structures, materials A and B must necessarily display complementary functionalities. These functionalities can be cationic/anionic groups, thus enabling electrostatic interactions; H-bonding sites, such as hydroxyl, ether, amine, amide, pyridine, and son on; charge-transfer complex; host-guest interaction; biologically specific; coordination chemistry; and covalent bonding [87]. Therefore, a myriad of sensing

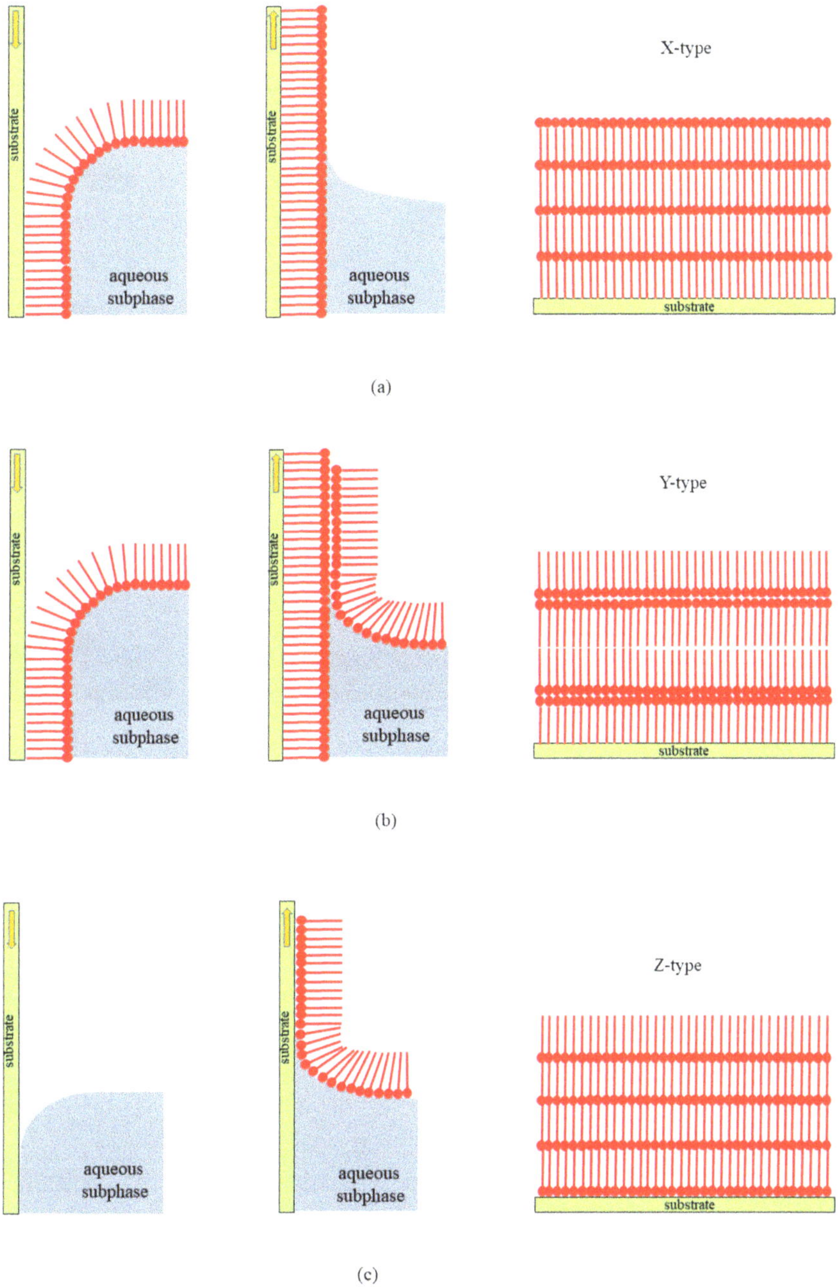

Figure 8.16. Types of LB films: (a) X-type, (b) Y-type and (c) Z-type.

materials for e-tongues can be assembled via LbL, including semiconducting polymers and small organic semiconductors, MNP and MONP, G, GO, RGO, SWCNT, MWCNT, biologically-active materials, transition metal dichalcogenides, to name but a few.

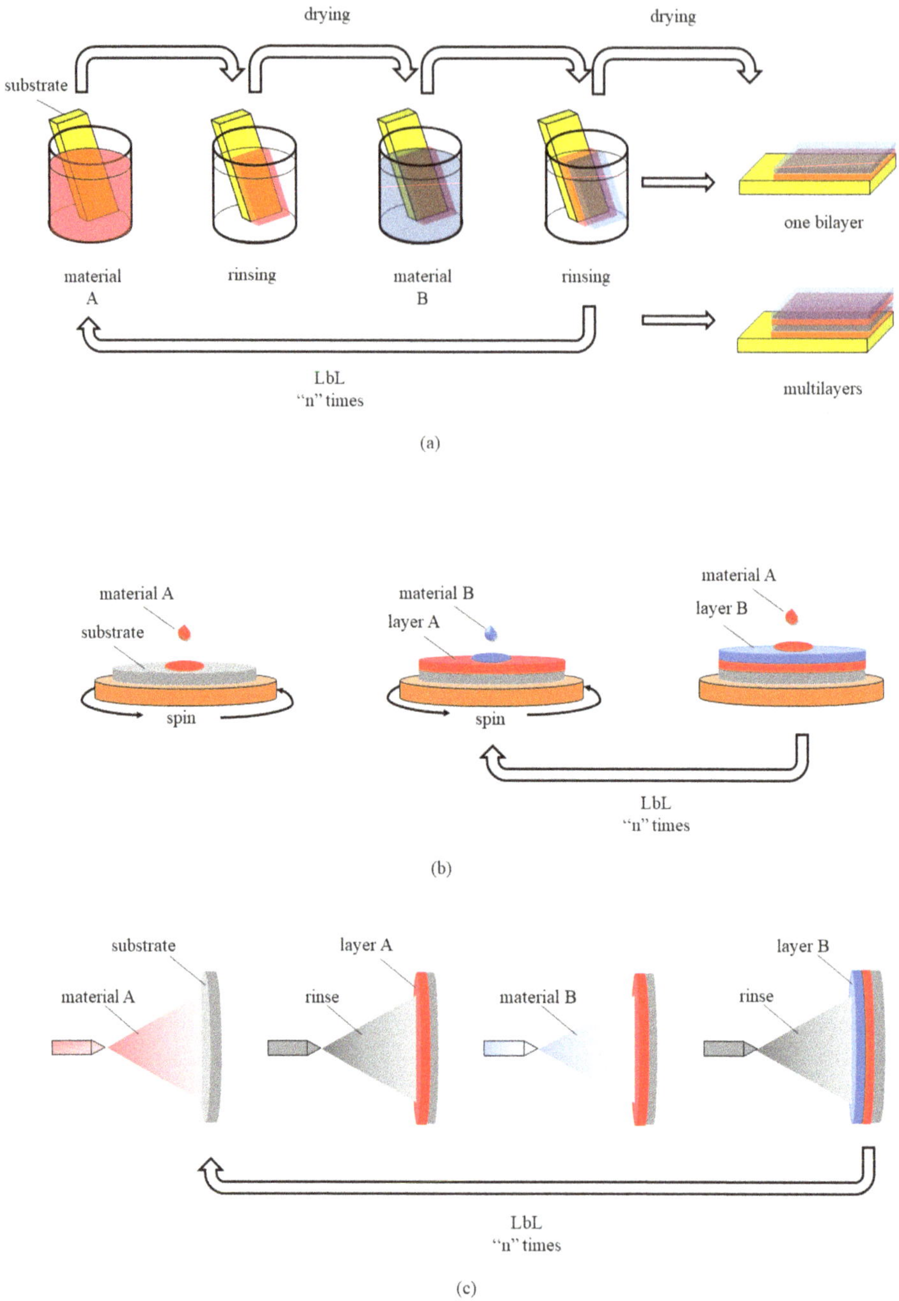

Figure 8.17. Different modes of LbL deposition: (a) immersive, (b) spin-assisted, (c) spray, (d) electrically-driven, and (e) microfluidic.

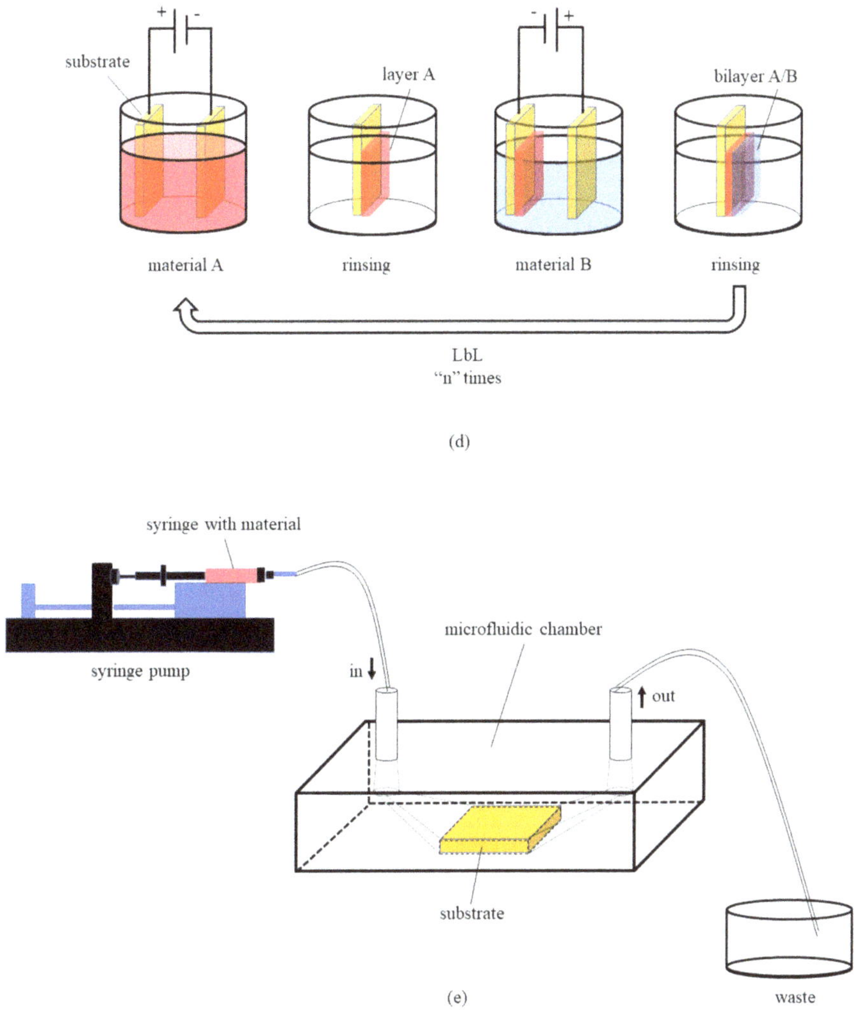

Figure 8.17. (Continued.)

The most widespread way of producing LbL films is that using mutual electrostatic interactions between materials A and B. For that purpose, the substrate must undergo some physical or chemical treatment to exhibit surface-charged groups prior to film deposition. When the substrate is immersed into a solution containing an oppositely charged material, the electrostatic interaction between them drives the adsorption and formation of a thin layer. The adsorption equilibrium is reached in a few minutes and ceases at its own. This occurs because charges from the substrate are overcompensated by the oppositely charged upcoming material, so that the substrate net charge is reversed and begins to repel alike species in solution. Also known as overcompensation, this mechanism occurs in the same way on the deposition of subsequent layers and is responsible for ensuring equal amounts of materials adsorbed in every deposition cycle [88]. Despite that there is a heat demand for replacing adsorbed water molecules by incoming materials, the adsorption is still spontaneous, since there is a much larger increase of the total entropy as the counter-ions from adsorbing materials are released to the supernatant [89].

Among the advantages offered by the LbL technique is its low operational cost. The successive immersion, rinsing and drying steps can be manually performed, and the entire process is conducted with simple glassware. Nonetheless, LbL films with reproducible properties are hardly achieved through hand-made deposition [90]. In this sense, automated LbL deposition machines have been made available, thus overcoming artifacts that are inherent to processes performed by different operators. Besides that, it is worth mentioning that immersive LbL is a time-dependent process since adsorbing materials must diffuse through the solution towards the substrate. Although this time is significantly shorter than that spent in the LB deposition, it is still considerably large whenever films with multiple bilayers are concerned. Thus, diffusion can be replaced by faster and high output processing, such as spray LbL (figure 8.17(b)) [91]. In this alternative LbL mode, solutions of different materials are sprayed against the substrate, thereby eliminating the lengthy diffusion. Moreover, spraying can be automated, and the film characteristics can thus be regulated by external operational parameters, including spraying time, nozzle dimension, spray–substrate distance, and so on. The LbL technique can also be performed in combination with other techniques such as spin-coating and electrochemical deposition [91]. For example, spin-assisted LbL consists of casting the material solution followed by spinning the substrate (figure 8.17(c)). Besides the fact that spinning the substrate accelerates solvent evaporation, a rather stratified film is attained in comparison to a fuzzy and highly interpenetrated network usually produced by immersive LbL. Alternatively, DC electrical field can be applied to a simple circuit containing two conducting substrates immersed into the deposition solution (figure 8.17(d)). Electrical polarization will drive charged species toward the substrate of opposite polarity, thus providing an additional force to move them instead of waiting on molecular diffusion. Finally, LbL deposition can be undertaken inside microfluidic channels (figure 8.17(e)). Since the solutions are pumped inside the channels with plunge syringes, adsorption is dictated by high flow rates rather than diffusion. Thus, the multilayer fabrication is faster than that made by immersive LbL. Moreover, the microfluidic-assisted deposition can be made at a selected region, something that is hardly achieved by conventional LbL. [92, 93].

LbL films are mostly built upon water-soluble materials. Besides water-soluble organic semiconductors (figure 8.3), carbon nanotubes (figure 8.10), GO and rGO (figure 8.7), this wide group is complemented with synthetic and natural-based polyelectrolytes, whose molecular formulae are given in figure 8.18.

8.9 Overview of materials for sensing in e-tongues

The large variety of sensing materials available for fabrication of sensing units has been reflected in an even greater number of possible applications of e-tongues. As such, tables 8.1–8.4 collect some of them, focusing on the type of sensing material

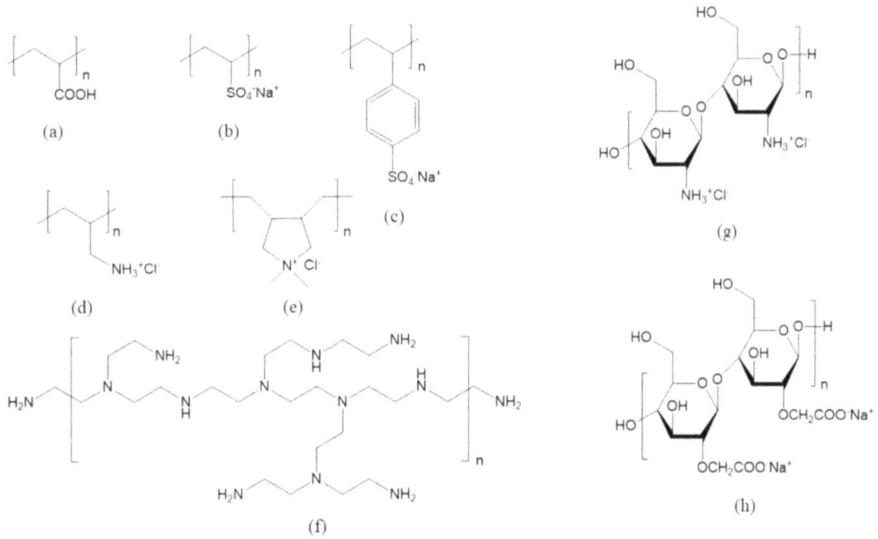

Figure 8.18. Molecular formulae of some common polyelectrolytes employed in the LbL deposition: (a) poly (acrylic acid), (b) poly(vinyl sulfate), (c) poly(sodium-4-styrene sulfonate), (d) poly(allylamine hydrochloride), (e) poly(diallyldimethyl ammonium chloride), (f) poly(ethylene imine), (g) chitosan, and (f) carboxymethyl cellulose.

Table 8.1. E-tongues made with PVC membranes doped with lipids and ionophores.

Sensing material	Deposition method	Transduction	Target analyte	References
PVC/lipids	Drop-casting	Potentiometry	Basic tastes Black tea Green tea Milk Ham Rice Pork Table salt Ginseng	[2, 3][94][95] [96][97] [98][99] [100][101]
PPy/ionophore	Electropolymerization	EIS	K^+, Na+, NH_4^+	[102]
PVC/ionophores	Drop-casting	Potentiometry	Beer and wine	[103]
PVC/ion-chromoionophores	Spin-coating	Image scanning	Na^+ and K^+	[104]
PVC + ionophores	Drop-casting	Potentiometry	Aminoacids	[105]
PVC + ionophores	Drop-casting	Potentiometry	Methane fermentation	[106]

PPy-poly(pyrrole), EIS-electrochemical impedance spectroscopy

Table 8.2. E-tongues made with organic semiconductors, carbon-nanomaterials, MNPs and MONPs.

Sensing material	Deposition method	Transduction	Target analyte	References
PEDOT:PSS, PPy, rGO, AuNPs	Drop-casting	IS	Organophosphate pesticides	[107]
MWCNT, carbon black, PTh	Drop-casting	IS	Basic tastes	[108]
SWCNT, MWCNT, rGO, AuNPs	Drop-casting	Resistance	Petroleum hydrocarbons in water	[109]
MWCNT, MOHNPs (Pd, Au, Cu, Ni, Co)	Drop-casting and electrodeposition	CV	Glucose, xylose, galactose, and mannose	[110]
PAH, CuTsPc, NiTsPc, AuNPs	LbL	IS	Milk	[111]
SWCNT, AuNPs	DEP	FET	Malignant tissues and cells	[112]
MWCNT and PLA nanofiber	Solution-blow spinning	IS	Water	[113]
PA nanofiber, cellulose nanowhisker, AgNPs	Electrospinning and dip-coating	IS	Pb^{2+}, Cu^{2+}, Ni^{2+}, Cd^{2+}	[114]
rGO, PDAC	LbL	IS	Basic tastes and VOC's	[115]
PANI, PPy, PSS, PAH, SL	LbL	IS	Methylisoborneol and geosmin	[116]
PA nanofibers, POEA, PPy, PSS	Electrospinning and LbL	IS	Paraoxon	[117]
PAH, PEDOT:PSS, PPy, CuTsPc	Microfluidic LbL	IS	Gliadin	[118]
MoS_2 and GO	Drop-casting	IS	Cloxacillin benzathine, erythromycin, streptomycin sulfate, and tetracycline hydrochloride	[119]
16-mer PANI, PPy, Ru(II) complex, azo-polymer	LB and LbL	IS	Basic tastes	[120]
SA, 16-mer PANI, PPy	LB	IS	Mineral water, tea, coffee	[121]
PDAC, CuTsPc, MMt-K, PEDOT: PSS	Microfluidic LbL	IS	Sucralose-based sweetners	[122]
PEI, GO, CHI, PAZO, PAH	LbL	IS	Triclosan	[123]
POMA, POEA, SL, PSS	LbL	IS	Sweetener formulations	[124]

PEDOT:PSS: poly(3,4-ethylenodioxy thiophene):sulfonated poly(styrene); PPy: poly(pyrrole); rGO: reduced graphene oxide; AuNPs: gold nanoparticles; MWCNT: multiwalled carbon nanotubes; PTh: poly(thiophene); SWCNT: single wall carbon nanotubes; MOHNPs: metal hydroxyde nanoparticles; PAH: poly(allylamine hydrochloride); CuTsPc: copper(II) tetrasulfonated phthalocyanine; NiTsPc: nickel(II) tetrasulfonated phthalocyanine; PLA: poly(lactic acid); PA: poly(amide); AgNPs: silver nanoparticles; PDAC: poly(diallyldimethyl ammonium chloride); PANI: poly(aniline); PSS: poly(sodium-4-styrene sulfonate); SL: sulfonated lignina; POEA: poly (o-ethoxy aniline); GO: graphene oxide; SA: stearic acid; MMt-K: potassium montmorrillonite; PEI: poly(ethylene imine); CHI: chitosan; PAZO: azo polymer; POMA: poly(o-methoxy aniline); IS: impedance spectroscopy; CV: cyclic voltammetry; DEP: dielectrophoretic deposition; FET: field-effect transistor

Table 8.3. E-tongues made with metallic and chalcogenide glass electrodes.

Sensing material	Deposition method	Transduction	Target analyte	References
Noble metals		CV, IS	TNT	[125]
Pt, Au		Pulse voltammetry	Fruit juices, still drink, milk	[126]
Doped chalcogenide glass		Potentiometry	Ni^{2+}, Zn^{2+}, Cd^{2+}, Pb^{2+}, Cu^{2+}	[127]
Doped chalcogenide glass		Potentiometry	Paracetamol formulations	[128]
Doped chalcogenide glass, PVC + ionophore, Mn(TPP)Cl	Drop-casting	Potentiometry	MIB/GEO/ISO	[129]
Doped chalcogenide glass		Potentiometry	Italian wine	[130]
PVC + metallophorphyrins	Glued onto electrodes	Potentiometry and QCM	Urine and milk	[131]
Doped Chalcogenide glass and ion-selective electrodes		Potentiometry	Cu^{2+}, Cd^{2+}, Fe^{2+}, Cr^{3+}, Zn^{2+}, Cl^-, SO_4^{2-} and H^+	[132]

Table 8.4. E-tongues made biologically-active and molecularly-imprinted materials.

Sensing material	Deposition method	Transduction	Target analyte	References
Carbon paste, tyrosinase, glucose oxidase, CoPc, CuPc, ZnPc, LuPc2	Mixing with carbon paste and drop-casting	CV	Red grapes	[133]
Tyrosinase, laccase, LuPc2, arachidic acid	LB	CV	Grapes	[134]
Cholinesterases, tyrosinase, peroxidases and cellobiase dehydrogenase	Drop-casting and crosslinking with glutaldehyde	amperometry	Organophosphates and carbamates pesticides	[135]
Taste ephitelium	Cells culture onto electrodes	LAPS	Basic tastes and ATP	[136]
PAMAM, PAH, PVS, antigens, proteolipossomes	LbL	IS	Anti-pasteurollosis antibodies	[137]

(*Continued*)

Table 8.4. (*Continued*)

Sensing material	Deposition method	Transduction	Target analyte	References
PAMAM, proteolipossomes, antigens	LbL	IS	Chaga's disease and leishimaniasis	[138]
PAM MIP	Drop-casting	CV	Haemoglobin, equine myoglobin, cytochrome C, and bovine serum albumin	[139]
PAM MIP	Drop-casting	OA	Arylamines	[140]
AA, EGMA MIP	Photopolymerization	OA	Atrazine, 2,4-D, chlorpyrifos, diazinon, prometryn, and bisphenol A	[141]
DNA aptamer	Crosslinking with glutaraldehyde	LAPS	ATP	[142]

CoPc: cobalt(II) phthalocyanine; CuPc: copper(II) phthalocyanine; ZnPc: zinc(II) phthalocyanine; LuPc2: lutetium(III) bis-phthalocyanine; PAMAM: poly(amidoamine) dendrimer; PVS: poly(vinyl sulfate); PAM MIP: poly(acrylamide) molecularly-imprinted polymer; AA: acrylic acid; EGMA: ethylene glycol mono-acetate; LAPS: light-addressable potentiometric sensor; OA: optical absorption

and the deposition method. Although many other possibilities have been investigated, those reported herein provide a good evolution picture of the field. It also points out that making a rational choice of sensing material, deposition method and signal transduction would enable the use of e-tongues in almost all types of routine processes carried out in food, beverage, and pharmaceutical industries, as well as on monitoring of pollutants and human diseases.

8.10 Final remarks and outlook

The previous sections have clearly shown that e-tongues can be built on different sensing materials platforms, ranging from polymeric membranes doped with lipids and ionophores to organic semiconductors, carbon-based nanomaterials, metal and metal oxide nanoparticles, biologically-active, and molecularly-imprinted materials. The right choice for sensing materials is mainly driven by the type of target analyte, even though availability and ease of processing have also to be concerned. This latter aspect is a major issue when one considers that e-tongues are envisaged as lower-cost alternatives to more robust analytical instrumentation. Nonetheless, most sensing materials described herein are both commercially available and readily attained by low-cost synthetic procedures. In addition, they can be solubilized in common solvents and thus processed into thin films by low-cost deposition techniques. In summary, a rational choice for sensing material, deposition method and signal transduction has enabled the use of e-tongues in almost all types of routine processes carried out in food, beverage, and pharmaceutical industries, as well as on monitoring of pollutants and human diseases. It is, therefore, expected that e-tongues will become an ordinary technology assisting professionals in different fields where rapid and reliable chemical/biochemical analyses are needed.

References

[1] Nelson D L and Cox M M 2005 *Lehninger Principles of Biochemistry* (New York: W. H. Freeman) p 343

[2] Tahara Y and Toko K 2013 *IEEE Sensors J.* **13** 3001–11 .

[3] Hayashi K, Yamanaka M, Toko K and Yamafuji K 1990 *Sens. Actuators* B *Chem.* **2** 205–13

[4] Bowman W C 1990 *Pharmacology of Neuromuscular Function* (London: Butterworth-Heinemann) pp 8–35

[5] Bakker E, Buhlmann P and Pretsch E 1997 *Chem. Rev.* **97** 3083–132

[6] Ciosek P, Augustyniak E and Wroblewski W 2004 *Analyst* **129** 639–44

[7] Morf W E 1981 *The Principles of Ion-Selective Electrodes and of Membrane Transport* (New York: Elsevier)

[8] Vlasov Y G and Bychkov E A 1987 *Ion-selective Electrode Rev.* **9** 5–93

[9] Vassilev V and Boychevea S 2005 *Talanta* **67** 20–7

[10] Vlasov Y G 1989 *Fresenius* Z *Anal Chem.* **335** 92–9

[11] Vlasov Y G, Legin A and Rudnitskaya A 1997 *Sens. Actuators* B *Chem.* **44** 532–37

[12] Hadziioannou G and Malliaras G G 2006 *Semiconducting Polymers: Chemistry, Physics and Engineering* (Weinheim: Wiley-VCH)

[13] Faria R M and Oliveira O N Jr 1999 *Braz. J. Phys.* **29** 360–70

[14] Rothemund P 1935 *J. Am. Chem. Soc.* **57** 2010–011

[15] Sakamoto K and Ohno-Okumura E 2009 *Materials* **2** 1127–179

[16] Zhu Y, Murali S, Cai W, Li X, Suk J W, Potts J R and Ruoff R S 2010 *Adv. Mater.* **22** 3906–924

[17] Novoselov K S, Geim A K, Morozov S V, Jiang D, Zhang Y, Dubonos S V, Grigorieva I V and Firsov A A 2004 *Science* **306** 666–69

[18] Dreyer D R, Park S, Bielawski C W and Ruoff R S 2010 *Chem. Soc. Rev.* **39** 228–40

[19] Bansal T, Mohite A D, Shah H M, Galande C, Srivastava A, Jasinski J B, Ajayan P M and Alphenaar B W 2012 *Carbon* **50** 808–14

[20] Somani P R, Somani S P and Umeno M 2006 *Chem. Phys. Lett.* **430** 56–9

[21] Hernandez Y, Nicolosi V, Lotya M, Blighe F M, Sun Z Y, De S, McGovern I T, Holland B, Byrne M and Gun'ko Y K 2008 *Nat. Nanotechnol.* **3** 563–68

[22] Brodie B C 1859 *Philos. Trans.* R. *Soc. London* **149** 249–59

[23] Staudenmaier L 1898 *Ber. Dtsch. Chem. Ges.* **31** 1481–487

[24] Hummers W S and Offeman R E 1958 *J. Am. Chem. Soc.* **80** 1339

[25] Koch K R and Krause P F 1982 *J. Chem. Ed.* **59** 973–74

[26] Konkena B and Vasudevan S 2012 *J. Phys. Chem. Lett.* **3** 867∓72

[27] Pei S F and Cheng H M 2012 *Carbon* **50** 3210–228

[28] Stankovich S, Dikin D A, Piner R D, Kohlhaas K A, Kleinhammes A, Jia Y, Wu Y, Nguyen S T and Ruoff R S 2007 *Carbon* **45** 1558–565

[29] Popov V N 2004 *Mater. Sci. Eng.* R *Rep.* **43** 61–102

[30] Saito R, Fujita M, Dresselhaus G and Dresselhaus M S 1992 *Phys. Rev.* B **45** 6234

[31] White C T and Todorov T N 1998 *Nature* **393** 240–42

[32] Dai H, Wong E W and Lieber C M 1996 *Science* **272** 523-26

[33] Weisman R B and Bachilo S M 2003 *Nano Lett.* **3** 1235–238

[34] Kataura H, Kumazawa Y, Maniwa Y, Umezu I, Suzuki S, Ohtsuka Y and Achiba Y 1999 *Synth. Metals* **103** 2555–558

[35] Iijima S 1991 *Nature* **354** 56–8

[36] Iijima S and Ichihashi T 1993 *Nature* **363** 603–05

[37] Guo T, Nikolaev P, Thess A, Colbert D T and Smalley R E 1995 *Chem. Phys. Lett.* **243** 49–54

[38] Cheng H M, Li F, Su G, Pan H Y, He L L, Sun X and Dresselhaus M S 1998 *Appl. Phys. Lett.* **72** 3282–284

[39] Meyyappan M, Delzeit L, Cassell A and Hash D 2003 *Plasma Sources Sci. Technol.* **12** 205–16

[40] Guo T, Nikolaev B, Rinzler A G, Tomanek D, Colbert D T and Smalley R E 1995 *J. Phys. Chem.* **99** 10694–97

[41] Gavillet J, Loiseau A, Ducastelle F, Thair S, Bernier P, Stephan O, Thibault J and Charlier J C 2002 *Carbon* **40** 1649–63

[42] Sinnott S B, Andrews R, Qian D, Rao A M, Mao Z, Dickey E C and Derbyshire F 1999 *Chem. Phys. Lett.* **315** 25–30

[43] Hilding J, Grulke E A, Zhang Z G and Lockwood F 2003 *J. Dispers. Sci. Technol.* **24** 1–41

[44] Tasis D, Tagmatarchis N, Bianco A and Prato M 2006 *Chem. Rev.* **106** 1105–36

[45] Ivanov V, Fonseca A, Nagy J B, Lucas A, Lambin P, Bernaerts D and Zhang X B 1995 *Carbon* **33** 1727–38

[46] Andrews R, Jacques D, Qian D and Rantell T 2002 *Acc. Chem. Res.* **35** 1008–17

[47] Zhang J, Lee J K, Wu Y and Murray R W 2003 *Nano Lett.* **3** 403–7

[48] Moore V C, Strano M S, Haroz E H, Hauge R H and Smalley R H 2003 *Nano Lett.* **3** 1379–82

[49] Katz E and Wilner I 2004 *ChemPhysChem.* **5** 1084–104

[50] Xu H, Shang H Y, Wang C and Du Y K 2020 *Adv. Funct. Mater.* **30** 2006317

[51] Li M, Cushing S K and Wu N 2015 *Analyst* **140** 386–406

[52] Krishnan S and Goud K Y 2019 *Magnetochemistry* **5** 64

[53] LaMer V K 1952 *Ind. Eng. Chem.* **44** 1270–77

[54] Voorhees P W 1985 *J. Stat. Phys.* **38** 231–52

[55] Hunter R J 1986 *Foundations of Colloid Science* (Oxford: Claredon Press)

[56] Fedlheim D L and Foss C A 2001 *Metal Nanoparticles Synthesis, Characterization, and Applications* (Boca Raton: CRC Press)

[57] Pillai Z S and Kamat P V 2004 *J. Phys. Chem.* **108** 945–51

[58] Turkevich J, Stevenson P C and Hillier J 1951 *Discuss. Faraday Soc.* **11** 55–75

[59] Brust M, Walker M, Bethell D, Schiffrin D J and Whyman R 1994 *J. Chem. Soc., Chem. Commun.* **1994** 801–2

[60] Ofir Y, Samanta B and Rotello V M 2008 *Chem. Soc. Rev.* **37** 1814–23

[61] Katz E and Willner I 2004 *Angew. Chem. Int.* **43** 6042–108

[62] Wu B, Kuang Y, Zhang X and Chen J 2011 *Nano Today* **6** 75–90

[63] Zhou X, Huang X, Qi X, Wu S, Xue C, Boey F Y C, Yan Q, Chen P and Zhang H 2009 *J. Phys. Chem.* **113** 10842–6

[64] Jolivet J P, Cassaignon S, Chaneac C, Chiche D, Durupthy O and Portehault D 2010 *Comptes Rend. Chim.* **13** 40–51

[65] Henry M, Jolivet J P and Livage J 1992 Aqueous chemistry of metal cations: hydrolysis, condensation and complexation *Chemistry, Spectroscopy and Applications of Sol-Gel Glasses. Structure and Bonding* ed R Reisfeld and C K JJørgensen (Berlin: Springer)
[66] Jolivet J P 2000 *Metal Oxide Chemistry and Synthesis: From Solution to Solid State* (Hoboken, NJ: Wiley)
[67] Kwon S G and Hyeon T 2008 *Acc. Chem. Res.* **41** 1696–709
[68] Ciosek P and Wroblewski W 2007 *Talanta* **71** 738–46
[69] Ha D, Sun Q, Su K, Wan H, Li H, Xu N, Sun F, Zhuang L, Hu N and Wang P 2015 *Sens. Actuators* B *Chem.* **207** 1136–46
[70] Siqueira J R Jr, Caseli L, Crespilho F N, Zucolotto V and Oliveira O N Jr 2010 *Biosens. Bioelectron.* **25** 1254–63
[71] Berg J, Tymoczko J L, Gatto G J Jr and Stryer L 2015 *Biochemistry* (New York: W. H. Freeman)
[72] Liu J, Cao Z and Lu Y 2009 *Chem. Rev.* **109** 1948–98
[73] Del Valle M, Cetó X and Gutiérrez-Capitán M 2014 Bioelectronic tongues: when the sensor array incorporates biosensors *Multisensor Systems for Chemical Analysis: Materials and Sensors* ed L Lvova, D Kirsanov, A Legin and C Di Natale (Boca Raton, FL: CRC Press) pp 211–46
[74] BelBruno J J 2019 *Chem. Rev.* **119** 94–119
[75] Huynh T P, Pietrzyk-Le A, Bikram K c c, Noworyta K, Sobczak J W, Sharma P S, D'Souza F and Kutner W 2013 *Biosens. Bioelectron.* **41** 634–41
[76] Huynh T P and Kutner W 2015 *Biosens. Bioelectron.* **74** 856–64
[77] Rangappa S M, Parameswaranpillai J and Siengchin S 2020 *Polymer Coatings: Technologies and Applications* (Boca Raton: CRC Press)
[78] Petty M C 1996 *Langmuir–Blodgett Films: An Introduction* (Cambridge: Cambridge University Press)
[79] Blodgett K B 1935 *J. Am. Chem. Soc.* **57** 1007–22
[80] Goncalves D, Bulhoes L O S, Mello S V, Mattoso L H C, Faria R M and Oliveira O N 1994 *Thin Solid Films* **243** 544–46
[81] Antunes P A, Santana C M, Aroca R F, Oliveira O N, Constantino C J L and Riul A 2005 *Synth. Metals* **148** 21–4
[82] Paul S, Pearson C, Molloy A, Cousins M A, Green M, Kolliopoulou S, Dimitrakis P, Normand P, Tsoukalas D and Petty M C 2003 *Nano Lett.* **3** 533–6
[83] Li X, Zhang G, Bai X, Sun X, Wang X, Wang E and Dai H 2008 *Nat. Nanotechnol.* **3** 538–42
[84] Mangadlao J D, Santos C M, Felipe M J L, de Leon A c c, Rodrigues D F and Advincula R C 2015 *Chem. Commun.* **51** 2886–89
[85] Zhang H, Jeon K W and Seo D K 2016 *ACS Appl. Mater. Interfaces* **8** 21539–44
[86] Decher G 1997 *Science* **277** 1232–37
[87] Borges J and Mano J F 2014 *Chem. Rev.* **114** 8883∓942
[88] Schelenoff J B, Ly H and Li M 1998 *J. Am. Chem. Soc.* **120** 7626–34
[89] Raposo M and Oliveira O N Jr 2000 *Langmuir* **16** 2839–44
[90] Braga G S, Paterno L G, Lima J P H, Fonseca F J and de Andrade A M 2008 *Mater. Sci. Eng.* C **28** 555–62
[91] Richardson J J, Malm M B and Caruso F 2015 *Science* **348** 2491
[92] Kim H J, Lee K, Kumar S and Kim J 2005 *Langmuir* **21** 8532–8

[93] Daikuzono C M, Dantas C A R, Volpati D, Constantino C J L, Piazzetta M H O, Gobbie A L, Taylor D M, Oliveira O N Jr and Riul A Jr 2015 *Sens. Actuators* B *Chem.* **207** 1129–35
[94] Hayashi N, Chen R, Ikezaki H, Ujihara T, Kitajima H and Mizukami Y 2007 *Biosci. Biotechnol. Biochem.* **71** 587–9
[95] Hayashi N, Chen R, Ikezaki H and Ujihara T 2008 *J. Agri. Food Chem.* **56** 7384–7
[96] Mizota Y, Matsui H, Ikeda M, Ichihashi N, Iwatsuki K and Toko K 2009 *Milchwissenschaft* **64** 143–6
[97] Kobayashi Y, Habara M, Ikezazki H, Chen R, Naito Y and Toko K 2010 *Sensors* **10** 3411–43
[98] Tran T U, Suzuki K, Okadome H, Homma S and Ohtsubo K 2004 *Food Chem.* **88** 557–66
[99] Sasaki K, Tani F, Sato K, Ikezaki H, Taniguchi A and Emori T 2005 *Sens. Mater.* **17** 397–404
[100] Chen R, Hidekazu I and Toko K 2010 *Sens. Mater.* **22** 313–25
[101] Cui S, Wang J, Geng L, Wei Z and Tian X 2013 *Sens. Mater.* **25** 241–55
[102] Cortina-Puig M, Munoz-Berbel X, Alonso-Lomillo M A, Munoz-Pascual F J and del Valle M 2007 *Talanta* **72** 774–79
[103] Nery E W and Kubota L T 2016 *Anal. Chim. Acta* **918** 60–8
[104] Erenas M M, Pegalajar M C, Cuellar M P, de Orbe-Paya I and Capitan-Vallvey L F 2011 *Sens. Actuators* B *Chem.* **156** 976–82
[105] Janczyk M, Kutyla-Olesiuk A, Ceto X, del Valle M, Ciosek P and Wroblewski W 2013 *Sens. Actuators* B *Chem.* **189** 179–86
[106] Witkowska E, Buczkowska A, Zamojska A, Szewczyk K W and Ciosek P 2010 *Bioelectrochem.* **80** 87–93
[107] Facure M H M, Mercante L A, Mattoso L H C and Correa D S 2017 *Talanta* **167** 59–66
[108] Pioggia G, Di Francesco F, Marchetti A, Ferro M and Ahluwalia A 2007 *Bionsens. Bioelectron.* **22** 20078–2623
[109] Cooper J S, Myers M, Chow E, Hubble L J, Cairney J M, Pejcic B, Müller K H, Wieczorek L and Raguse B 2014 *J. Nanoparticle Res.* **16** 2173
[110] de Sa A C, Cipri A, Gonzalez-Calabuig A, Stradiotto N R and del Valle M 2016 *Sens. Actuators* B *Chem.* **222** 645–53
[111] Mercante L A, Scagion V P, Pavinatto A, Sanfelice R C, Mattoso L H C and Correa D S 2015 J. *Nanomater.* **2015** 890637
[112] Silva G O, Michael Z P, Bian L, Shurin G V, Mulato M, Shurin M R and Star A 2017 *ACS Sens.* **2** 1128–32
[113] Oliveira J E, Grassi V, Scagion V P, Mattoso L H C, Glenn G M and Medeiros E S 2013 *IEEE Sens. J.* **13** 759–66 .
[114] Teodoro K B R, Kelcilene B R, Shimizu F M, Scagion V P and Correa D S 2019 *Sens. Actuators* B *Chem.* **290** 387–95
[115] Gross M A, Sales M J A, Soler M A G, Pereira-da-Silva M A, da Silva M F P and Paterno L G 2014 *RSC Adv.* **4** 17917–24
[116] Braga G S, Paterno L G and Fonseca F J 2012 *Sens. Actuators* B *Chem.* **171–2** 181–9
[117] Oliveira J E, Scagion V P, Grassi V, Correa D S and Mattoso L H C 2012 *Sens. Actuators* B *Chem.* **171–2** 249–55
[118] Daikuzono C M, Shimizu F M, Manzoli A, Riul A Jr, Piazzetta M H O, Gobbi A L, Correa D S, Paulovich F V and Oliveira O N Jr 2017 *ACS Appl. Mater. Interfaces* **9** 19646–52
[119] Facure M H M, Schneider R, dos Santos D M and Correa D S 2020 *Talanta* **217** 121039

[120] Riul A, dos Santos D S, Wohnrath K, Di Tommazo R, Carvalho A C P L F, Fonseca F J, Oliveira O N, Taylor D M and Mattoso L H C 2002 *Langmuir* **18** 239–45

[121] Riul A, Soto A M G, Mello S V, Bone S, Taylor D M and Mattoso L H C 2003 *Synth. Metals* **132** 109–16

[122] Braunger M L, Fier I, Shimizu F M, de Barros A, Rodrigues V and Riul A Jr 2020 *Sensors* **20** 6194

[123] Marques I, Magalhaes-Mota G, Pires F, Serio S, Ribeiro P A and Raposo M 2017 *Appl. Surf. Sci.* **421** 142–47

[124] Dyminski D S, Paterno L G, Takeda H H, Bolini H M A, Mattoso L H C and Candido L M B 2006 *Sensor Lett.* **4** 403–8

[125] Garcia-Breijo E, Peris R M, Pinatti C O, Fillol M A, Civera J I and Prats R B 2013 *IEEE Trans. Instrum. Meas.* **62** 424–31

[126] Winquist F, Wide P and Lundstrom I 1997 *Anal. Chim. Acta* **357** 21–31

[127] Vlasov Y, Legin A and Rudnitskaya A 1997 *Sens. Actuators* B *Chem.* **44** 532–7

[128] Belugina R B, Monakhova Y B, Rubtsova E, Becht A, Schollmayer C, Holzgrabe U, Legin A V and Kirsanov D O 2020 *J. Pharm. Biomed. Anal.* **188** 113457

[129] Lvova L, Jahatspanian I, Mattoso L H C, Correa D S, Oleneva E, Legin A, Di Natale C and Paolesse R 2020 *Sensors* **20** 821

[130] Legin A, Rudnitskaya A, Lvova L, Vlasov Y, Di Natale C and D'Amico A 2003 *Anal. Chim. Acta* **484** 33–44

[131] Di Natale C, Paolesse R, Macagnano A, Mantini A, D'Amico A, Legin A, Lvova L, Rudnitskaya A and Vlasov Y 2000 *Sens. Actuators* B *Chem.* **64** 15–21

[132] Di Natale C, Macagnano A, Davide F, D'Amico A, Legin A, Vlasov Y, Rudnitskaya A and Selezenev B 1997 *Sens. Actuators* B *Chem.* **44** 423–28

[133] Medina-Plaza C, Revilla G, Munoz R, Fernandez-Escudero J A, Barajas E, Medrano G, de Saja J A and Rodriguez-Mendez M L 2014 *J. Porphyr. Phthalocyanines* **18** 76–86

[134] Medina-Plaza C, de Saja J A and Rodriguez-Mendez M L 2014 *Biosens. Bioelectron.* **57** 276–83

[135] Solná R, Dock E, Christenson A, Winther-Nielsen M, Carlsson C, Emnéus J, Ruzgas T and Skládal 2005 *Anal. Chim. Acta* **528** 9–19

[136] Chen P, Wang B, Cheng G and Wang P 2009 *Biosens. Bioelectron.* **25** 228–33

[137] Zucolotto V, Daghastanli K R P, Hayasaka C O, Riul A Jr, Ciancaglini P and Oliveira O N Jr 2007 *Anal. Chem.* **79** 2163–67

[138] Perinoto A C *et al* 2010 *Anal. Chem.* **82** 9763–8

[139] Bueno L, El-Sharif H F, Salles M O, Boehm R D, Narayan R J, Paixão T R L C and Reddy S M 2014 *Sens. Actuators* B *Chem.* **204** 88–95

[140] Greene N T, Morgan S L and Shimizu K D 2004 *Chem. Commun.* 1172–3

[141] Xu D, ZhuW, Wang C, Tian T, Cui J, Li J, Wang H and Li G 2014 *Chem. Eur. J.* **20** 16620–5

[142] Wu C, Du L, Zou L, Zhao L and Wang P 2012 *Biomed. Microdevices* **14** 1047–53

Chapter 9

Hybrid and optical multisensory systems for liquid analysis: theoretical basis, trends and applications

Larisa Lvova, Corrado Di Natale and Roberto Paolesse

Dr L Lvova and Professor R Paolesse would like to dedicate their contributions to this chapter to their respected co-author Professor Corrado Di Natale on the occasion of his 60th birthday and in recognition of his outstanding contributions to multisensory analysis and chemometrics.

In this chapter the main aspects of optical and hybrid sensor arrays employed for liquid system analysis will be discussed. The operating principles, advantages and challenges of multisensory optical systems regarding electroanalytical ones will be considered. Moreover, the examples of hybrid and multitransduction multisensory systems applications, including the recent advances from our team, will be reported to demonstrate the effectiveness of such systems for a wide range of analytical tasks, such as environmental monitoring, foodstuff quality assessment, medical and health care applications.

9.1 Introduction

The second half of the last century until nowadays was characterized by considerable attention devoted to the development of chemical sensors. This specific interest arises from the emerging demand for fast and inexpensive chemical analyses in various fields, such as industrial control, environmental monitoring, medicine, foodstuff production and testing, etc. Since sensors do not require costly hardware, are portable and can be easily applied on-site for the direct monitoring of several parameters, spectacular progress in chemical sensors developments and tremendous growth in their application have been occurring in the recent years. Modern chemical sensors are characterized by high sensitivity, low detection limits

doi:10.1088/978-0-7503-3687-1ch9

(few ppb) and are developed for many analytes. Nevertheless, the development of novel sensing materials and analytical sensory systems still remains a big challenge. Various material structures and transduction principles may be used in the chemical sensors; these characteristics often are determined by the nature of the analytical quantity to be measured.

Among different types of chemical sensors, the optical sensors, due to the fast response time, simplicity of preparation, and signal acquisition gain continuously growing interest finding applications in plenty of analytical tasks [1–3]. Moreover, the construction of most modern optical sensors requires neither wire connections with the detector nor sophisticated and energy-consuming hardware, enabling wearable and wireless sensors development [4, 5]. The analytical signal of an optical sensor can be registered with consumer optoelectronic devices such as digital scanners, digital cameras (alone-staying or embedded in commonly used PCs), tablets and smartphones [6–8], or even without any power supply in a so-called 'naked-eye' mode [9].

The response of an optical sensor originates from intermolecular interactions of an analyte with an active receptor (chromophore or fluorophore) [10]. Such an interaction is tuned to be specific for only one determined target analyte in a selective optical sensor. In contrast, in multisensor approach the series of non-specific interactions will provide an individual analyte pattern, composed by the simultaneous responses of several cross-sensitive elements of the optical sensor array [11, 12]. Recent technological advances resulted in a notable employment expansion of artificial intelligent sensing systems based on optical transduction [13]. Such systems, due to the similarity of their working principle to the 'chemical' mammalian senses, olfaction and taste, by analogy are often called optical e-nose and optical e-tongue (or optoelectronic nose and optoelectronic tongue) and employed as the fast and non-invasive diagnostic instruments for multicomponent analysis [12]. In fact, the ability of mammalian olfactory and gustatory systems to distinguish specific smells and tastes is not provided by specific lock-and-key receptors. Instead, the receptors are not selective towards single species and they may interact with a large family of chemicals, although in different intensity, with the brain in charge of extracting the information derived from the senses, classifying and discriminating among different odors or tastes. When applied to the sensor field, these concepts lead to a different idea of chemical sensors, where it is no more necessary the high selectivity of the device, but it should be able to interact with almost all the analytes present in the environment, although with different sensitivities. This concept is especially well-fitting to the optical transduction, since the same sensing ligand may produce a large number of spectral variations in the presence of various, sometimes very similar structurally, but having different affinity to optically active ligand analytes [13]. Tasks on discriminations, classifications, and even precise quantitative assessment of various complex chemical environment components can be solved combining together in an array the optical chemical sensors based on different optically active ligands. Sometimes, analyzed samples have tens, hundreds or thousands of different compounds that can be afforded and sufficiently assessed, as shown recently in works of several research groups.

This chapter will discuss the main aspects of optical and hybrid sensor arrays employed for liquid system analysis. We will consider the operating principles, advantages and challenges of multisensory optical systems. Moreover, the examples of hybrid and multi-transduction multisensory systems applications, including the recent advances from our team, will be reported in order to demonstrate the effectiveness of such systems for a wide range of analytical tasks, such as environmental monitoring, foodstuff quality assessment, medical and health care applications, and industrial processes control.

9.2 Optical sensors definition and classification

Optical devices transform changes of optical phenomena, as the result of the interaction between the analyte and the receptor part [14], and optical chemical sensors detect the intensity of photon radiation that arrives at a sensor. The photon irradiation intensity can be modulated either by ***absorption*** (in a transparent medium) caused by the analyte absorptivity or its interaction with selected indicator, or by the variation of the wave-propagation direction. The wavelength of emitted irradiation can be modulated by ***scattering*** (caused by particles of different sizes present in the analyzed sample), ***refraction*** (expressed through the variation of refraction index upon the variation of analyzed sample composition), or ***reflectance*** (measured in non-transparent media) [15]. Photons detected by optical sensors can be emitted by ***fluorescence*** and ***phosphorescence*** (both measured as a positive emission effect caused by irradiation) or ***chemi-/bioluminescence*** (intensity of light emitted by a bio/chemical reaction in the receptor system). Each of these photon sources has a known relationship to a given chemical reaction, facilitating chemical sensing with high sensitivity.

Optical chemical sensors may be subdivided according to the recognition element, the type of optical technique employed and the sensor construction (including transductor and reagents involved), as shown in figure 9.1.

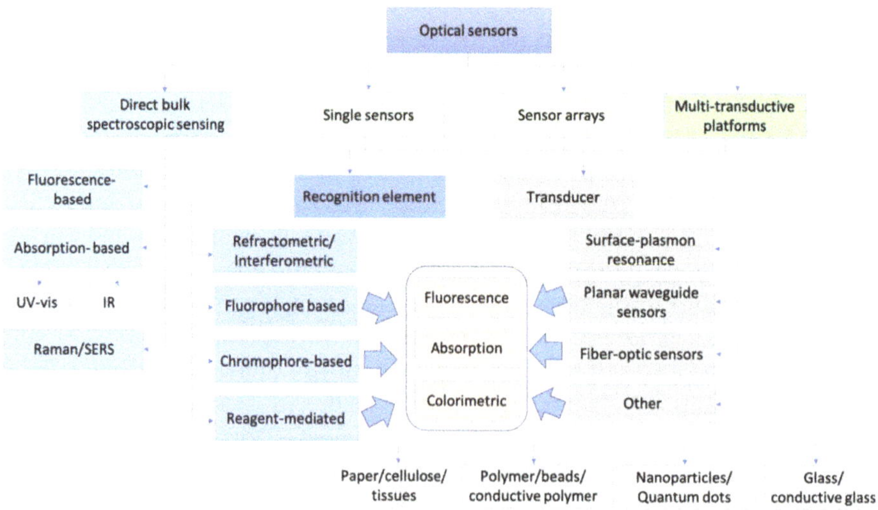

Figure 9.1. Classification of optical sensors. In the figure the following abbreviations are applied: infrared (IR) and ultraviolet-visible (UV–vis) absorption.

In direct spectroscopic sensing, the analyte is detected via its intrinsic optical property. In reagent-mediated sensing systems, a change in the optical response of an intermediate agent, usually an analyte-sensitive ligand or dye molecule, is used to monitor analyte concentration. The most widely used techniques employed in optical chemical sensors are fluorescence, absorption, and luminescence; sensors based on other spectroscopies and optical parameters, such as refractive index and reflectivity, have also been developed. Such devices have also been called optodes; the general characteristics [16] and the sensing materials [17] applied in the bulk optodes were summarized previously by Pretsch and colleagues.

A comprehensive review on modern optical chemical sensors was previously published by McDonagh *et al* [10]; optical biosensors [18], wearable optical sensors [5], and optical sensor arrays with different transduction types [19–22] were also reviewed. Nowadays, optical sensor arrays are wildly used in many analytical applications. Due to the huge amount of research in this area, in this chapter we will limit our discussion to the recent achievements in developing multisensory systems with optical transduction, particularly to optical chemical sensor arrays for complex liquid phase analysis, optical e-tongues. We divide this chapter into two main sections according to the sensing principles, construction/transduction and applications. Primary attention is given to colorimetric and luminescent (fluorometric and chemiluminescent) e-tongues used for the assessment of various analytes; also, hybrid (multitransduction) multisensory systems are discussed, and the basics of optical signal output interpretation and chemometric data analysis are introduced.

9.3 Colorimetric sensors, arrays and e-tongues

We uncommonly start this chapter by describing colorimetric sensors and sensor arrays due to the popularity of these devices nowadays, attracting immense research interest thanks to the low cost, ease of application and no need for sophisticated instrumentation for signal readout. By definition, colorimetric sensors demonstrate different color change of their sensing element/sensing material under external stimuli. In the case of chemical sensor array, the change in chemical environment composition is the main registered stimulus. Generally, optical signal intensity changes at a specific wavelength within visible range (400–700 nm) can be determined using special instrumentation. The simple qualitative or semi-quantitative estimation of the analyte presence can be performed through naked-eye visualization of the colorimetric sensor response, or, through an application of familiar devices (smartphones, tablets, etc), accomplished with particular colorimetric output interpretation app(lication)s.

There are two main approaches employed in colorimetric sensor design: the first consists of direct binding of an analyte to the chromophore (so-called lock and key principle), the second is based on the competitive binding of an analyte on ionophore in the presence of the indicator (chromophore). In both cases, the chromophore color variation is registered as a functional sensor readout [13], figure 9.2.

Colorimetric sensor arrays combine several cross-sensitive sensing elements (or sensing spots), providing a multi-color output to be interpreted through the pattern recognition methods. In this way, colorimetric sensor arrays may provide sensitive

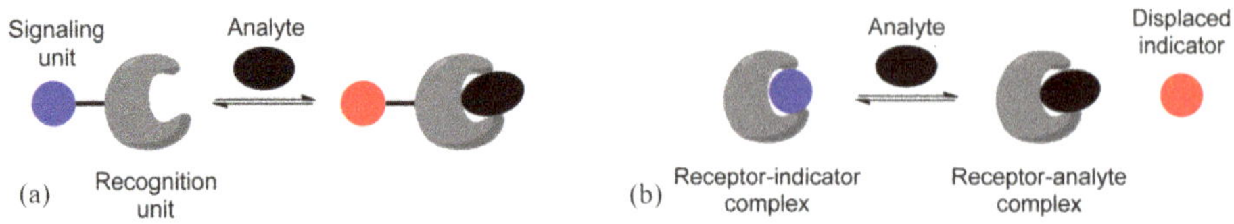

Figure 9.2. Representation of the direct (a) and competitive (b) binding-based approaches for colorimetric sensing. Reprinted with permission from [13], copyright (2015) American Chemical Society.

and efficient detection of several analytes under interest or perform samples identification [9–12].

Colorimetric e-tongues employed for liquid phase analysis require particular construction. Depending on the signal readout method, they may be composed of single physically separated sensors employing optical fibers in various configurations, or a set of microtubes, or flow-through cells containing sensing spots prepared with one or more active chromophores, or ionophore/chromophore mixture deposited on the same support. They can be transparent (glass, polymer, and microwell plate), non-transparent (TLC plate, paper, and other cellulose matrices). Moreover, the sensing ligands can be suspended in the solution or adsorbed on the surface of nanoparticles (NPs), quantum dots (QDs), and different composite materials.

Among colored binding ligands, porphyrinoids and their metallic complexes have found a particular interest in colorimetric sensors and sensor arrays development due to their peculiar chemistry and rich optical properties. Porphyrins represent versatile lock-to-key sensing materials and are successfully applied to detect analytes in either liquid or gaseous phases, operating in a wide range of different sensing mechanisms [23]. For example, metalloporphyrins can exploit metal coordination to bind gases and volatile organic compounds, or anions in the liquid phase, whereas free-base porphyrins can reversibly bind to metal ions or protons. These interactions often induce color variation of porphyrin-based sensing materials that have been successfully used as ionophores and chromophores in optical sensors [24]. After the pioneering work of Professor K S Suslick's group from the University of Illinois [25], the colorimetric metalloporphyrin (and other chromophores)-based optoelectronic nose and e-tongue systems have received significant attention and were widely employed for discrimination of pollutants in water [26], and in soil [27], pathogenic bacteria [28], taste assessment of soft drinks, beers and green tea [29–31], proteins discrimination [32], and monitoring of milk adulteration [33]. Moreover, the versatility of porphyrins as sensing materials enables the combination of different transduction mechanisms for the same sensing layer, thus obtaining hyphenated sensing systems. Previously our group has reported the porphyrin-based opto-electrochemical electronic tongue system [34]; the applications of this hyphenated sensing system for several analytical tasks are discussed in detail in section 9.6 of the present chapter.

When a binding ligand (ionophore) is colorless, incorporating an additional colored component (often lipophilic pH-indicator, such as ETH2439, ETH5350, etc), that competes with the analyte to ionophore binding, is required to develop efficient colorimetric optodes. Such systems were also called 'indicator displacement assays' [35], since the free and bound indicators have different optical (colorimetric or fluorescent) properties that are responsible for the effective color signal change; this definition is also extended to the fluorescent indicator-bearing optodes. Theory [16, 36, 37] and methods of digital color analysis for optochemical sensor arrays [38] have been previously developed, and their various applications also reported.

The requirements of practical utility and ease in the application and the necessity of fast, accurate readout of colorimetric e-tongues output determine in a high degree the construction of such systems, and in particular, define the choice of appropriate solid support used for sensing molecules immobilization. Several different strategies of sensing spots immobilization on a solid support are used for colorimetric sensor arrays developments [19]. The most common constructions recently reported in the literature are e-tongues: printed on silica gel plates, incorporated in molecularly imprinted polymers (MIPs), or immobilized on different polymeric matrices deposited on NPs, on paper, and other cellulose-based supports. More details can be found in several comprehensive reviews on colorimetric sensors and sensor arrays published previously. Among them are the works by You, Zha and Anslyn [13], by Suslick's group [11, 12], by Pohanka [6], by Ebralidze *et al* [19].

9.3.1 Colorimetric output interpretation

The colorimetric e-tongue array visual output interpretation and data analysis have particular importance. As previously mentioned, preliminary naked eye analysis can be done without additional equipment, simply through registering the presence/absence of the colored readout of the system and comparing the system response with a provided colored calibration scale. This last condition is more valid for single selective colorimetric sensors, as reported in works from Meyerhoff's group on single selective optodes development [39, 40], and our recent works on the selective colorimetric assessment of fluoride [41] and cyanide [42] anions. In the case of a more complex system comprising, for instance, the matrix of the sensing spots of different color and intensity, the digitalization of such a system output (including the subtraction of a signal part, provided by only the color/luminescence/fluorescence variation of the employed solid substrate) is required. A detailed discussion on visual output interpretation from colorimetric sensor arrays is provided in the literature [19, 43]. The most common approaches will be briefly described here for readers' convenience.

The colorimetric sensor array output can be registered in the form of a digital image, which is further analyzed with an appropriate (selected from the wide-user available or home-designed) image editing/processing app(lications)s. The RGB color space is the most used to reproduce the image colors and to interpret their intensity in respect to the three main visible spectrum colors: Red (630 nm), Green

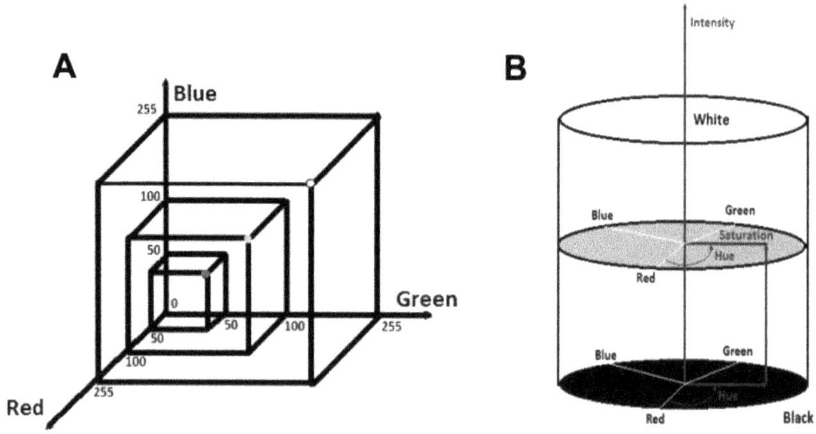

Figure 9.3. (A) The RGB and (B) HSI color spaces representation. Reprinted with permission from [43], copyright (2014) American Chemical Society

(530 nm), and Blue (480 nm) [44]. The color components of an 8-bit RGB image are integers in the range [0, 255], and the image intensity is given by equation (9.1):

$$I = \int_{\lambda} [R(\lambda) + G(\lambda) + B(\lambda)] \cdot S(\lambda) \cdot d\lambda \tag{9.1}$$

Where *R, G*, and *B* represent the color intensity of a sensing spot at wavelength λ, and S(λ) accounts for the substances' transmittance and emission.

The RGB color space can be represented as a cube with three perpendicular axes R, G, and B, each with the axis range from 0 to 255 [43], figure 9.3(A). Other two-color spaces often employed for image digitalization are CIELAB (referred to also as L*a*b*) and HSI. The L*a*b* scale is detector independent and expresses color according to the three dimensions with the values: *L** for perceptual lightness, and *a** and *b** for the four unique human vision color opponents, red–green and blue–yellow, respectively [45]. The values of hue (H), saturation (S) and intensity (I) in HSI color space are normally obtained from the RGB component values, and HSI space has a cylinder shape, figure 9.3(B). Other components, Value or Lightness are used in similar models, HSV and HSL, which have pyramid and double pyramid shapes, respectively [43].

Color differences of the colorimetric sensor array image before and after a sensing event can be determined as differences in selected color space coordinates. This procedure consists of several steps, such as: (i) image acquisition; (ii) inversion to the appropriate color space; (iii) background substruction (or differential measurement); (iv) selection of the ROI—region of interest; (v) averaging of ROI signals; (vi) if necessary, integration of the grey pixel value for each ROI. The obtained color variation values are used for analyzed samples identification in the case of qualitative analysis tasks. For quantitative analysis, the digitalized optical e-tongue color outputs are employed for the construction of the linear or polynomial calibration curves and/or for the appropriate regression models (in the coordinates real–found) development. A more detailed discussion on the main multivariate data analysis (chemometric) techniques will be given in section 9.7 of the present chapter.

9.3.2 Applications of colorimetric e-tongues

Frequently, colorimetric sensor arrays have been reported recently for rapid screening of the various environmental pollutants, such as transition and heavy metals [19, 27, 46–48], anions [45, 46, 49–53], pesticides [54], for recognition of different biomolecules [51–55] and drugs [56], for foodstuffs analysis, as for instance, grape ripening prediction [61], etc.

In fact, since the standard detection methods of various contaminations and toxic chemicals are time-consuming, often expensive, and require trained personnel and computing devices to capture and analyze the data, the rapid, cost-effective, and user-friendly pollutants assessment with colorimetric sensors became a ubiquitous detection technique with acceptable sensitivity.

Thus, the colorimetric assessment of Hg^{2+}-ions on the aptamer conjugated gold nanoparticles (AuNPs) loaded to the lightweight 3D material was reported in [46]. The interaction between the aptamer and the analytes led to the solution color change due to aggregation of AuNPs and could be registered with a smartphone employing the novel RGB image processing protocol. Multiple linear regression analysis was used for response interpretation, which permitted a high sensitivity range from 1 nmol L^{-1} to 0.2 ppb. In [48] a portable transmission densitometer was employed for direct quantification of the optical density of a paper-based colorimetric array for the assessment of heavy metals in aqueous systems. The 1,10-phenanthroline, bathocuproine, and dimethylglyoxime indicators for detection of Fe(II), Cu(II) and Ni(II) ions, respectively, were added to the circular reservoirs obtained on paper support by screen-printing technique of hydrophobic barriers, figure 9.4.

The reading of transmitted light through the assay spot results in a significant improvement on the accuracy and sensitivity of high analyte concentrations in comparison to the common reflection method, expanding the linear quantification range to 0.5–500 mg L^{-1} for Cu(II), and Fe(II), and to 2–500 mg L^{-1} for Ni(II)

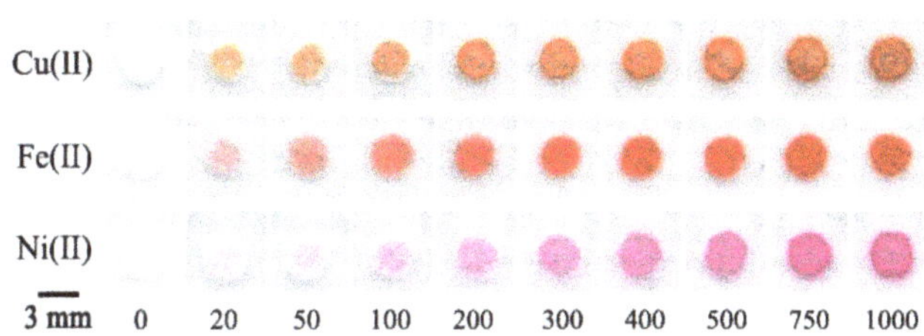

Figure 9.4. Photogram showing the on-paper patterned colorimetric Cu(II), Fe(II) and Ni(II) array response with the concentrations from 0–1000 mg L^{-1}. Reprinted from [48] with permission of the Royal Society of Chemistry.

and suggesting the utility of the proposed readout for the future development of paper-based analytical sensors.

We have recently reported colorimetric optical sensors based on two novel ZnTPP porphyrin ligands, functionalized with 1,3-bis(2-pyridylimino)isoindoline, ZnTPP-BPI, and Zn(II)TPP-BPI-crown (ZnPC), functionalized with two dibenzocrown-ether moieties as cation-sensitive ionophores [27]. The ligands were incorporated inside PVC-based polymeric membranes, which then were tested on paper or glass solid support for cations assessment in soil extracts. The visibly (naked eye) observed color change of the sensing material from green to red demonstrated the suitability of the ZnPC-based optodes to perform fast monitoring of Cu(II) ions in the aqueous soil extracts; the results obtained were in good agreement with a standard ASS method. Moreover, the utility of on-paper strip sensor array in the identification of soils in terms of pollution was shown. The developed optical sensor array was able to rapidly asses the presence of copper ions above the established borderline concentration of 1×10^{-3} mol L^{-1} (the higher copper levels in soils become hazardous for many species), providing different visual response for soils contaminated with other metal, and sampled close to the airport and fuel stations. Also, a high copper concentration was detected in a soil sample, collected in the 'Tor Vergata' University garden (placed in proximity to the highway).

A low-cost colorimetric sensors array based on pyridylazo- and porphyrin indicators has been developed by Han *et al* in [47] and employed for the quantitative prediction of Pb, Cd and Hg toxic heavy metal residues in forty-eight fish samples (*Carassius carassius*) retailed in Zhenjiang, China. The best results were obtained for extreme learning machine (ELM) data treatment technique and provided the possibility to predict Pb, Cd, and Hg with correlation coefficients of 0.854, 0.83, and 0.845, respectively; the corresponding root mean squares errors were 0.102, 0.026, and 0.016 mg kg^{-1}, thus demonstrating a great potential of the developed system for fish quality assessment.

Among different chromophores employed for the colorimetric anion sensing, the Schiff base compounds have been the most widely applied. In [50] the comprehensive review of 104 papers reporting the Schiff bases agent's role in the detection of anions is provided, and the main applications over the last two decades are listed. The interactions between polar groups of Schiff bases and analyzed ionic species result in the evident color change, which is attributed to changing the sensing system electronic structure due to the process of charge transfer, electron transfer and/or hydrogen bonding between Schiff bases and analytes. Naked eye detection of several anions, such as cyanide, fluoride and hydroxide [51, 52], acetate and hydrocarbonate [53] in tap water by Schiff base chemosensors has been reported recently. In [51] Uahengo *et al* have synthesized and tested the new naphthyl derivative bearing hydrazone receptors, which were uploaded on paper strips and used for colorimetric detection of cyanide and hydroxide in tap water. The same authors have demonstrated the recognition properties of naphthoquinone ligand bearing hydrazone receptors towards several anions, namely F^-, OH^-, CN^- and AcO^- [52]. Among these species, naked eye observable color changes, from green to light blue (for F^-, CN^- and AcO^-) and to violet (for OH^-) were registered. The fluorides in commercially available kinds of toothpaste, cyanide and hydroxide in tap water were tested with colorimetric paper-based test strips. In [53] the azo-derivatives of 1-naphthylamine, 2-((E)-((4-chloro-3-(trifluoromethyl)phenyl)imino)methyl)-4-((E)-naphthalene-1-yldiazenyl)-phenol(2) receptor, was synthesized and employed for colorimetric detection of sub-millimolar concentrations of HCO_3^- and AcO^- anions.

A rapid and straightforward method for colorimetric organophosphate and carbamate pesticides determination in water based on microfluidic paper biosensor is presented in [54]. The acetylcholine esterase (AChE) and acetylcholine chloride (AChCl) together with bromocresol purple (BCP) pH indicator were immobilized by the sol–gel method on paper support. The color variation upon enzyme inhibition was used, and the rate of the inhibited reaction was measured with a digital camera by quantifying the appropriate color coordinate. Authors show the possibility of determining carbaryl and chlorpyrifos pesticides with the detection limits 0.24 and 2.00 μg L^{-1}, respectively, and with reproducibility of 4.2%–5.5% in water samples, with no sample preparation needed.

Mahmoudi *et al* have reported an interesting application of the colorimetric sensor array to identify various nanoparticles, NPs, in aqueous solution [62]. The colorimetric sensor array was prepared with five dyes, bromocresol green (BCG), bromophenol blue (BPB), bromophenol red (BPR), bromopyrogallol red (BGR), and acridine orange base (APB), dissolved and then seven nanoparticles types (spherical Au NPs of four different sizes – 2,8,20 and 40 nm; Au nanorods with two different aspect ratios – 2.4 and 3.5; and a multifunctional carbon/iron oxide NPs) and three controls were added. The non-specific interactions of dyes with NPs were registered at elected wavelengths (i.e., 480, 590, and 620 nm) assigned to RGB values, and the color pattern changes for all dyes provided a unique molecular fingerprint for each type of NP tested. The developed colorimetric sensor array was able to discriminate among different sizes, shapes, and compositions of NPs. Moreover, a semi-quantitative determination of concentration of each type of NP was demonstrated with excellent sensitivity (< 100 ng mL^{-1}).

Previously, the term electronic eye (e-eye) was introduced for the employment of optical sensor arrays. An elegant method to predict parameters related to grape phenolic ripening based on colorimetric electronic eye (EE) application has been developed by Orlandi *et al* in [61]. For the analysis, different grape varieties were collected at different harvesting periods, from unripened to maturity, and the must from the berries was deposited on a white sheet of absorbent paper to obtain a sort of paper chromatography. The RGB images of the must spots were registered with a flatbed scanner and converted into one-dimensional colorgrams, which were then employed to build calibration models relating the color of the images with the phenolic composition of the samples, determined by reference analytical methods. Satisfactory calibration models were obtained for the prediction of several parameters related to phenolic ripening of grapes: total anthocyanins content, malvidin-3-O-glucoside and petunidin-3-O-glucoside, color index and grape tonality.

9.3.3 Colorimetric sensors support material

Different optical reporters and solid supports have been used for the colorimetric arrays employed for organic and biomolecules analysis, i.e. polymers [27, 55, 63], paper and other cellulose fibers [64–67], Ag and Au nanoparticles [56, 57, 64, 65], nanoparticles deposited on paper [58], molecularly imprinted photonic crystals [59] or other nanomaterials and substrates [70].

A poly(aniline-co-anthranilic acid) (ANI-co-AA) composite enzymatic film biosensor doped with horseradish peroxidase (HRP), glucose oxidase (GOx), horseradish peroxidase-glucose oxidase (GOx-HRP) and tyrosinase (Tyr) enzymes and coupled with smartphone detecting was developed in [55]. The enzymes were immobilized on the composite polymer film by adsorption; they catalyzed a reversible redox color change of the host polymer from light green to dark blue in the presence of substrates. The developed sensing platform was used for hydrogen peroxide, glucose, and catechol assessment in food and pharmaceutical analysis. Thus, in particular, the tests on catechol content in several commercially available red and white wine, pear juice, fresh pomegranate juice and pharmaceutical products have shown good recoveries.

The unique optical properties of nanoparticles, NPs, make them promising candidates for application in optical sensing arrays [68, 69]. Recently the possibility of protein detection through the application of colorimetric sensor array based on DNA-protected gold nanoparticles (AuNPs) as non-specific receptors was shown in [56]. It was demonstrated that different proteins could trigger the DNA–AuNPs aggregation resulting in various color changes. A multi-protein discrimination array prepared with two ssDNA strands (15A and 30A bases) was employed for naked eye discrimination of 12 different proteins at the 50 nM level and tested for assessment of proteins in human serum. Linear discriminant analysis (LDA) and hierarchical cluster analysis (HCA) were employed with 100% accuracy of discrimination of similar mixtures of proteins. The same research group has reported an application of colorimetric AuNPs modified with tree single-strand oligonucleotides of A15, C15, and T15 for analysis of proteins (BSA, HRP, HAS, Pap, TRF, Hem, Try) in human urine [57].

Sheini *et al* reported a paper-based colorimetric assay for analysis of urea in biological samples [58]. Ag NPs and two different capping agents, thiomalic acid and maltol, were used as colored sensing elements. Through the urea hydrolysis in the presence of urease, ammonia and carbon dioxide products were formed and interacted with nanoparticles that caused their aggregation and consecutive color change from yellow to brown and red, respectively. The resulting colorations were captured by a scanner, and permitted the individual detection of ammonia, carbon dioxide, and urea with a reliable performance in saliva and blood samples. A colorimetric array discriminating 20 food antioxidants of natural, synthetic and biological groups was reported by Bordbar *et al* in [71]. The array was formed with 12 types of Au and Ag NPs synthesized using six different reducing and/or capping agents. Upon the antioxidants interactions with the nanoparticles the aggregation or morphological changes occurred, which caused a change in the sensors' colors.

The array absorbances were registered at wavelengths of 405, 450, 490 and 630 nm, or by capturing the images with a digital camera, and then treated with principal component analysis (PCA) and hierarchical cluster analysis (HCA). The method was applied to the quantitative assay of gallic acid, catechin, butylated hydroxytoluene, dopamine, and ascorbic acid, and the respective root mean square errors for prediction of the test set were 0.0650, 0.0782, 0.811, 0.0206 and 0.135 nmol L^{-1}. The developed method was approved for the analysis of antioxidants in tea and lemon juice.

Lu *et al* have developed a colorimetric sensor array for nitroaromatic explosives detection based on molecularly imprinted colloidal particles (MICs), templated with 2,4,6-trinitrotoluene (TNT), 2,6-dinitrotoluene (2,6-DNT), 2,4-dinitrotoluene (2,4-DNT) and 4-nitrotoluene (4-MNT) [59]. The 3D close-packed molecularly imprinted photonic crystal (MIPC) was then received through the self-assembly of MICs on glass slides, pressed onto the adhesive tape, and then peeled off for sensing purposes. The individual MIPC sensors were tested and compared to the colorimetric sensor array performance. The array was composed of TNT, 2,6-DNT,2,4-DNT and 4-MNT, respectively, templated MIPC sensors and the cross-sensitivity of array was investigated in methanol/water solutions (3/2, v/v, 30 mM KDP buffer, pH 7.0) containing nitro aromatic explosives in varying concentrations. The red-shift diffractions of MIPC films to every sample under saturated adsorption were recorded, with the data treated with PCA. The developed colorimetric MIPC sensor array showed the possibility of precise identification of nitroaromatics with improved selectivity compared with the individual sensors

9.3.4 Quality assessment and medical applications of colorimetric e-tongues

The application of colorimetric e-tongues was also reported for potable water valuation [67], food brands discrimination, and quality assessment [43, 72]. Thus, the simultaneous naked-eye detection of microcystin-LR, lead(II), chloramphenicol, testosterone, and chlorothalonil in drinkable water by immunochromatographic assay was reported by Xing *et al* in [67]. For this, five different antigens were immobilized on separated test lines of nitrocellulose membrane. The recognition was accomplished by specific monoclonal antibodies, which recognized the corresponding antigens and permitted a fast semi-quantitative detection of all pollutants at low concentrations.

Jia *et al* recently reported an optical colorimetric sensor array for the discrimination of Chinese teas [72]. The array was based on the covalent combination of phenylboronic acids with different substituents as the receptors and tea polyphenols. In particular, 3-aminophenylboronic acid (APA), phenylboronic acid (PA) and 3-nitrophenylboronic acid (NPA) were utilized as receptors, while the catechol-containing indicators alizarin red S (ARS), bromopyrogallol red (BR) and pyrocatechol violet (PV) were chosen for the construction of the nine-channel (3 indicators × 3 receptors) colorimetric sensor array. The application of partial least-squares discriminant analysis (PLS-DA) technique permitted the successful distinction of tea samples within different categories, quality grades, and geographical origins.

A colorimetric sensor consisting of series of carbonyl-sensitive dyes, pH indicators, metal chromogens and metalloporphyrins, immobilized onto a reverse silica thin-layer-chromatography plate for the identification of volatile aldehydes and ketones was developed by the Suslick group in [66]. The possibility of discriminating aliphatic or aromatic aldehydes and ketones up to 2 ppm levels with high accuracy of classification (> 99%) within 2 min by applying a handheld reader was shown. The system was then employed for differentiation among complex odors of ten liquor samples (whiskies, brandy, vodka) and ethanol controls, demonstrating its potential in the beverage industry for quality control and assurance. In [73] Ghanem *et al* have reported the cross-reactive indicator displacement colorimetric sensor array composed of short peptides, divalent copper or nickel metal ions, and one of three colorimetric indicators: free chrome azurol S (CAS), pyrocatechol violet (PCV), and bromopyrogallol red (BPR), respectively, for discriminating cachaça extracts from oak and Brazilian woods based on their tannin composition. The discrimination pattern was reproducible for oaks obtained from different countries and for regional Brazilian woods; 62.5% of the samples were correctly identified. Finally, the array was used to classify unknown wood extracts based on their clustering pattern with 60%–100% accuracy.

Medical applications of colorimetric sensors and sensor arrays were focused recently on the possibility of the fast and reliable detection of bacteria [63], viruses [74], and point-of-care devices development [64, 65]. The rapid detection of pathogenic species, bacteria, and viruses became crucial for addressing serious public health issues in clinical, environmental, and food settings, especially nowadays, during the worldwide COVID-19 pandemic situation. For instance, Ranjbar *et al* have developed a novel electrochromic sensor assay based on converting an electrochemical signal into optical readout to visualize *Escherichia coli* detection [63]. The specific antibodies were adsorbed on the surface of polyaniline (PANI) film, electropolymerized on an indium tin oxide screen-printed electrode (ITO SPE). Through the application of a constant potential to the PANI-modified ITO SPE, a color variation upon the change of electropolymer oxidation state was registered. In the presence of *E. coli*, electrode surface resistance increases, thus affecting the PANI oxidation states, and producing a different electrochromic response, figure 9.5. The developed sensing platform has permitted the naked eye detection of *E. coli* spanning 4 orders of magnitude with a limit of detection of 10^2 colonies forming unit per 1 mL (CFU mL^{-1}); LD of 10^1 CFU mL^{-1} was achieved using ImageJ software.

Recent achievements in the development of on-paper-based, rapid, low-cost and time effective POCs devices are discussed in [64, 65]. The utility of vertical flow assay (VFA) for high sensitivity C-Reactive Protein (hsCRP) testing, commonly used for assessing risk of cardio-vascular disease (CVD) in combination with neural-networks machine learning framework as a use-case is discussed in [65]. The possibility of the correct hsCRP prediction for CVD testing was demonstrated. Thus, using a custom-designed handheld VFA reader for a clinical study of 85 human samples a competitive coefficient-of-variation of 11.2% and linearity of $R^2 = 0.95$ among blindly-tested VFAs in the hsCRP range (i.e., 0–10 mg L^{-1}) were obtained.

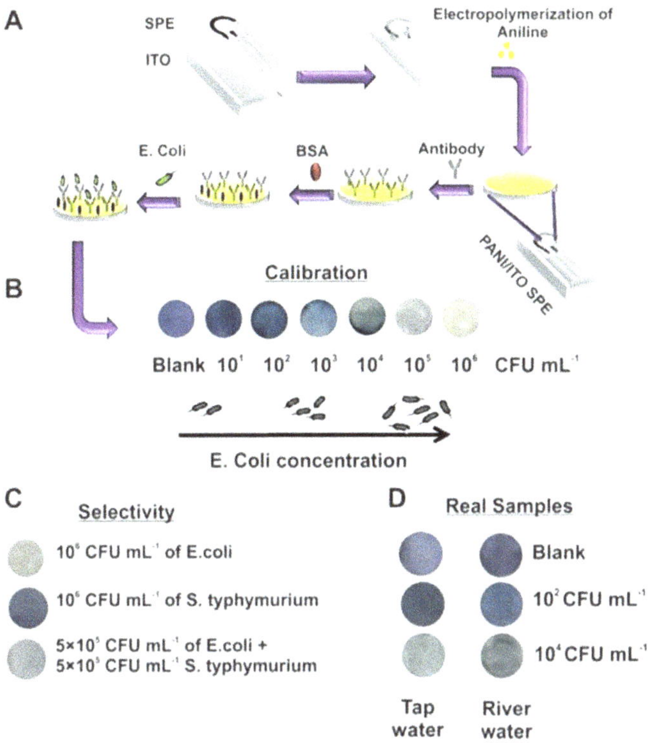

Figure 9.5. Schematic representation of electrochromic *E. coli* sensor (A) fabrication, (B) calibration, (C) selectivity using versus S. typhimurium as the interfering agent. (D) application for tap and river water spiked with *E. coli* analysis. Reprinted with permission from [63], copyright (2019) American Chemical Society.

9.4 Fluorescent sensors, sensor arrays and e-tongues

9.4.1 Principles of fluorometric sensing

Fluorescent sensors are the most popular and commonly used optical sensors due to their high sensitivity in a tunable emission range and the possibility to modulate the luminescent response through the variation of excitation wavelength. The specific ligands employed in such sensors are called fluorescent probes or fluorophores. There are several criteria of fluorophore selection for effective optical sensor development, among them:

- (i) the presence in the fluorophore structure of a specific (often macrocyclic) chelating site that is linked to (or functions as) the optically active fluorogenic fragment;
- (ii) the strong absorption in the selected region (often near-UV region) and high luminescence in easily readable wavelength diapason (often visible diapason);
- (iii) depending on the task, enhanced fluorophore sensitivity or cross-sensitivity properties, i.e., a strong binding ability toward several analytes;
- (iv) commercial availability or simple synthetic procedures required for fluorophore synthesis.

Fluorescence occurs when an electron from a fluorophore molecule is first excited and then relaxes to the ground state by emitting a photon from an excited singlet state. For phosphorescence, the excited singlet state of fluorophore first passes to the forbidden excited triplet state transition, followed by subsequent light emission upon relaxation to the ground state [11]. Both fluorescence and phosphorescence are measured as a positive emission effect caused by irradiation, and analyte interaction with fluorophore binding site modulates (enhancing or quenching) the fluorescence emission during sensing event. Finally, chemi- and bioluminescence represent the intensity of light emitted by a bio/chemical reaction in the receptor system leading to a product in electronic exited state. The mechanisms involved in relaxation of the electronically excited state are the same as in fluorescence.

The main mechanisms involved to translate the analyte binding into fluorophore signal variation are: the fluorescence increase through photoinduced electron transfer (PET) mechanism; photoinduced proton transfer (PPT); a chelation enhanced fluorescence emission (CHEF effect); intermolecular and intramolecular electron transfer (IEI) and intramolecular charge transfer (ICT) processes, aggregation-induced emission, and others. The emission wavelength shift upon complexation is often contributed by the ICT process inside the fluorophore molecule.

Moreover, in optodes and optical sensor arrays the intensity of luminescence is often a subject of their construction and influenced by the position of excitation light source, the sensor material, and the substrate optical properties. As previously mentioned, the optode fluorescence intensity can be tuned with the excitation wavelength selection; moreover, quantum yield, lifetime, anisotropy, fluorescence decay, etc can be employed for data analysis in order to enrich the sensor output. Also, the simultaneous acquisition of several parameters may transform the single optical sensor in a multiparameter array providing multivariate readout upon sensing.

In the pioneering works of the Suslick [11, 26] and Anslyn [13, 75, 77] groups, two main concepts to construct a functional sensor array were established. Suslick *et al* state that ligands chemical diversity is necessary to construct effective optoelectronic tongues, while Anslyn *et al* apply the idea of a 'relaxed lock-and-key principle' for the discrimination of complex analytes by optical sensor arrays. Finally, the concept of the 'hypothesis-free sensor array' was proposed by Rotello *et al* [76], suggesting that for the development of optical sensor arrays, the structural prerequisites to sensing material can be quite generalized, thus permitting one to sense 'everything' with any fluorescent dye. Several small sensor arrays on the basis of fluorescence modulation through fluorescence quenching or enhancement can be simultaneously or subsequently employed, as will be illustrated in the next sections.

9.4.2 Fluorometric sensors arrays applications

The fluorophore-based optical chemical sensors and sensor arrays, or fluorometric e-tongues coupled with chemometric approach have been applied for the identification and quantitation of various analytes, such as metal ions [77–79], anions [80], toxins [81], pesticides and other organic compounds [82, 83], explosives [84–86], biomolecules (such as amino acids, proteins, etc) [87–90] and biomarkers [91], drugs [92], for beverages analysis [93–95], in foodstuffs identification and control [96], in point-of-care-testing (POCT) systems [97, 98], in medical applications [98, 99], and to taste the taste [100]. In [100] the review of 341 research papers, dedicated to the application of fluorescent optical biosensors to specific tastes assessment and/or to screen new agonists/antagonists for different taste modalities were systematically described; with new approaches for studying taste physiology and taste responsible compound screening proposed. In this chapter we leave out the discussion of biological optical sensors applications, sending the readers to many comprehensive reviews reported in literature to date.

9.4.3 Fluorometric e-tongues for metal ions detection

The huge release of metal pollutants into the environment occurs due to the growing industrial impact and incorrect preservation of waste, and that has become a serious problem in many countries and entire regions. The high multi-metal contamination of soil, air, and drinking water cause their consecutive accumulation in humans and this phenomenon has been connected to the insurgence of allergies, tumors, and other pathologies. This emphasizes the necessity of careful and frequent assessment of metals in the environment, with optical sensor arrays actively used recently for this task.

For instance, an artificial tongue using DNA and commercially available dyes, namely N-methylmesoporphyrin IX (NMM) and thiazole orange (TO), for metal ions discrimination has been reported by Qui *et al* in [78]. The fluorescence intensities of the sensor array elements measured at 610 and 532 nm corresponding to NMM and TO, respectively, varied in different degrees in respect to the different interactions between dye-bound DNA and metal ions, since DNA strands with different lengths and sequences display various binding ability to the target analytes. The possibility to distinguish fourteen metals (K^+, Ag^+, Mg^{2+}, Ca^{2+}, Ba^{2+}, Cu^{2+}, Zn^{2+}, Mn^{2+}, Ni^{2+}, Ce^{3+}, Eu^{3+}, Fe^{3+}, Al^{3+}, Tb^{3+}) by PCA and HCA was shown. A highly sensitive discrimination of metal ions with 92.8% identification accuracy was achieved. The non-covalent combination between DNA and dyes has provided the advantages of simple design, convenient operation, and low cost for the developed optical e-tongue system. Moreover, commercially available dyes can be selected to interact with the DNA strands, thus extending e-tongue applications. In [79] the ratiometric fluorescent sensor array (NRFSA) prepared with flexible and transparent Bacterial Cellulose (BC) nanopaper with immobilized on it photoluminescent (PL) ratiometric probe (Carbon Dots-Rhodamine B (CDs-RhB) was used for visual discrimination analysis of heavy metals. To create the test zones with the desired patterns and hydrophobic walls, a pattering method using an office laser printer was employed via direct-printing a hydrophobic toner layer on the dried BC nanopaper films. Through registration of the color variation of NRFSA platform with a smartphone, upon UV-irradiation the heavy metal ions, Hg(II), Pb(II), Cd(II), Fe(III) and Cu(II), have been successfully distinguished with RGB analysis by producing characteristic fingerprint patterns, figure 9.6. Besides, the developed optical sensor array was used to identify heavy metal ions in water and fish samples.

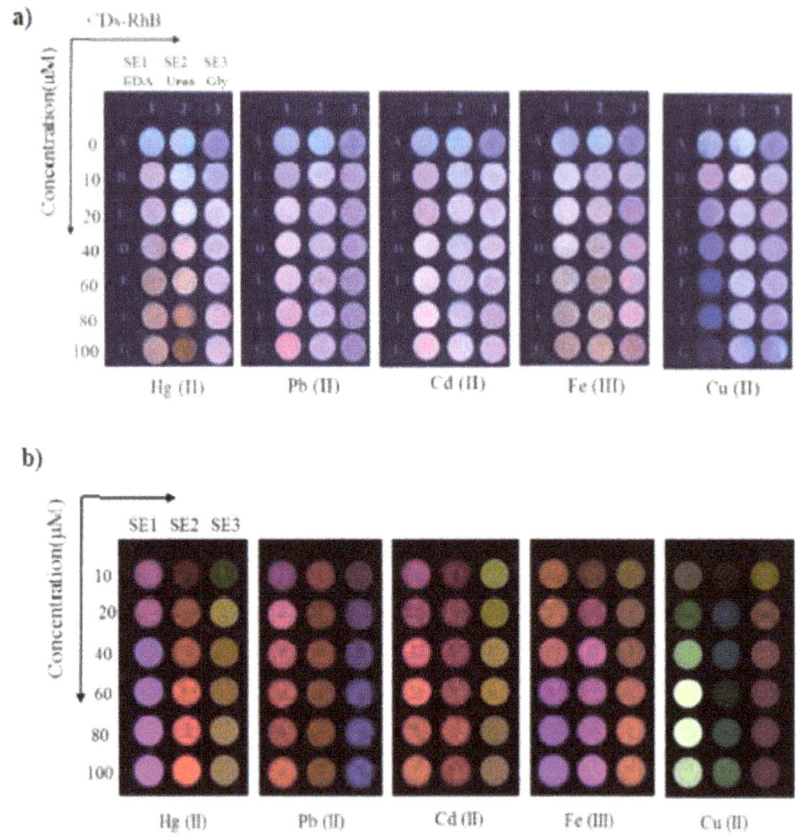

Figure 9.6. (a) Visual color variation of the NRFSA for various concentration of heavy metal ions at pH 7 and (b) the corresponding color difference map. Reprinted from [79] with permission from The Royal Society of Chemistry.

The NRFSA spectra recorded by a spectrofluorometer were exploited for discrimination applications: signal outputs were also analyzed with pattern recognition techniques such as HCA and LDA; the latter provided a 100% with a classification accuracy of all metal ions.

Recently in [101] two linear 1,4-distyrylbenzenes and five star-shaped 1,3,5-tristyrylbenzene derivatives (L_{2a} and L_{2b}, Y_0–Y_3 and Y_{NBu}) were synthesized and employed for optical sensing of acids and metal ions. The possibility of discriminating protons in different acids in THF and metal cations, such as Al^{3+}, Mn^{2+}, Fe^{3+}, Fe^{2+}, Cd^{2+}, Ag^{+} and Hg^{2+} in DCM was shown. The interaction of novel all styrylbenzene (SBs) fluorophores with metal ion could be observed by naked-eye upon the excitation at 365nm. The picoline, quinoline, and series of BODIPY fluorophores, were used for selective multicomponent detection of cations Zn^{2+}, Cu^{2+}, Hg^{2+}, Pb^{2+}, Cd^{2+}, Fe^{3+}, and Cr^{3+} in work of Xu *et al* [102]. A PCA analysis of these fluorophore responses in PBS upon addition of metals measured with Spectra Max M2 plate reader has shown a possibility of quantitative identification of each heavy metal ion.

Our research group has previously reported several applications of optical sensor arrays for monitoring metal ions [77, 103–107]. Thus, e-tongue arrays, composed of PVC-based solvent polymeric membranes based on two different ligands, 5,10,15,20- tetraferrocenyl porphyrin (H_2TFcP) and 5-(7-methoxycoumarin-4-methyl)–2, 8-dithia-5-aza-2,6-pyridinophane (L3, the abbreviation employed during our studies), respectively, were employed for detection of transition metal ions, Pb(II), Cu(II) and Cd(II) in [103] and Hg(II) in [104]. While the H_2TFcP fluorescence quenches under exposure to growing amounts of metal ions, L3 increased the fluorescence intensity with the mercury concentration growth. A Computer Screen Photoassisted Technique, CSPT, complementing the image sensor device with a computer controlled light source, namely, computer screen, first reported in 2003 by Fillipini *et al* [108] was employed for the above-mentioned measurements. In CSPT measurements the e-tongue system was backside illuminated by a sequence of 50 colours, generated by TFT-LCD computer monitor screen upon the analyte exposure and the output luminescence was registered with a webcam connected to the same computer. CSPT measurements were made using in-house written codes running in MATLAB (MathWorks, Inc., Natick, USA). PCA and partial least-squares regression (PLS) methods were employed for sensor array data treatment. CSPT multisensor array application permitted obtaining a linear response towards Hg (II) ion in the concentration range from 9.0×10^{-9} to 3.1×10^{-3} mol L^{-1} with a low detection limit of 9.1×10^{-9} mol L^{-1}, lower, then maximal permissible concentrations of mercury in potable water of 0.001 and 0.002 mg L^{-1} recommended, respectively, by WHO (World Health Organization) and EPA (United States Environmental Protection Agency). The e-tongue applicability for a fast and cheap analysis of Hg(II) in natural water samples was also demonstrated. The significant improvement of H_2TFcP-based e-tongue response toward lead, copper and cadmium presence in natural waters was shown upon the fusion of CSPT and potentiometric data and has permitted to clearly discriminate between all tested metal ions and to observe their concentration gradients [103]. More details of this investigation will be given in section 9.6 of the present chapter. In [105] we have performed a systematic study of a series of diaza-18-crown-6 8-hydroxyquinoline (DCHQ) fluorophores, incorporated in PVC-based polymeric films and deposited on glass slides, devoted to Mg(II) ion detection. The variations in increase of the luminescence intensity were registered upon optodes exposure to the growing concentrations of Mg(II) ions; optodes emission signal was sufficiently brilliant to be captured by a low-cost computer webcam. The photometric measurements were carried out using a modified CSPT setup, by substitution of the computer monitor with two different external monochromic light sources: a UV lamp ($\lambda_{ex} = 367$ nm) and a blue-colored light emitting diode, InGaN LED ($\lambda_{ex} = 380$ nm). The use of an e-tongue array composed of five optodes (doped with different amounts of the lipophilic sites and different plasticizers) allowed the reduction of the relative standard deviation (RSD, $n = 6$) of magnesium determination in commercial fertilizer samples (down to 3.7% with respect to 5.5% for single optodes), improvement in the analysis recoveries, and to achieve a detection limit (estimated by s/n = 3 method) down to 4.6×10^{-7} mol L^{-1}.

The binding properties of two 2,8-dithia-5-aza-2,6-pyridinophane-based ligands, bearing respectively coumarin (HNCum) and naphthol-benzoxaazole (HNBO) optically active units as pendant arms, have been investigated as fluorophores inside PVC-based polymeric film with the aim of transition metal ions, cadmium(II) and zinc(II), detection in [106]. An InGaN LED ($\lambda_{ex} = 380$ nm) was used as a light source and a digital camera as optodes luminescent signal detector. The results obtained demonstrated the suitability of the developed optodes to perform fast and low-cost simultaneous monitoring of Zn^{2+} and Cd^{2+} ions using familiar devices and chemometrics. Finally, in [107] diaza-crown ethers, pyridinophans and metallocorroles, previously investigated heteroatomic macrocyclic fluorophores, were combined in a unique portable analytical system for fast and low-cost detection of polluting species in real samples. Satisfactory PLS1 regression models between sensor array optical response and analyte concentration were obtained for Cd^{2+}, Cu^{2+}, Zn^{2+}, and NO_2^- ions in all the range of tested concentrations. Our attention was focused on the detection of cadmium and nitrite ions, for which the detection limits, DL, estimated by 3s method were found 0.0013 mg L^{-1} and 0.21 mg L^{-1}, respectively, and these values were lower than the corresponding WHO guideline values of 0.003 mg L^{-1} (Cd^{2+}) and 2 mg L^{-1} (NO_2^-). The suitability of the developed optical e-tongue for the accurate monitoring of polluting species, was shown in five natural water samples: three samples from Rome city fountains, one sample of surface water from

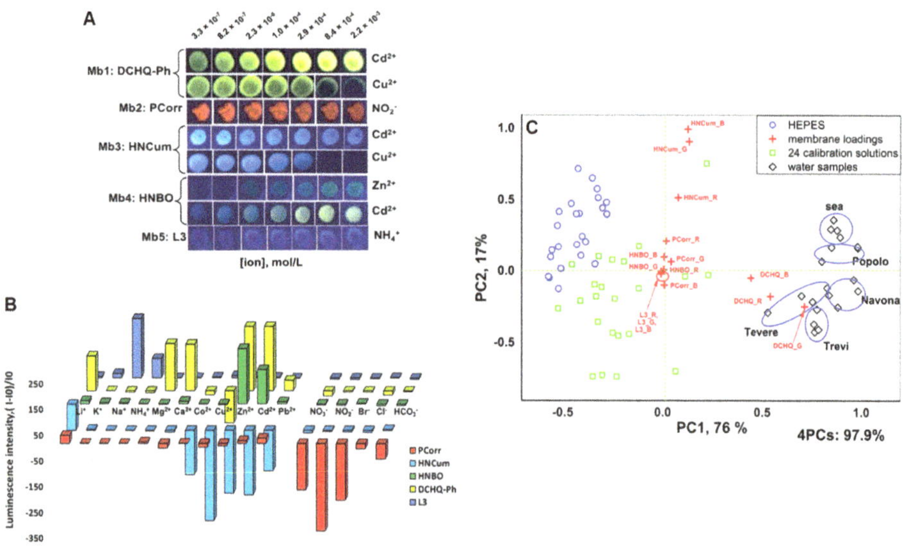

Figure 9.7. Fluorescence responses of membranes Mb1–Mb5 toward several analytes in individual solutions evaluated with PT at $\lambda_{ex} = 380$ nm LED illumination in 0.01 mol L^{-1} HEPES buffer at pH 7.5: (A) Photograms of sensors PT responses with the highest registered luminescence changes; (B) pattern of the relative intensity changes of membranes Mb1–Mb5 toward all tested analytes at a concentration of $2.9 \times 10{-}4$ mol L^{-1}. (C) PCA scores and loadings plot of sensor array fluorescence responses in 24 multicomponent calibration solutions and real water samples. Reproduced with permission from [107], copyright 2018 Lvova, Caroleo, Garau, Lippolis, Giorgi, Fusi, Zaccheroni, Lombardo, Prodi, Di Natale and Paolesse (CC BY 4.0).

the Tiber river and a marine water sample, figure 9.7. The samples were spiked with 1.0×10^{-8} mol L^{-1} and 1.7×10^{-7} mol L^{-1} for Cd^{2+}, and with 3.2×10^{-5} mol L^{-1} and 9.6×10^{-5} mol L^{-1} for NO_2^-, respectively. The luminescence variance of sensors array before and after the addition of pollutants was registered with PT technique, and the previously calculated PLS1 regression models were used to evaluate the added pollutant amounts. The obtained recoveries were in the range from 90.5 to 106.6% for Cd^{2+} and in the range from 95.1 to 105.1% for NO_2^- with the mean RSD of 3.7% (Cd^{2+}) and 2.61% (NO_2^-), respectively, thus demonstrating the system utility for inexpensive, rapid and accurate monitoring of heavy metals and anions pollution of natural environments that can be performed by untrained personnel.

9.4.4 Anions detection with fluorometric e-tongues

Fewer examples of multisensory fluorescent systems for the analysis of anions have been reported. Anzenbacher group have employed six biguanide derivatives, S1-S6, as receptors for various anions (halides, carboxylates, phosphates) binding in paper-based fluorescence-based sensor array [109]. The array was prepared by printing hydrophobic barriers on paper, and the solutions of sensors S1-S6 in DMSO were applied on the obtained microzone plates. The sensor fluorescence changes upon 365 nm excitation were recorded using a UV-scanner. The aqueous solutions of 11 different analytes were correctly identified with LDA analysis. In [110] an anion-sensitive array based on a ratiometric calix[4]pyrrole introduced in poly (ether-urethane) hydrogel matrices with varying comonomer proportions was reported to discriminate eight different anions (acetate, benzoate, fluoride, chloride, phosphate, pyrophosphate, hydrogen sulphide, and cyanide) in urine samples and anti-inflammatory drugs with 100% classification accuracy.

Lin *et al* in [80] reported a fluorescent sensory array based on supramolecular metallo-gels for the identification of inorganic anions in water, namely CN^-, SCN^-, S^{2-}, and I^-. The anionic response of the sensor array was due to the competitive coordination of the special gelator compound with different metal ions and anions, figure 9.8.

Sun *et al* reported five anionic sulfonate surfactants satisfactory distinguished using two luminescent metal–organic frameworks (LMOFs), based on NH_2–UiO-66 and NH_2–MIL-88 materials [111]. The LMOFs were placed in a black 96-well plate, in 5 columns (for 5 surfactants) and 5 rows (5 replicates), respectively, after the

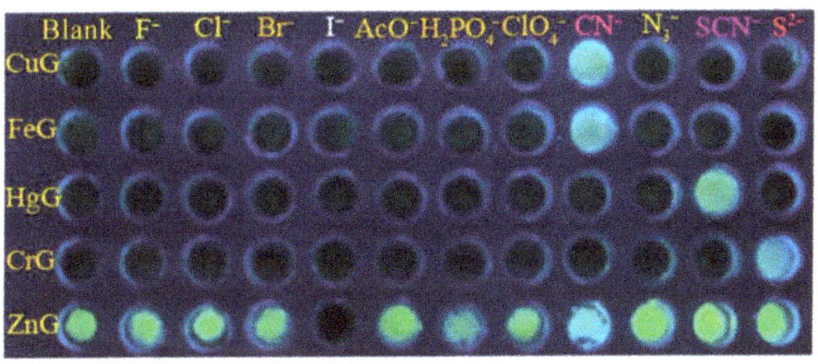

Figure 9.8. The fluorometric response of sensor array based on supramolecular metallo-gels toward one equivalent of various anions. Reprinted from [80] with permisson of The Royal Society of Chemistry.

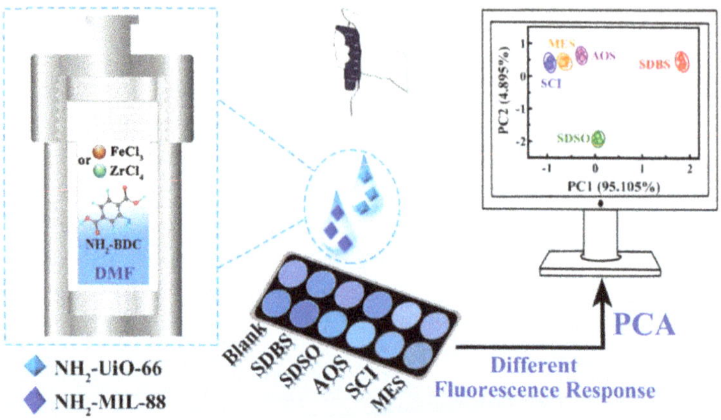

Figure 9.9. Schematic diagram of two LMOFs-based sensor array for anionic sulfonate surfactants discrimination. Reprinted with permission from [111], copyright (2020) American Chemical Society.

known concentration of anionic sulfonate surfactants was added into their corresponding columns, followed by the recording of fluorescence intensity (λ_{ex} = 320 nm, λ_{em} = 430 nm) of the mixtures by a microplate reader, figure 9.9. The output data matrix was treated with PCA. It was demonstrated that the stacking interaction of aromatic rings and dynamic quenching play essential roles in the generation of diverse fluorescence responses and unique fingerprint maps for individual anionic sulfonate surfactants and their mixtures. The latter satisfactorily distinguished in environmental water samples, demonstrating the practicability of the developed sensor array.

9.4.5 Fluorescent polymers as sensitive materials for optical e-tongue

Bunz and co-workers have published a series of research works investigating several polymer-based fluorometric multisensing platforms for a wide range of analytical applications [84, 85, 92–95, 112]. In [112] authors discussed poly (aryleneethynylene)s (PAE) water-soluble polymers that are easily synthesizable and possess stable fluorescent properties such as super quenching and high quantum yield. Their multivalency and molecular wire effect have been used as cores in optical chemical sensors for specific tasks. An application of hyperbranched conjugated polymers (HCP) obtained with truxene core units and benzene, bearing several alkyl- and alkyloxy long chain substitutes comonomers for fluorometric sensor development is presented in [84]. The small sensor array formed from two HCPs and two poly(para-phenyleneethyny-lene)s (PAE) polymers was tested for nitroaromatic explosives detection. The quenching efficiencies of polymers was dependent on the spectral overlap between the absorbance of the explosive analytes and the emission of the fluorescent polymer. LDA analysis was carried out on the micelles fluorescences quenched by the analytes. 100% accuracy was achieved in the discrimination of all the nitroaromatic analytes, namely picric acid (PA), nitrobenzene (NB), dinitrobenzene (DNB), dinitrotoluenes (DNT), trinitrotoluenes (TNT), nitrophenol (NP), 2-nitroaniline (2-NA), 3-nitroaniline

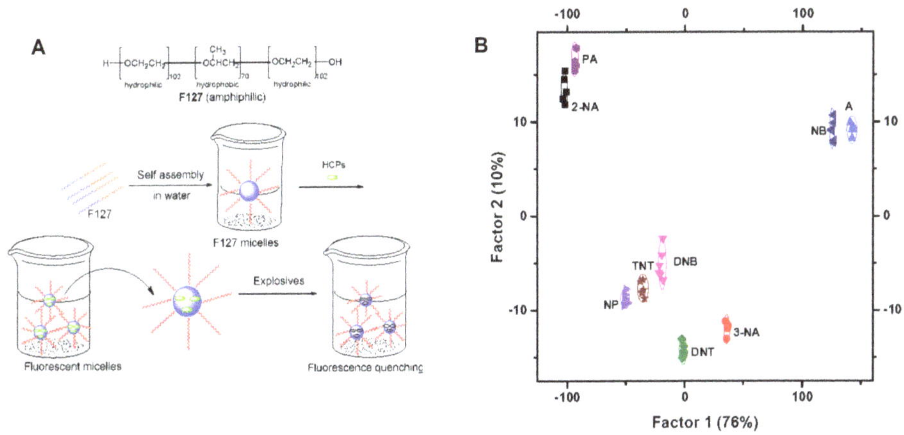

Figure 9.10. (A) The schematic representation of sensing micelles formation and fluorescence quenching process in water. (B) LDA score plot of explosives (95% confidence) by fluorescent micelles array. Reprinted with permission from [84], copyright (2017) American Chemical Society.

(3-NA) and aniline (A). Moreover, fluorescent micelles formed by addition of a concentrated THF solution of the hydrophobic conjugated polymers into an aqueous solution of amphiphilic polymer F127 (a polyethylene glycol-polypropylene glycol block copolymer) efficiently detected and discriminated nitroarenes in water, figure 9.10.

One year later the same authors extended their research to discriminate 14 different nitroaromatic explosives on tetraphenylethene-based polymers (TPEPs), showing the aggregate-induced emission (AIE) in water [85]. All analytes were discriminated by the four-element sensor array and LDA, moreover, the possibility to distinguish the regioisomeric nitroarenes with very similar physical and chemical properties was also demonstrated. Moreover, the same research group reported several applications of conjugated polyelectrolyte-based chemical tongues for beverages discrimination, such as white wines [93], fruit juices [94], whiskies [95], honeys and syrups identification [96], assessment of non-steroidal anti-inflammatory drugs [92] and amino acids [89].

A different approach for sensing nitroaromatic compounds was employed by Behera, Mohanty and De in [86]. Gold nanoclusters (Au NCs), MoS_2 quantum dots (MQD) and WS_2 quantum dots (WQD) were used as fluorescent receptors with different quenching efficiency. A sequential on–off strategy was used to sense nitroaromatic explosives. For this, Cu^{2+}/EDTA was used as a quencher/masker pair for AuNCs, while the Fe^{3+}/F^- pair served the purpose for MQD and WQD. In this way, nine different nitroaromatics, and their binary mixtures at 0.1 mmol L^{-1} concentration were detected with 100% classification accuracy.

9.4.6 Drugs assessment

The optical sensor array based on three poly(p-phenyleneethynylene)s (PPEs), and their complexes electrostatically modified with Fe^{2+}, Cu^{2+} and Co^{2+} cations was employed to identify phenylthiohydantoin (PTH) amino acids—the degradation products of oligopeptide through Edman sequencing, thus demonstrating a powerful tool to determinate the primary structure of oligopeptide and proteins [87]. A paper-based e-tongue obtained with fluorescent CTAB-encapsulated CuNCs for specific detection of dithiocarbamates, DTS, was developed by Cheng, Wang and Feng in [82]. The modified CuNCs displayed orange-reddish fluorescence in solid phase and its emission intensity could gradually decrease upon the addition of CS_2 and DTCs pesticides. By deposition the as-synthesized CuNCs on the paper with PVA-PEG immobilizing system, a versatile sensing platform was established for specific detection and discrimination of four different DTCs: ferbam, metham, dazomet and nabam). In addition, the as-fabricated sensor was used to evaluate the total added amounts of dithiocarbamates, in apple, pear and cherry tomato fruits with recoveries from 91% to 94%.

The optical array system assembled by multiple cholinesterases (ChE) and thiocholine (TCh) sensors for low-cost and accurate pesticides screening has been described in [83]. The inhibition of ChE activity by pesticides reduced the TCh produced by the hydrolysis of butyryl/acetylthiocholine iodide, thus changing the fluorescence intensity of TCh sensor. These diverse responses were collected in pesticide database, which was then used for the screening of unknown real contaminated samples and compared with high-performance liquid chromatography estimations. The pesticides assessment possibility was verified in real samples, such as soil, groundwater, spinach and soil extracts with two anonymous pesticides with unknown concentration, which were successfully identified.

9.4.7 Biomolecules testing

The biomolecules testing using optical sensors arrays has been actively investigated recently. Tomita *et al* have reported the fabrication of a pattern-recognition-based multichannel sensor using a set of fluorophore-conjugated single-stranded DNAs (ssDNAs) [87] and DNA/graphene oxide conjugates [113]. In [87] an optical sensing system was prepared in microplate wells, where ssDNAs were placed together with a target protein. It was used to discriminate eight protease analytes and to identify the composition of protease/inhibitor mixtures. Additionally, the sensing platform could perform the inverse task—identify proteins (which are potential targets for disease diagnosis) in protease and inhibitor mixture of different composition ratios. One year before, the same research group reported a fingerprint-based sensor using an array of single-stranded DNAs (ssDNAs) labeled with environment-responsive 3′-carboxytetra-methylrhodamine (TAMRA) for the identification of 8 proteases and 12 different mixtures of trypsin and its inhibitor protein (α_1-antitrypsin) by multivariate analysis [88].

Nano-graphene oxide (nGO) and suitable fluorophores, 7-hydroxy coumarin (7-HC), fluorescein (Fl), rhodamine 6G (Rh-6G), rhodamine B base (Rh-B) and methylene blue (MB), as sensor elements to construct a multichannel sensor array for the detection of protein analytes was reported by Behera and De in [114]. The possibility to detect different proteins, such as cytochrome C (Cyt C), β galactosidase (β gal), chymotrypsin (ChT), bovine serum albumin (BSA), lipase (Lip), pepsin (Pep), xylanase (Xyl), macerozyme (Mac), and mixtures of proteins with 100% classification accuracy at sub-nanomolar concentration was shown.

Bacterial cellulose (BC) was used as solid support in a nanopaper analytical device (NAD) or 'lab-on-nanopaper' for visual sensing of human serum albumin (HSA) reported by Naghdi *et al* in [115]. Color changes of curcumin embedded in BC (CEBC) were visually monitored either by naked eye using a smartphone camera, or spectroscopically, and both were linearly proportional to the HSA concentration from 10–300 μmol L^{-1} and 25–400 μmol L^{-1}, respectively. The developed NAD/CEBC as a novel albumin assay kit was successfully utilized in the determination of HSA in human blood serum samples with satisfactory results.

Zhang *et al* have reported in [116] a sensor array based on ionic liquids and aggregation-induced emission (AIE) luminogens for the detection and discrimination of several biologically important analytes, and multi-analyte systems, such as 20 natural amino acids, nine important phosphate derivatives, ten metal ions and three pairs of enantiomers. Further, a novel customizable sensing system for multi-analytes based on a single AIE-doped poly(ionic liquid) photonic sphere was developed, and it was successfully applied for discriminating between 20 amino acids in human urine and their 26 unprecedented complex mixtures, figure 9.11.

Shen *et al* have developed a fluorescence sensor array consisting of three sensing units (two commercially available small molecular fluorescent dyes, Nile Red and ANS, and one fluorescent conjugated polymer, PPE2) to discriminate different urinary proteins, including human serum albumin (HSA), transferrin (Trf), lysozyme (Lys), immunoglobulin G (IgG), myoglobin (Myo), and acid phosphatase (ACP) [90]. The 100% accuracy in HCA differentiation and LDA quantitative discrimination of every single protein and their complex mixtures was achieved. Also, protein profiles of urine samples from healthy people and patients with different urinary diseases, including tubular injury (TI), diabetic nephropathy (DN), and nephropathy (Nep) were successfully distinguished both with LDA and HCA,

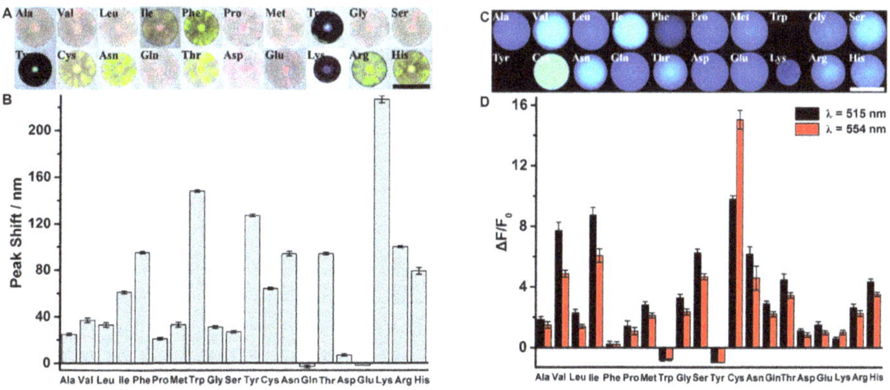

Figure 9.11. Responses of AIE-doped poly(ionic liquid) photonic spheres to 20 natural amino acids, 10 mmol L^{-1}. (A) Optical images. (B) Histogram of reflection peak shifts. (C) Fluorescence images (false color, exposure time 30 ms). (D) Histogram of the folds of fluorescence enhancement. Reprinted from [116] with permission of The Royal Society of Chemistry.

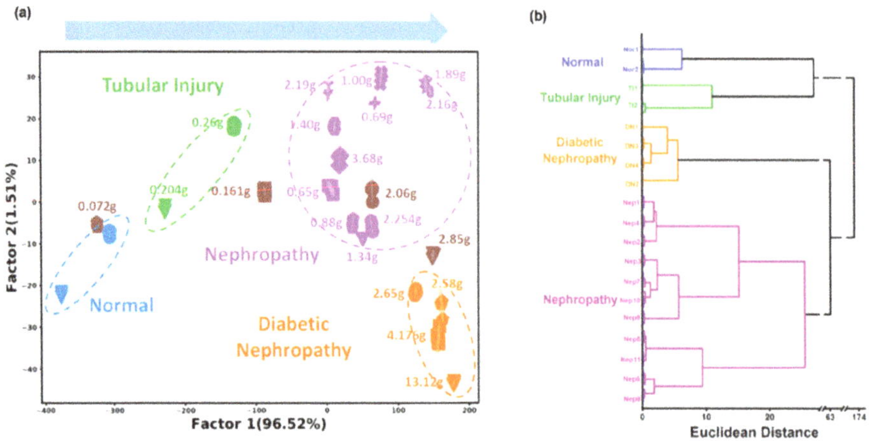

Figure 9.12. Discrimination of urine samples from patients with different urinary system diseases: (a) LDA score plot and (b) HCA dendrogram derived from the fluorescence change profile of the optimized sensor array upon addition of different urine samples. Reprinted with permission from [90], copyright (2020) American Chemical Society.

demonstrating the potential application of developed sensing array in biomedical diagnosis and monitoring, figure 9.12. In [91] Jafarinejad *et al* developed an optical sensor array using three fluorescent dyes with distinct emission bands, monitoring variations in their emission profiles upon the addition of catecholamine neurotransmitters, specifically, dopamine (DA), epinephrine (EP), and norepinephrine (NE), in the presence of gold ions. Neurotransmitters, as substantial indicators of various neurological diseases, require the development of effective and rapid detection methods. The formation of differently sized gold nanoparticles (GNPs) having distinct aggregation levels was generated *in situ*, because of different reducing power of catecholamines, resulting in different amounts of spectral overlap between the absorption band of GNPs and the emission bands of the fluorescent dyes. These energy-transfer-based fingerprint profiles were used to discriminate neurotransmitters by applying pattern recognition methods, such as linear discriminant analysis (LDA) and artificial neural networks (ANN), and to determine the analyte concentrations using multiple linear regression (MLR). Finally, the developed e-tongue showed good performance in the discrimination of DA, EP, and NE in human urine.

Sun *et al* employed a different strategy for the effective discrimination of DA analogues with the help of hyperbranched polyethyleneimine (hPEI)-based fluorescent sensor array [117]. The self-polymerization formation of fluorescent polymer nanoparticles (FPNs) with diverse emission profiles occurred in the presence of different analytes. The fluorescence patterns of the formed FPNs were treated with LDA for differentiation of DA analogues. An excellent discrimination of four DA analogues at 1.0 μmol L^{-1} was demonstrated when LDA and HCA analyses were combined. The possibility to distinguish binary, tertiary and quaternary mixtures of DA analogues was also shown with the proposed sensor array; for 88 blind samples containing a single analogue or mixtures, presenting 100% accuracy of evaluation. In [118] Ghasemi, Hormozi-Nezhad, and Mahmoudi have developed a thioglycolic acid functionalized cadmium telluride (CdTe) quantum dots-based sensor array for identification and discrimination of catecholamine neurotransmitters based on optical properties of their oxidation products under alkaline conditions. The spectral changes were analyzed by HCA and PCA, and the effective discrimination of individual catecholamines, DA, norepinephrine, and l-DOPA and their mixtures in the concentration range 0.25–30 μmol L^{-1}. The system was successfully applied to identify the various catecholamines in urine. In [119] Han *et al* showed the possibility to distinguish original and counterfeit nonsteroidal anti-inflammatory drugs (aspirin, ibuprofen, diclofenac, or naproxen) by poly(para-phenylene ethynylene)-based sensor array and LDA.

9.4.8 Medical applications

The development of optical point-of-care-testing (POCT) systems has received vast amounts of research interest recently. POCT systems are wearable sensors attached to a part of the human body and employed to measure/monitor one or more physical/chemical parameters, or even represent contactless non-invasive devices monitoring biomarkers in different biofluids, such as sweat, interstitial fluid, blood, tears, saliva, tissues, etc. For instance, Ardalan *et al* recently reported a smart wearable sweat patch (SWSP) sensor based on highly fluorescent sensing probes embedded in paper substrates. Microfluidic channels consisted of cotton threads to harvest sweat from the skin surface and transport it to paper-based sensing probes [97]. The simultaneous non-invasive and *in situ* multi-sensing of sweat biomarkers, including glucose, lactate, pH, chloride, and sweat volume, could be detected with an SWSP system upon UV-LED lamps illumination, figure 9.13. Finally, the designed smartphone app was used to quantify the concentration of the biomarkers using a developed detection algorithm.

Bogomolov *et al* have reported an LED-based near-infrared sensor system for kidney oncological pathology diagnostics [98]. The LED sensor concept is based on the rapid scanning of a few (typically, from two to seven) selected wavelengths. LEDs applied at the chosen wavelengths assured an alternating illumination of the sample. The reflected light is then measured by a photodiode. The developed system used four LEDs at selected wavelengths and one detecting photodiode. Two sensor prototypes were tested using 14 *in vitro* clinical samples of seven different patients.

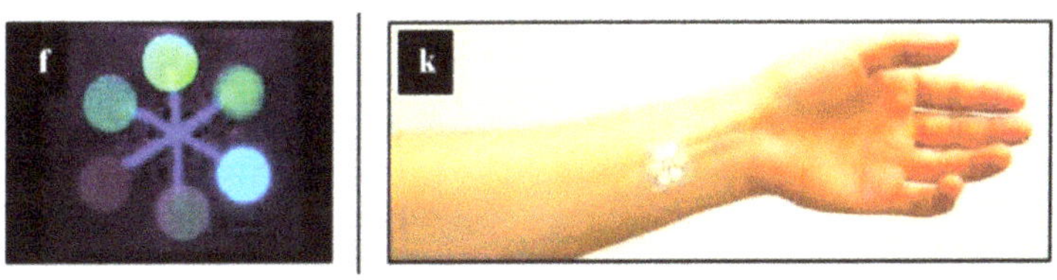

Figure 9.13. The POCT sweat patch under UV illumination and mounted on the forearm. Reprinted from [97], copyright (2020) with permission from Elsevier.

Statistical data evaluation using PCA and partial least-squares discriminant analysis (PLS-DA) statistical evaluation of results using PCA and PCA-DA confirmed the general applicability of the LED-based sensing approach to kidney tumor detection. Additional validation of the results was performed through sample permutation.

Anslyn lab reported an array of semi-specific peptides conjugated to thiazole orange to act as both a fluorescence reporter and a DNA intercalator capable of self-assembly [99]. In contrast to static nanoparticle-based arrays, these receptors allowed a dynamic arrangement of intercalator-conjugated peptides expected to bind to cancer cells. The chemometric fingerprints for eight cancer cell lines were obtained from fluorescence data for the peptides exposed to cells, with and without exogenous double-stranded DNA (dsDNA). The correct classification from 60% to 100% was found for the developed sensing platform depending upon the cell line Comparing to the 'hypothesis non-specific sensor array', it was demonstrated that the use of a limited level of specificity, according to the 'relaxed lock-and-key principle', generated high dimensional and linear discriminant score plots, with several PCs carrying significant variance. Finally, to further test the cross-reactivity of the sensing array, it was exposed to the unknown cellular analyte for which it had not been designed, and to an LDA model generated for the original pattern data. The array successfully discriminated this cell line from the other eight cell lines tested.

9.4.9 Environmental monitoring and food quality assessment applications

Applications of optical sensors and e-tongues for environmental monitoring purposes have received recent growing attention. Simple, fast, low-cost optical sensors handled by any person, instead of qualified personnel only, are the main benefits of optical e-tongues employed in environmental control. Since water is an essential natural resource for humanity, the appropriate maintenance of water resources and protection of supplies, sanitation, and reasonable quality control are the main actions for safer, broader water access to everyone [120]. Water contamination with toxic pollutants may damage flora and fauna, and result in serious health risks and environmental problems. Optical sensor arrays are one of the most promising opportunities to carry out inexpensive and real-time monitoring of water quality, and to control both the presence of abnormal alarm situations or detection of a high level of some components. Thus, in our previous research we have demonstrated the possibility of Microcystin-LR (MC-LR) assessment in potable water with non-enzymatic optical sensor array [81]. MC-LR is one of the most poisonous among toxins released during cyanobacteria proliferation. It belongs to a family of hepatotoxic cyclic heptapeptides, being the inhibitor of protein phosphatases (PP1A and PP2A) MC-LR and considered a tumor promoter. Also, MC-LR affects the normal liver functioning and is responsible for acute and chronic poisoning phenomena in animals and humans, even at nanomolar concentrations. In our work, we developed an all-solid-state optical sensor system based on phenyl-substituted diaza-18-crown-6 hydroxyquinoline (DCHQ-Ph), previously investigated as Mg(II)-sensitive fluorophore [105], for the indirect selective detection of MC-LR. The inverse principle of sensor functioning was based on the fact that MC-LR in a basic medium carries two negative charges due to the deprotonation of the carboxylic groups present in its structure. These negatively charged species, MC-LR^{2-} may either electrostatically interact with the cationic species present in the solution or complex in the microcystin cavity where inner core amide groups serve as electronic density donors. Consequently, measuring variations of free Mg^{2+} cation concentrations upon exposure to different MC-LR contents permits an indirect analysis of the toxin through the measured fluorescence intensity variation of DCHQ-Ph- based PVC membrane upon UV-light illumination, figure 9.14. The application of an optical array composed of sensing membrane spots with different amounts of DCHQ-Ph and deposited on glass or paper support permitted differentiatiation of surface water samples and samples spiked with *M. aeruginosa* culture filtrates. The PLS regression method was employed to correlate the data obtained from the optode array and the concentrations of different hepatotoxins in filtrates of toxic *M. aeruginosa*, determined with UHPLC-DAD method. We obtained an Root Mean Square Error of Validation (RMSEV) of 0.219 $\mu g\ L^{-1}$ 1.5 times lower than the WHO reference value of 1 $\mu g\ L^{-1}$ in potable water. The array was used for assaying the amount of MC-LR spiked in natural water samples one from the surface of Lake

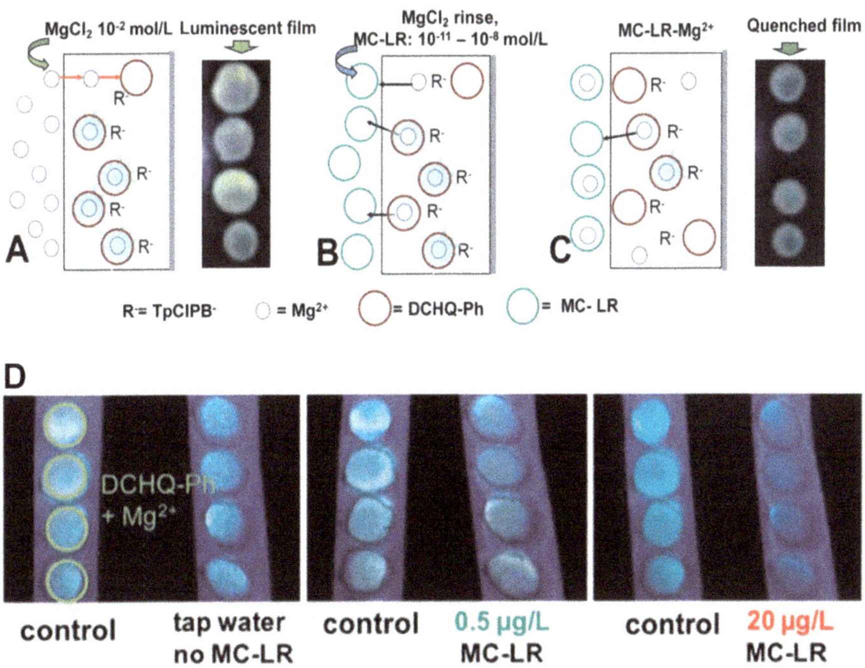

Figure 9.14. Schematic presentation of the indirect MC optical detection: (A) membrane stabilization in 0.01 mol L^{-1} TRIS pH 8.6 containing 0.01 mol L^{-1} $MgCl_2$; (B) addition of MC-LR; (C) Mg^{2+} flux from the membrane phase resulting in a film luminescence decrease; and (D) assessment of MC-LR in tap water upon the addition of 0.5 and 20 ug L^{-1} of MC-LR, λ_{ex} = 365 nm; a digital camera of common smartphone was employed for photogram registration. Reprinted from [81] with permission from The Royal Society of Chemistry.

Albano (Rome, Italy) and another from a public fountain located in the Rome area, thus demonstrating its potential applicability.

Recently, we have described the synthesis and coordination properties towards Zn^{2+} and Cd^{2+} ions of the N_2S_2- type ligands L1 and L2 containing the 2,8-dithia-5-aza-2,6-pyridinophane (L) macrocyclic receptor unit and the 2-(20-hydroxy-30-naphthyl) –4-methylbenzoxazole (HNBO) or 7-(2-ethylamino)–4-methylcoumarin [106]. The fluorophores were dispersed inside PVC membranes and employed in an optical sensor array based on the photoassisted technique (PT) for signal acquisition for a quantitative determination of Cd^{2+} in soils [77]. Soil samples from three Italian sites with different anthropogenic influences (from a site close to the landing strip of Ciampino Airport at Rome; from Rome 'Tor Vergata' university garden; and from waste dump close to Naples) were analyzed with the sensor array, and the results were compared with standard AAS method. An application of the optical sensor array to detect the amount of spiked Cd^{2+} ions in two soil samples gave recoveries lying in a narrow range from 98.5% to 100.5%, and mean RSD of 0.96%, while the detection of Cd^{2+} amount spiked in a sample from the airport was not possible due to the high initial concentration of the analyte. The PLS calibration model was fitted with 3PCs, representing 99.3% and 98.1% of the total explained variance at calibration and validation steps, and the correlation coefficients, R^2, were 0.993 and 0.983, respectively. A detection limit of 1.8×10^{-7} mol L^{-1} for Cd^{2+} ion was three orders of magnitude lower than the USEPA (U.S. Environmental Protection Agency) value of 20 mg kg^{-1} (corresponding to 1.8×10^{-4} mol L^{-1}) for Cd^{2+} in the agricultural topsoils, indicating good analytical performance of the developed optical sensor array.

In [121] the optical sensor array of chemoresponsive dyes consisting of metalloporphyrins and pH indicators differently blended with ionic exchanger salts and between them, was used for the identification of toxic compounds from different origins such as intermediate products of the chemical industry (dimethyl methylphosphonate and cyclohexanone), drugs (5-fluorouracil and piperazine) and pesticides (imidacloprid and paraoxon). The light reflected by the sensor array upon illumination with red, green, and blue LED light emitters was registered with a high resolution Philips SPC900NC/00 digital camera connected via USB to the computer. The obtained luminescence signals were interpreted in RGB scale, and PCA was employed for e-tongue data treatment. Results showed that all tested sensing materials were sensitive, although scarcely selective to the tested compounds from 10^{-7} to 10^{-4} mol L^{-1}. However, the obtained selectivity patterns were sufficiently different to enable multivariate identification of compounds, independently of the concentration in water. A simple highly sensitive chemiluminescent sensor array for the discrimination of organophosphate and carbamate pesticides is reported in [122]. The sensor array is based on simultaneous utilization of the triple-channel properties of the luminol-functionalized Ag nanoparticle and a H_2O_2 chemiluminescent system. The triple-channel properties were simultaneously altered after interaction with pesticides, producing distinct response patterns as 'fingerprints' related to each specific pesticide. The 'fingerprints' were treated with PCA to generate a clustering map. Five organophosphate and carbamate pesticides, including dimethoate, dipterex, carbaryl, chlorpyrifos and carbofuran were identified at 24 μg mL^{-1}. Moreover, twenty unknown pesticides have been successfully identified with 95% accuracy. The conjugated polymer sensor array constructed using three anionic conjugated polyelectrolytes was used for a rapid quantitative profiling of Red and Orange Aza Dyes pollutants in [123]. The LDA was used to treat the fluorescence response patterns, enabling the individual identification of 12 closely related Azo dyes; the practical application of the system was shown through differentiation between Aza dyes at nanomolar concentrations in water. Seo *et al* have developed a spilled oil point-of-testing device (SOPD) for immediate on-site fluorescence monitoring of oil spills [124]. The device employing only two LED light sources, five optical lenses, and a CMOS sensor was able to analyze quantitatively fluorescence intensities at both low (0–1 ppm) and high (1–800 ppm) crude oil concentrations in marine and inland environments, respectively. The system utility for five different kinds of crude oil monitoring and the possibility of system remote control by smartphone was demonstrated.

9.4.10 Foodstuffs analysis

Several tasks of food quality assessment and tasting have been satisfactorily solved with fluorometric e-tongues. Among the most popular recently reported applications are the analysis of beverages as fruit juices [92], beer and wines [93, 125, 126], whiskies [95, 127], quality and authenticity assessment of olive oils [128, 129], food safety assessment [130], etc.

For instance, in [95] the selection of fluorescent poly(p-aryleneethynylene)s (PAEs) polyelectrolytes or chimeric green fluorescent proteins (GFPs) sensing materials for discriminating 33 different whiskies according to their production origin, brand, blend status (single malt or blend) and taste (rich or light) was performed following the above-mentioned hypothesis-free sensor array construction principle. A library of 22 PAEs, and GFP protein tongue consisted of three elements: conventional GFP with a net charge of −7, a highly positively charged variant (GFP-K36), and a highly negatively charged variant (GFP-E36), figure 9.15.

The arrays were used without any pre-treatment; with the analyte pipetted to the solution of the fluorescent dyes in a standard plate reader having a 96-well plate. The interactions of e-tongues materials with colorants naturally present in whiskies in different amounts (vanillic acid, vanillin, oak lactones, tannins, etc) provided sensitive fluorescence quenching patterns. These patterns were analyzed by linear discriminant analysis, LDA, demonstrating their ability to discriminate any soluble analyte, apparently regardless of its structure, function, or origin.

A reusable optical tongue device made with multiplexed Au and Al nano-arrays applied to differentiate seven types of off-the-shelf whiskies and three controls (water, 40% ethanol and absolute ethanol) with > 99.7% accuracy using LDA was reported by Macias *et al* in [127]. The developed system had three sensing regions, each of them capable of obtaining two partially selective responses from a single measurement through the orthogonal chemical-functionalization of superimposed arrays via thiol (for Au) and silane (for Al) chemistries, providing two resonance peak-shifts for a single optical measurement.

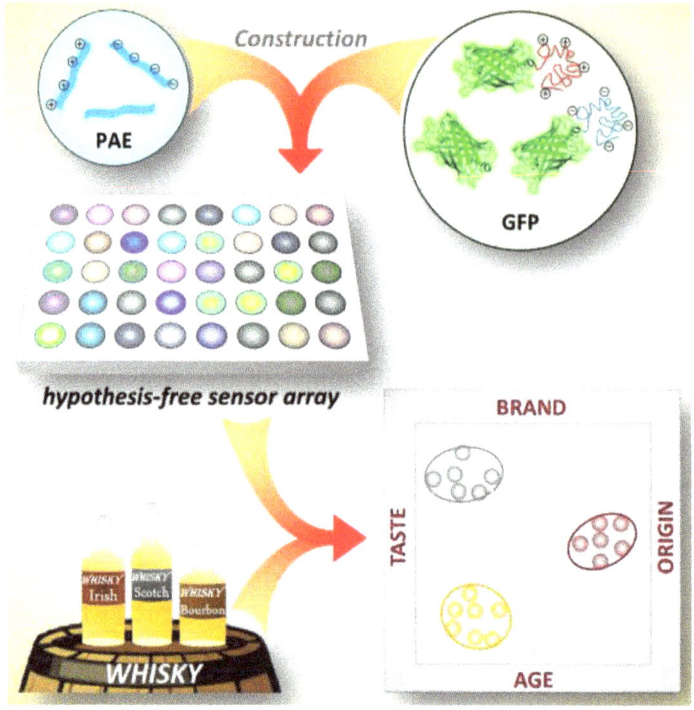

Figure 9.15. The schematic presentation of hypothesis-free sensor array based on fluorescent poly(p-aryleneethynylene)s (PAEs) polyelectrolytes and chimeric green fluorescent proteins (GFPs) for whiskies discrimination. Reprinted from [95], copyright (2017) with permission from Elsevier.

9.4.11 Sensing materials and supports employed in fluorometric e-tongues

As can be noticed from the previous sections, various fluorophore probes and supports were employed in fluorometric e-tongues. Among them there are metallic nanoparticles [122], quantum dots [80, 84, 86, 118, 131], water soluble conjugated polymers [84, 85, 92, 95, 112], hydrogels [80], fluorophores in solution and fluorophores incorporated in solid support—either polymeric [77, 105–107] or cellulose matrix [79, 115], and others in order to both modulate the fluorescence readout and improve the analytical performance of developed artificial sensing systems.

9.5 Single staying optical sensors as multisensor systems applications

The employment of single optical cross-sensitive sensors working in multisensory mode was also reported. The sensor response at different wavelengths represents a sort of multisensor array output UV–vis absorbance or fluorescent emission intensities, which can be further used to generate a recognition pattern of the analyzed sample [132]. The recently developed optical single sensor-based discriminative systems functioning through multisensory principle were comprehensively reviewed by Fan and Ding in [133]. Various construction strategies, such as the application of cross-reactive conjugated polymers, fluorophore/surfactant aggregates, combinatorial molecular sensors, etc, were employed for single optical cross-sensitive sensory systems, while the colorimetric and mainly fluorometric outputs were used for discrimination tasks solving [134]. In [134] a mixture of two amphiphiles bearing the fluorescent head groups were used for the pattern recognition of aminoglycoside antibiotics. The fluorescent spectra were recorded in a buffered aqueous solution, where the amphiphiles form micellar aggregates of different compositions in the absence and the presence of analyzed antibiotics. Since the re-equilibration of the system was analyte-specific, the characteristic fluorescence spectra were obtained for different tested aminoglycosides and the possibility to differentiate analytes in micromolar concentration range with PCA was shown. The possibility of different amyloid beta (Aβ) aggregates discrimination with a cis-amino proline scaffold appended with three fluorescent reporters: thioflavin T (ThT), sulforhodamine B (SRB), and sulfo-Cy5 (sCy5) and functioning as a combinatorial fluorescent molecular sensor was shown in [135]. The unique optical fingerprints were generated by this unimolecular analytical device, providing a simple way to track the dynamic changes occurring as a result of molecular aggregates formation.

A chemical tongue employing a single fluorescent sensor based on pyrene-based fluorophore (PB) containing a metal ion receptor group together with an anionic surfactant sodium dodecyl sulfate (SDS) assembly was developed in [136]. The binary PB/SDS ensembles have exhibited multiple emission bands and provided wavelength-based cross-reactive responses to a series of metal ions. This chemical tongue device could differentiate 13 heavy and transition metal ions and was applied for identification of several brands of drinking water, figure 9.16. The same research group has used a similar principle for the detection of proteins [137–139] and biothiols [140] in serum and urine, while the opposite task was solved in [141] through an application of the dual-chromophore-based cross-reactive fluorescent sensor for the discrimination of anionic surfactants. Thus, in [137] a cholic

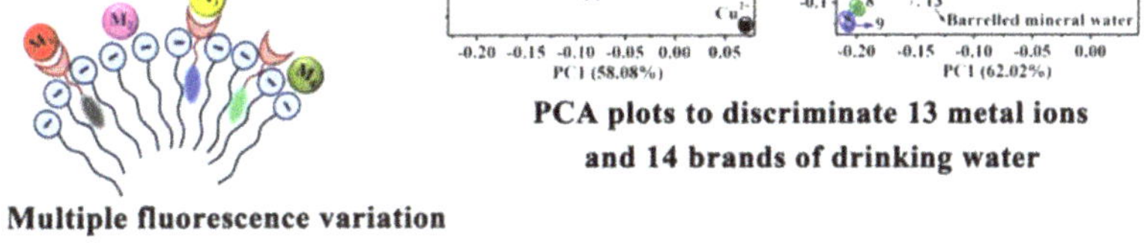

Figure 9.16. Schematic representation of chemical tongue based single PB/SDS fluorescent sensor for 13 metal ions and mineral waters brands discrimination. Reprinted with permission from [136], copyright (2019) American Chemical Society.

acid-modified pyrene derivate fluorophore was assembled with sodium dodecylbenzene sulfonate (SDBS) surfactant to form supramolecular aggregates in aqueous solution. These aggregates exhibited cross-reactive sensing behavior, and the possibility to discriminate seven different metalloproteins in biological fluids such as serum or urine was shown.

In [138] a bispyrene-modified perylene derivative (PEPBI) as a fluorophore with multiple emission bands that was used together with a cationic surfactant cetyltrimethylammonium bromide (CTAB) has shown an enhanced discrimination via ratiometric response to non-metalloproteins and turn-off response to metalloproteins. The single probe fluorescent sensor array for the identification of four different biothiols employing a dual-fluorophore containing pyrene–pyronine and varied concentrations of CTAB surfactant was developed in [140]. The array application for biothiols discrimination in serum and urine was shown. This study demonstrated that the surfactant modulation strategy associated with dual-fluorophore use can offer an opportunity to build a simple and effective fingerprinting sensing platform for multiple biothiols. In [141] the authors used a dual-chromophore-based fluorescent probe with pyrene and rhodamine signaling units and oligoethoxyl group used as a linker. Fluorescence of the probe varied in the presence of different anionic surfactants, including SDS, SDBS, SDeS, SDSO, SDeSO, and AOT. The combination of the two ratio signals of pyrene monomer and excimer to rhodamine permitted generating a specific recognition pattern and distinguished anionic surfactants in a two-dimensional PCA scattering plot. The same group has reported the water-soluble squaraine dyes (SQOH and SQPY) as sensitive colorimetric and fluorescent chemosensors for aldehydes in [142].

9.6 Hybrid multisensory systems for analysis of complex liquid media

9.6.1 Multitransduction e-tongues

The acquisition of different outputs from the same sensing layer, or combining several sensors with different transduction principles, may improve the analytical performance of the resulting hybrid sensing device in terms of selectivity and sensitivity [34]. The fast and simple monitoring of analytes under investigation may be performed through the increase of the obtained chemical information. Recently, the application of hyphenated sensing platforms and the multitransduction approach became more and more popular [126, 143]. Due to the universality and acquisition simplicity, optical transduction is one of the favorite sensory outputs employed in the development of multitransduction sensor systems. The simultaneous registration of fluorescence (either enhancement, or quenching, or both of them), UV–vis absorption and colorimetric output, light-scattering, electrochemical or photoelectrochemical response (including voltage, current, and photocurrent), conductivity, resistance, and other values have been reported previously for hybrid and multitransduction multisensory systems development.

For instance, in [144] Liu *et al* proposed a competitive host-guest fluorescent sensor array for the selective discrimination of several metal ions, including lanthanide and

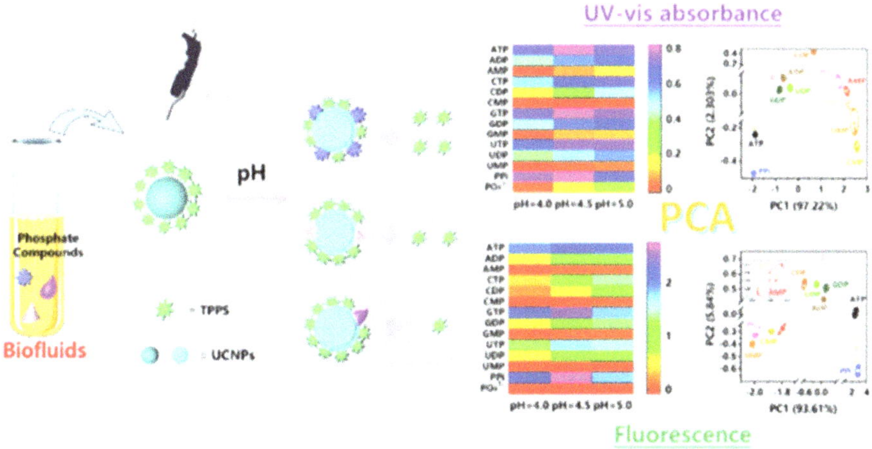

Figure 9.17. The schematic illustration of the dual-mode optical sensor array based on TPPS/Ln-UCNPs functioning Phosphate Compounds discrimination. Figure is reprinted with permission from [145], copyright (2018) American Chemical Society.

actinide salts in aqueous solution. The host–metal interactions resulted in both a fluorescence enhancement and quenching, improving in this way the system discriminatory properties. The fluorescence assay was performed on 96-well plates and the fluorophore signal was recorded in a microplate reader. Yan *et al* [145] developed a dual-mode optical assay based on lanthanide-doped up-conversion nanoparticles (Ln-UCNPs), which upon the interaction with tetraphenyl porphyrin tetrasulfonic acid hydrate, TPPS, has demonstrated a new absorption band and quenched the up-conversion fluorescence of Ln-UCNPs due to fluorescence resonance energy transfer (FRET). The optical behavior of TPPS/Ln-UCNPs organic/inorganic hybrid was characteristically correlated with the pH in aqueous solution and employed for dual-mode optical sensing (UV–vis absorption and fluorescence) of 14 phosphate-containing compounds (PCs), nucleotides and pyrophosphate. In particular, the TPPS/Ln-UCNPs prepared at three different pH (4, 4.5 and 5) were prepared to form an optical sensing array that provided individual absorbance-fluorescence optical patterns upon addition of various PCs analytes through a competitive mechanism between TPPS and PCs. The application of pattern recognition methods (PCA in particular) has permitted differentiating the different concentrations of all tested PCs, and a mixture of them, figure 9.17.

Liu *et al* have reported the synthesis and application of dicyanovinyl-substituted oligothiophene derivative (NA-3T-CN) employed with a zinc porphyrin derivative (Zn-TPP) inside a colorimetric and fluorescent sensor array for dual-mode detection of aliphatic amines and hydrazine [146]. The cost-effective TLC-supported sensor array employing drop-test method and supported with a smartphone-enabled readout system and RGB data treatment has demonstrated a satisfactory discrimination capability among hydrazine, aliphatic and aromatic amines, figure 9.18. The operation of the system both in selective single-sensor mode and in enhanced multisensory mode functioning was demonstrated, and the possibility for naked-eye

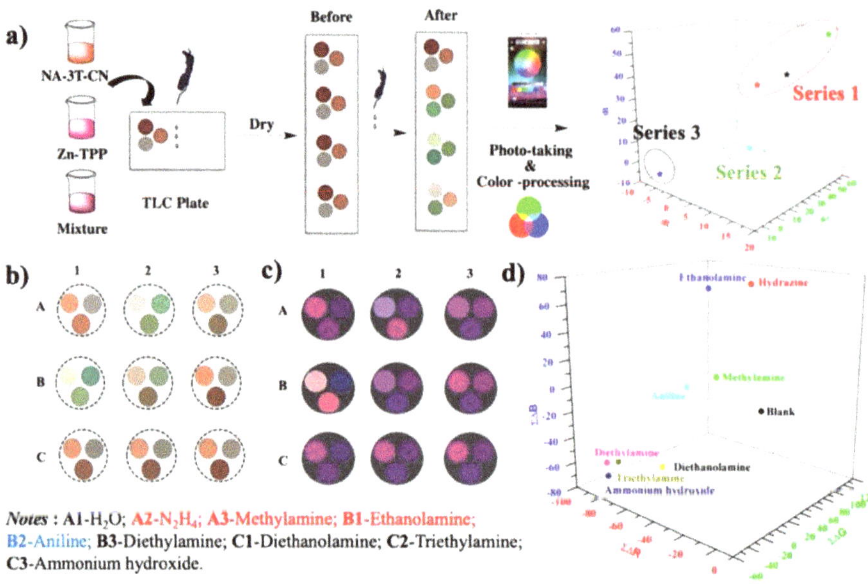

Figure 9.18. Schematic representation of preparation and application for amines discrimination of dual-mode colorimetric/fluorometric NA-3T-CN/ZnTPP-based sensor array. Reprinted with permission from [146], copyright (2020) American Chemical Society.

sensing of ethanolamine, anillin, and hydrazine was shown; for the latter, the colorimetric assessment was possible down to 0.5 wt% solutions.

The detection of the carboxylate drugs by supramolecular sensor array consisting of eight hydrogen-bonding based chemosensors embedded in hydrophilic polyurethane films was reported by Anzenbacher's group in [60]. The simultaneous fluorescence and color readout of the sensor array was treated with pattern recognition methods, namely by PCA and LDA. The 14 carboxylates were identified in water with 100% classification accuracy. The practical utility of the array for carboxylates analysis in human urine was also shown. Moreover, the simultaneous semi-quantitative detection for six NSAIDs was performed in a wide range of concentrations (0.5–100 ppm). Hong and Chang [147] developed a practical app for immediate point-of-care (POC) multi-analyte sensing by using only a smartphone. Their diagnosis results made by using the app of the contents of ascorbic acid, leukocyte, glucose, nitrite, ketones, protein, urobilinogen, bilirubin, and red blood cells in urine are consistent with those obtained by using a commercial urine analyzer. Mitra and co-workers [148] developed a rapid, low-cost water-monitoring sensor termed Mobile Water Kit (MWK). The MWK can simultaneously detect total coliform and *Escherichia coli (E. coli)* bacteria in contaminated drinking water within 30 min. For one of the field samples, the MWK was even able to detect the total coliform within 35 s.

Moczko *et al* have employed the cells as part of an optical sensing array and developed the innovative optical diagnostic assay, based on a mixture of human skin cells and environmental-sensitive fluorescent dyes that generate fluorescence spectra patterns distinctive for particular physicochemical and physiological conditions [149]. The physicochemical changes produced by cells in contact with different compounds reflected in changes of the fluorescent pattern of dyes that were treated with chemometric techniques provided qualitative information about analytical characteristics of samples. This integrated approach has been successfully applied (with a sensitivity of 93% and specificity of 97%) to assess chemical agents toxic for human skin.

In order to identify 10 types of Gram-negative bacteria (including three types of antibiotic-resistant strains), a novel hyphenated 3D nanocomposite array consisting of positively charged cetyltrimethylammonium bromide (CTAB)-capped gold nanoparticles (CTAB-AuNPs) and negatively charged vancomycin-templated gold nanoclusters (Van-AuNCs) was developed in [150]. The triple optical response of (+)AuNPs/AuNCs composite, including fluorescence, UV–vis absorption, and light-scattering channels, was analyzed by LDA and HCA. The highest impact on bacteria discrimination was found by LDA analysis transformation of the output data matrix into canonical factors for fluorescence (90.7%) and UV–vis absorption channel (7.0%). The possibility of the sensor array to discriminate accurately ten types of Gram-negative bacteria at low concentration level of $OD_{600} = 0.015$ was shown, figure 9.19.

Also, the discrimination of Gram-negative bacteria was accomplished in human urine, and all tested bacteria were well discriminated. One year before, the same research group had reported a 3D sensor array based on the simultaneous variation of the optical properties (fluorescence, light scattering, and UV–vis absorption) of gold–silver alloy nanocluster (AuAgNCs) gold nanoparticle (AuNPs) composite for rapid identification of 13 sulfur-containing species and sulfur-oxidizing bacteria [151]. Through the different affinities of sulfur-containing species toward AuAgNCs, the unique response patterns were generated, and the possibility to divide the analytes into three categories including organic sulfide, inorganic sulfide, and thiols 0.5 μM concentrations was shown by PCA analysis. Also, sulfur-oxidizing bacteria and non-sulfur bacteria were discriminated at a level of $OD_{600} = 0.005$.

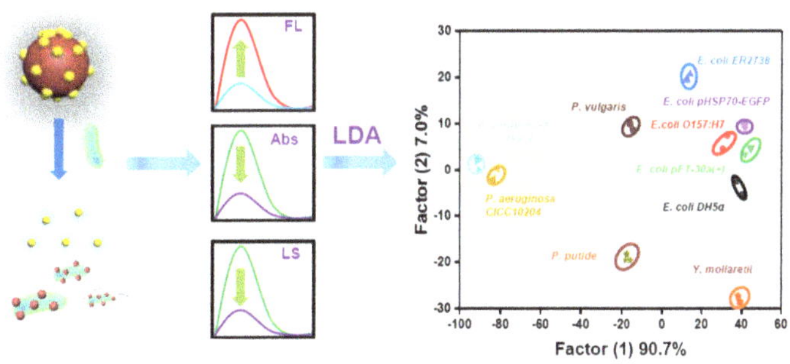

Figure 9.19. Schematic illustration of the sensor array based on the (+)AuNPs/AuNCs composite for the discrimination of Gram negative bacteria. Reprinted from [150] with permission of The Royal Society of Chemistry.

After the development and application of an electronic eye for grape phenolic ripening assessment in [61], Orlandi *et al* have extended their research to the complementary information analysis obtained from the hyphenated analytical system comprising an electronic eye (EE) and an electronic tongue (ET) to monitor grape ripening based on the must analysis [152]. Different data fusion strategies (low- and mid-level fusion approaches) were investigated in order to develop an integrated device able to quickly and easily quantify the physicochemical parameters that are used to assess phenolic ripeness. PLS regression was applied to the fused data, with the aim of relating the information brought by the two sensors having twelve physicochemical parameters measured on must samples by standard analytical methods. The results achieved with mid-level data fusion outperformed those obtained using EE and ET separately and highlighted that both artificial sensory systems have made a significant contribution in the prediction of each one of the considered physicochemical parameters. The possibility of sugars discrimination on a single-material two-signal 2D sensor, prepared with Mn-doped ZnS (Mn-ZnS) quantum dot (QD) possessing 2D optical properties (phosphorescence or fluorescence) and using boronic-acid-substituted viologens (BBV) as sugar receptors, was reported by Miao *et al* in [153]. The six sugars (D-fructose, D-glucose, D-galactose, D-mannose, L-sorbose and sucrose) were analyzed through photoinduced electron-transfer (PIET): initially, BBV quenched the phosphorescence and fluorescence of Mn-ZnS QDs, after sugars added into the quenching system, the differences in the phosphorescence and fluorescence signals were registered and differentiated using LDA. The 2D sensor array was used to detect four orange juice brands and was able to discriminate sugar concentrations, thus proving its feasibility for real analysis.

Our group has developed a hyphenated opto-electrochemical probe based on pyrrole-substituted electropolymerized porphyrins for the detection of redox analytes [154]. The simultaneous CSPT-optical and DPV (differential pulse voltammetry)-electrochemical response of singular electropolymer sensing layer was registered for the analysis of Red-Ox active diazo-conjugated dyes of Sudan family. CSPT response of porphyrin electropolymers significantly supplemented the voltammetric one. The Sudan family dyes were properly discriminated with the proposed hyphenated system, and its feasibility for food matrices analysis was established. Previously the multitransduction systems based on porphyrin sensing materials were reported for the discrimination of olive and seed oils [129], for assessment of several inorganic ions [103, 155] and natural waters and soil pollution monitoring [120, 143]. In [103] the fusion of CSPT and potentiometric data obtained from H_2TFCP-based sensor array permitted us to clearly discriminate between all the tested metal ion species and to observe their concentration gradients, figure 9.20.

The PLS regression model was constructed for lead in the range 2.7×10^{-7} to 3.0×10^{-3} mol L^{-1}. The correlation coefficient of the obtained calibration curve was $R^2 = 0.983$, with root mean square error of calibration (RMSEC) of 3.9·μmol L^{-1}. The detection limit of lead determination was 0.27 μmol L^{-1}, and it was low enough to perform the direct analysis of Pb^{2+} in natural waters.

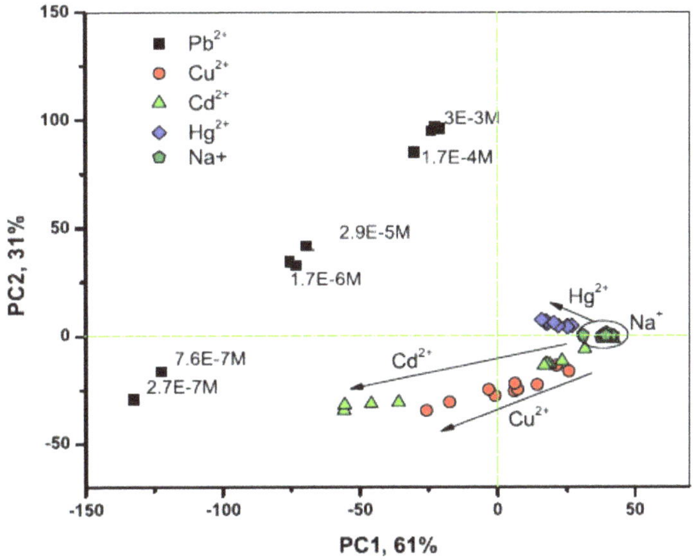

Figure 9.20. PCA scope plot of metal ions identification with opto-potentiometric system based on H_2TFcP. Reprinted from [103], copyright (2019) with permission from MDPI.

9.6.2 Other fusion types for optical e-tongues

Litvinenko *et al* presented an interesting hybrid e-tongue system approach based on low selective sensors, with the p-n junction on a silicon wafer having a porous silicon layer that produces a laser light-induced photovoltaic (PV) signal for the detection of various alcohol solutions [156]. The PCA of e-tongue PV signals permitted satisfactory discrimination of water, ethanol, isopropanol, and their mixtures.

Jansod *et al* [157] have recently described a novel approach consisting in the optical readout registration of a potentiometric sensor array of bipolar ion-selective electrodes (ISEs) based on PVC membranes. The novelty of the reported hybrid system consists in the physical separation of optical detection and sample compartments. In this manner, the drawbacks of sample turbidity and optical signal interpretation were overcome, which is not always possible through direct measurement, especially for colored samples. In the reported arrangement, the detection compartment contains a Prussian Blue (PB) film deposited on conductive indium-tin-oxide (ITO) glass and used as the optical redox indicator, which changes its color from colorless (Prussian White – PW, reduced form) to blue (PB, oxidized form) upon the ISE potential variation in the presence of analyte ion and keeping all the electrochemical cell at constant potential (through external voltage application from potentiostats), figure 9.21.

The colored orange, grape, multifruit and coconut fruit juices tested with the developed system, and the concentrations of sodium, potassium, and calcium determined in juices from optical readout registered by a convenient digital camera were in good correlation with a direct potentiometric measurement used as a reference method. The authors also mention the possibility to achieve self-powered sensing devices where the optical signal is directly generated from the potential change at the ion-selective electrode, without the need for additional electronics or power sources [158].

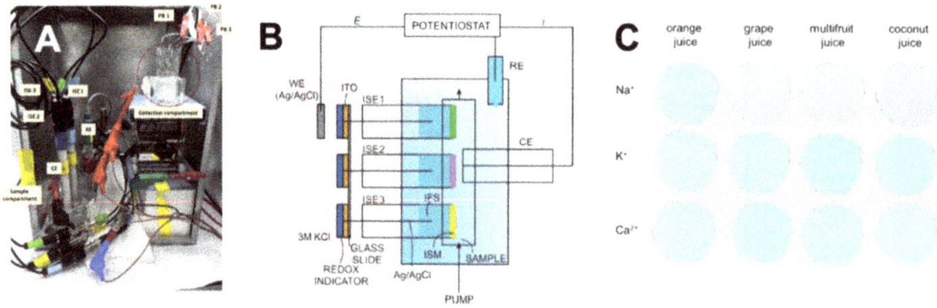

Figure 9.21. (A) The photograph and (B) the schematic presentation of the multiple ion-selective BPEs sensor array coupled to multiple PB films. (C) Optical readout of developed system in different colored fruit juices. Reprinted with permission from [157], copyright (2020) American Chemical Society.

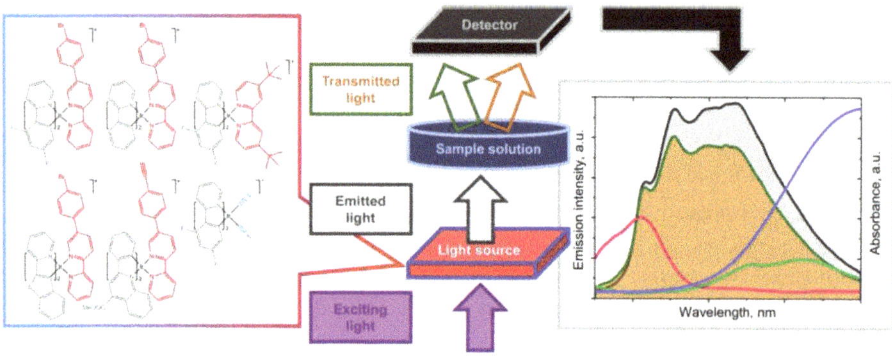

Figure 9.22. Schematic representation of the Ir(III)-complexes based optical multisensor system. Reprinted from [159], copyright (2020) with permission from Elsevier.

A novel approach to design an optical multisensor system was suggested by Kirsanov *et al* in [159]. Through a combination of molecular emitters—cyclometalated Ir(III) compounds—as tunable multiband light sources, the sample absorption spectra were registered, and the possibility of quantification of particular analytes in mixtures, namely cobalt, nickel, and copper metal ions after the chemometric processing of the resulted optical system output was demonstrated. In total eight Ir(III) complexes were used to prepare the multiband light source. The resulting mixture upon excitation with 365 nm laser diode has shown emission spectra in the visible range from 450 to 800 nm, figure 9.22. The emitted light was then sent to a sample and the obtained spectra were employed to construct PLS calibration models. As a result, the precise quantitative analysis of all three ions simultaneously was possible with root mean-square errors of prediction, RMSEP, below 0.007 M in the concentration range 0.01–0.1 mol L^{-1}.

A turbidimetric sensor array based on charged poly(N-isopropylacrylamide) (PNIPAM) nanogel receptors for proteins differentiation was reported by Culver *et al* in [160]. The PNIPAMs were copolymerized with methacrylic acid and modified via carbodiimide coupling to introduce sulfate, guanidinium, secondary amine, or primary amine groups. The modification of ionizable groups in the network changed the physicochemical and protein binding properties of the nanogels. The turbidity of the nanogel solution increased for high affinity protein-polymer interactions, while for low affinity interactions a minimal change in turbidity was observed. In order to improve the proteins differentiation, the turbidity of the solutions was measured at different pHs but keeping comparable ionic strength of background solution. It was possible to achieve 100% classification accuracy of eleven model protein biomarkers with only two nanogel receptors. The developed turbidimetric sensing platform was tested in simulated tear fluid to detect changes in lysozyme concentration.

9.7 Basics of optical sensor arrays data analysis

The detailed representations of the common pattern recognition protocols employed in multivariate analysis of chemical sensor arrays output, including optical sensor arrays, can be found in literature, for instance in [43, 161], and will be provided in the next chapters of the present book. Here, we briefly overview some common chemometric methods used for the interpretation of the optical sensor arrays previously discussed. There are two main groups of chemometric methods: supervised and unsupervised, that by turn, can be subdivided into linear and nonlinear techniques. An overview of the most utilized chemometric techniques and their advantages versus drawbacks are listed in table 9.1.

Table 9.1. Common chemometric techniques.

Method	Abbreviation	Supervised	Linear	Advantages versus drawbacks
Principal component analysis	**PCA**	No	Yes	Easy interpretation *versus* sensitive to the data drift
Hierarchical cluster analysis	**HCA**	Yes	No	Easy interpretation *versus* sensitive to the data drift
Linear discriminant analysis	**LDA**	Yes	Yes	Easy interpretation *versus A priori* knowledge on data structure
Multiple linear regression	**MLR**	Yes	Yes	Statistical results description *versus* Sensitive to the data drift
Principal component regression	**PCR**	Yes	Yes	Statistical results description *versus* Sensitive to the data drift
Partial least square regression	**PLS**	Yes	Yes	Statistical results description *versus* Sensitive to the data drift
PLS-discriminant analysis	**PLS-DA**	Yes	Yes	Statistical results description *versus A priori* knowledge on data structure

9.7.1 Principal component analysis

PCA is an unsupervised technique used to describe a dataset often in the two- or tri-dimensional graphical representation carrying most of the data variance [162]. The axes of the new representation space are given by a linear combination of the original axes and they are requested to be uncorrelated and orthogonal. The first axis, called first 'Principal Component', PC1, is chosen in the direction of the largest variance, and the second axis, called second 'Principal Component', PC2, is taken along the orthogonal direction to the first PC having the largest variance, and so on. This procedure is depicted in figure 9.23 and this technique allows visualizing multi-dimensional datasets for preliminary data exploration and study of the intrinsic capability of the system to discriminate the data in clusters [163]. PCA is one of a wide set of linear transformations that may be described as in equation (9.2):

$$S = W \cdot X \tag{9.2}$$

where X is the original data set, W is the transformation matrix, and S is the data matrix in the representation space. The peculiarity of the PCA is the faithful

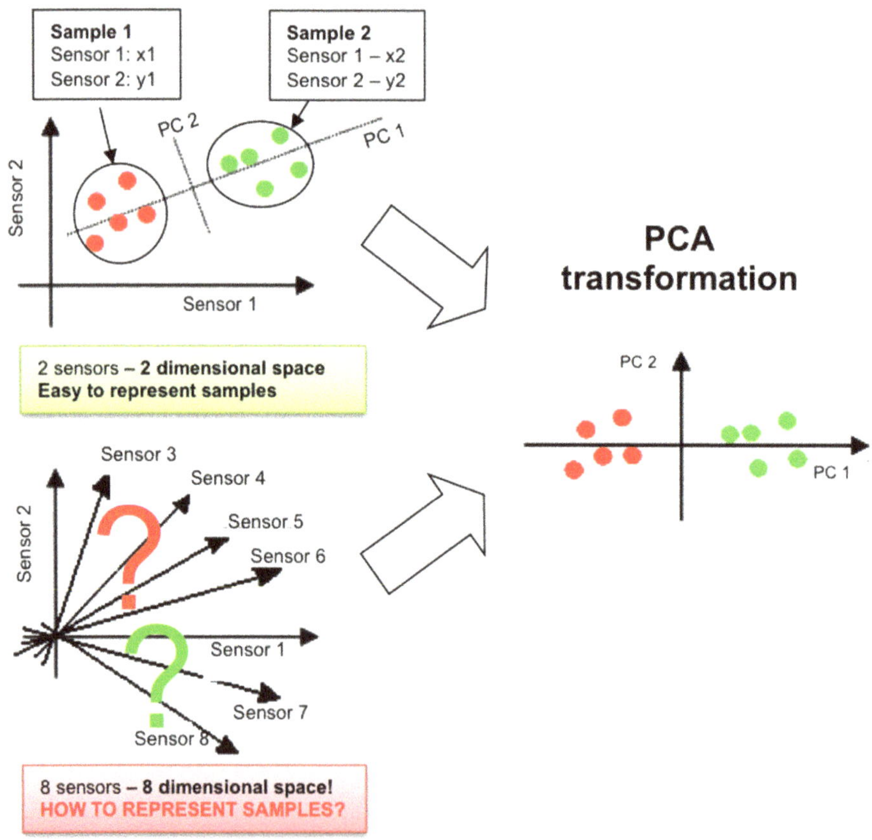

Figure 9.23. A sketch of PC procedure. PCA finds in the dataset the directions with maximum variance.

representation of the data set onto a subspace of reduced dimensionality. The term faithful here means preserving the statistical properties of the original data set. Chemical sensors always exhibit a certain degree of correlation among them, and then their data set may be represented in a subspace of reduced dimensionality. In order to preserve the statistical properties of a data set, it is necessary to make an assumption on the statistical distribution of the data.

In PCA the Gaussian distribution of the data is assumed. Indeed, PCA is calculated considering only the second momentum of the probability distribution of the data (covariance matrix). Only for normally distributed data the covariance matrix describes the data completely once they are zero-centered. Geometrically, the covariance matrix describes a hyper-ellipsoid in the N-dimensional space, and the PCA rotates the coordinates to represent this hyper-ellipsoid in canonical form. Mathematically, it is expressed as in equation (9.3):

$$C = X^T \cdot X \tag{9.3}$$

where C is the covariance matrix of matrix X of data. The ith principal component of X is $X^T\lambda(i)$, where $\lambda(i)$ is the *ith* normalized eigenvector of C corresponding to the *ith* largest eigenvalue.

PCA provides three kinds of information in a given matrix of data: scores, loadings, and eigenvalues of the covariance matrix. Scores are concerned with measurements and they are defined as the coordinates of each vector measurement (a row of matrix X) in the base of the principal components. Loadings are concerned with the sensors. They measure the contribution of each sensor to the array. A high loading for a sensor means that the principal component is aligned along the sensor direction. The eigenvalues are directly proportional to the variance explained by their correspondent eigenvector, so looking at the relative eigenvalues $\lambda(i)$ it is possible to confine the representation to those components carrying most of the information.

9.7.2 Linear discriminant analysis

The LDA method describes how well a group of variables supports *a priori* grouping of objects [164, 165]. The method is closely related to other linear methods such as PCA, MLR, etc. In LDA, a grouping variable is treated as the response variable and is expected to be categorical. Groupings should reflect *a priori* known or established information, such as belonging of samples to the same group, or reflect the results of a preliminary performed exploratory method such as cluster analysis (see next section). After discovering a significant difference among groups, LDA finds linear combinations of the explanatory variables that best discriminate between groups defined. The discriminant functions based on these combinations are further constructed and can be used to classify new objects described by the same explanatory variables used in the LDA. However, LDA itself is not a classification method. Further, the contribution of each of the explanatory variables to the discriminant functions may be examined allowing, for example, variables with high discriminatory power to be identified. An example of the LDA 2D score plot can be seen in figure 9.12(A) of section 9.4.6 of this chapter, where the discrimination of the urine samples of the patients with different urinary system diseases was accomplished.

9.7.3 Hierarchical cluster analysis

The main idea of HCA lies in the consecutive algorithmic approach to find discrete groups of samples having different degrees of similarity in a data set represented by a dissimilarity matrix [164]. These groups are organized hierarchically and represented in the form of a dendrogram graph, illustrated in figure 9.12(B). There are several agglomerative algorithms in HCA, among them: ***single-linkage algorithm*** computing the (dis)similarity between the nearest two members of each group; ***complete-linkage algorithm*** computing the (dis)similarity between the farthest two members of each new groups to be clustered; ***unweighted pair-group method using arithmetic averages (UPGMA) algorithm***, estimating the average (dis)similarity between all members in each group. The most robust algorithm used for HCA is ***Ward's method*** [164] that determines clusters to merge by evaluating the merge 'cost' against an objective function. At each stage, the algorithms that merge with the minimum cost are performed through the evaluation of the sum of squared deviations from cluster centroids. The merged values that yield the smallest increase in the sum of squared deviations are selected. Agglomerative algorithms end when all clusters have been merged into a single one. Implementations of Ward's method generally assume that individual entities form regular hyperellipsoids in the multidimensional space, but it almost flops in a case if irregular clusters occur.

9.7.4 Regression methods

All regression methods relate variations in one or several response variables (Y-variables) to variations of several predictors (X-variables), with explanatory or predictive purposes. Regression can be used to describe and interpret the relationship between X-variables and Y-variables, and to predict Y-values of new samples from values of the X-variables.

9.7.4.1 Multiple linear regression (MLR)

MLR is a linear and parametric method [164] based on ordinary least squares supposing that a complicated system response can be represented by a linear combination of individual system components (predictors) according to equation (9.4):

$$R_i = \beta_{1i} \cdot p_1 + \beta_{21} \cdot p_2 + \cdots + \beta_{ni} \cdot p_n + \varepsilon_i \tag{9.4}$$

where β_{ji} is a coefficient of the *i*-component influencing the signal R_i, p_j is the signal obtained only for component *j*, and ε_i is the residual error. The residual ε_i for a given sample and a given variable is computed as the difference between the observed and fitted (or projected or predicted) values on a sample. For n components of equation (9.4) the following data matrix can be constructed, equation (9.5):

$$R = P \cdot \beta + \varepsilon \tag{9.5}$$

In this matrix, P and R are known values, while β can be evaluated by the least square method. This operation involves a matrix inversion that leads to collinearity problems if variables are not linearly independent. The ability to vary independently from each other is a crucial requirement to variables used as predictors with this method. MLR also requires more samples than predictors, or the matrix cannot be inverted.

9.7.4.2 Partial least square regression (PLS) and principal component regression (PCR)

PLS regression is a technique combining characteristics from principal component analysis and multiple regression, figure 9.24. It is mostly useful when there is a need to predict a set of dependent variables from a large set of independent variables (predictors).

Considering M observations described by K dependent variables that are stored in a matrix denoted $Y = M{\cdot}K$, in which values of J predictors collected on these M observations are collected in the matrix $X = M{\cdot}J$. The goal of PLS regression is to predict Y from X and to describe their common structure. When Y is a vector and X is full rank, this goal is accomplished using ordinary multiple regression [167]. When the number of predictors is large compared to the number of observations, X can be singular and the regression approach is not feasible (i.e., because of multi-collinearity).

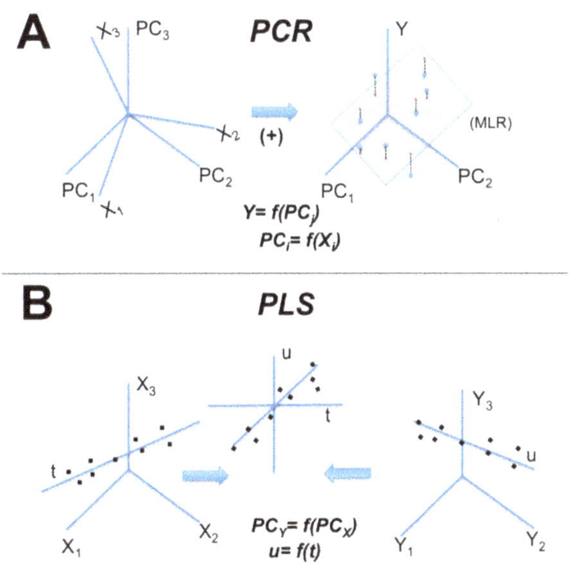

Figure 9.24. Schematic presentation of: (A) principal component regression and (B) partial least square regression methods principle.

Several approaches have been developed to resolve this problem. One approach is to eliminate some predictors (trying to avoid the multi-collinearity).

Another approach, called principal component regression, PCR, is to perform a principal component analysis of the *X* matrix and then fit an MLR model, using the PCs of *X* as regressors on *Y*, figure 9.24(A). The orthogonally of the principal components eliminates the multi-collinearity problem, but the problem of choosing an optimum subset of predictors remains. As an example, the choice of the first PCs of X cannot be relevant to find a correlation with the variables *Y*.

Instead, PLS regression finds components from *X* that are relevant for *Y*. In particular, PLS regression searches for a set of components, called latent vectors **t** and **u** that achieves a simultaneous decomposition of *X* and *Y* with the restriction that these components explain as much as possible of the covariance between *X* and *Y*, figure 9.24(B). Then PLS algorithm searches the subset of *X* that shows the maximum correlation with *Y*. The prediction error of the model is strongly dependent on the number of variables (features) and by the number of measurements. In the case of a small dataset, in order to have realistic error estimation a cross validation technique is applied. Then, latent variables that minimize the prediction error in the validation phase are used to build a model for the further test phase [168].

9.7.5 PLS discriminant analysis

The discriminant analysis, PLS-DA, approach consists in the assignment of a sample to the classes included in the analysis. The most common case is that of a binary discriminant variable: a question with a Yes/No answer. In this case the discriminant variable is coded 0/1 (Yes = 1, No = 0) as Y-variable in the model. With multivariable PLS2 technique this analysis can easily be extended to the case of more than two classes. Each class is represented by an indicator variable, i.e., a binary variable with value 1 for members of that class, 0 for non-members. By building a PLS model with all indicator variables as Y, it is possible to directly predict class membership from the X-variables describing the samples. The model is interpreted by viewing comparing the predicted values for each class indicator Y-variable, if $Y_{pred} > 0.5$ it means 'roughly 1' that is to say, 'class member'; when $Y_{pred} < 0.5$ means 'roughly 0' that is to say 'non-member'. Once the PLS model has been checked and validated it is ready for new samples classification [164].

9.8 Conclusions

This chapter describes the principles and recent advances in optical e-tongue devices development and applications. Specific examples have shown the obvious advantages of multisensory optical systems and multitransduction platforms employment to improve the efficiency of analysis. Taking into account the simplicity in construction and utilization, low-cost and sufficient sensitivity of most of the modern optical sensors and sensor arrays, this technology has a really great potential for the wide range of analytical applications, and for development of economical, portable detectors operating in real-time mode.

References

[1] Ferreira M F S *et al* 2017 *J. Optics* **19** 083001
[2] Di Natale C, Dini F and Scozzari A 2016 *Hdb. Env. Chem.* **40** 279–311
[3] Wolfbeis O S 1986 *Z. Anal. Chem.* **325** 387–92
[4] Kassal P, Horak E, Sigurnjak M, Steinberg M D and Murković Steinberg I 2018 *Rev. Anal. Chem.* **37** 20170024
[5] Ballard Z S and Ozcan A 2017 Wearable optical sensors *Mobile Health* ed J Rehg, S Murphy and S Kumar (Cham: Springer) https://doi.org/10.1007/978-3-319-51394-2_16
[6] Pohanka M 2020 *Rev. Anal. Chem.* **39** 20
[7] Geng Z, Zhang X, Fan Z, Lv X, Su Y and Chen H 2017 *Sensors* **17** 2449
[8] Li F, Bao Y, Wang D, Wang W and Niu L 2016 *Sci. Bull.* **61** 190–201
[9] Mc Cracken K E and Yoo J-Y 2016 *Anal. Methods* **8** 6591
[10] McDonagh C, Burke C S and MacCraith B D 2008 *Chem. Rev.* **108** 400–22
[11] Askim J R and Suslick K S 2017 *Compr. Supramolec. Chem. II* **8** 37–88
[12] Li Z, Askim J R and Suslick K S 2019 *Chem. Rev.* **119** 231–92
[13] You L, Zha D J and Anslyn E V 2015 *Chem. Rev.* **115** 7840–892
[14] Hulanichi A, Geab S and Ingman F 1991 *Pure Appl. Chem.* **63** 1247
[15] Riza M A, Go Y I, Harun S W and Maier R R J 2020 *IEEE Sens. J.* **20** 7614–27
[16] Bakker E, Bühlmann P and Pretsch E 1997 *Chem. Rev.* **97** 3083
[17] Bühlmann P, Pretsch E and Bakker E 1998 *Chem. Rev.* **98** 1593
[18] Borisov S M and Wolfbeis O S 2008 Optical biosensors *Chem. Rev.* **108** 423
[19] Ebralidze I I, Laschuk N O, Poisson J and Zenkina O V 2019 Colorimetric sensors and sensor arrays *Micro and Nano Technologies, Nanomaterials Design for Sensing Applications* ed O V Zenkina (Amsterdam: Elsevier) pp 1–39
[20] Lvova L, Kirsanov D, Legin A and Di Natale C (ed) 2014 Multisensor Systems for Chemical Analysis—Materials and Sensors (Singapore: Pan Stanford Publishing)
[21] Wang Q and Wang L 2020 *Nanoscale* 485–99
[22] Homola J 2008 *Chem. Rev.* **108** 462
[23] Paolesse R, Nardis S, Monti D, Stefanelli M and Di Natale C 2017 *Chem. Rev.* **117** 2517
[24] Ding Y, Zhu W H and Xie Y 2017 *Chem. Rev.* **117** 2203–56
[25] Rakow N A and Suslick K S 2000 *Nature* **406** 710–3
[26] Zhang C and Suslick K S 2005 *J. Am. Chem. Soc.* **127** 11548–9
[27] Lvova L, Acciari E, Mandoj F, Pomarico G and Paolesse R 2020 *ECS J. Solid State Sci. Technol.* **9** 061004
[28] Lonsdale C L, Taba B, Queralto N, Lukaszewski R A, Martino R A, Rhodes P A and Lim S H 2013 *PLoS One* **8** e62726
[29] Zhang C and Suslick K S 2007 *J. Agric. Food Chem.* **55** 237–42
[30] Zhang C, Bailey D P and Suslick K S 2006 *J. Agric. Food Chem.* **54** 4925–31
[31] Huo D, Wu Y, Yang M, Fa H, Luo X and Huo C 2014 *Food Chem.* **145** 639–45
[32] Hou C, Dong J, Zhang G, Lei Y, Yang M, Zhang Y, Liu Z, Zhang S and Huo D 2011 *Biosens. Bioelectron.* **26** 3981–6
[33] Yang L, Huo D, Jiang Y, Hou C and Zhang S 2013 *Food Addit. Contam.* A **30** 786–95
[34] Lvova L, Pudi R, Galloni P, Lippolis V, Di Natale C, Lundstrom I and Paolesse R 2015 *Sens. Actuators* B **207B** 1076–86
[35] Nguyen B T and Anslyn E V 2006 *Coord. Chem. Rev.* **250** 3118

[36] Kalinichev A V, Frosinyuk A, Peshkova M A and Mikhelson K N 2017 *Sens. Actuators* B **249** 123–30
[37] Dekina D I, Kalinichev A V, Pokhvishcheva N V, Peshkova M A and Mikhelson K N 2018 *Sens. Actuators* B **277** 535–43
[38] Kalinichev A V, Pokhvishcheva N V and Peshkova M A 2019 *Talanta* **197** 638–44
[39] Wang X, Zhang Q, Nam C, Hickner M, Mahoney M and Meyerhoff M E 2017 *Angew. Chem. Int. Ed.* **56** 11826–30
[40] Zhang Q, Wang X, Decker V and Meyerhoff M E 2020 *ACS Appl. Mater. Interfaces* **12** 25616–24
[41] D'Andrea A, Pomarico G, Nardis S, Paolesse R, Di Natale C and Lvova L 2019 *JPP* **23** 117–24
[42] Lvova L, Pomarico G, Mandoj F, Caroleo F, Di Natale C, Kadish K M and Nardis S 2020 *JPP* **24** 964–72
[43] Śliwińska M, Wiśniewska P, Dymerski T, Namieśnik J and Wardencki W 2014 *J. Agric. Food Chem.* **62** 1423–48
[44] IEC 61966-2-1:1999 1999 Multimedia systems and equipment, colour measurement and management. Part 2-1: colour management – default RGB colour space – sRGB
[45] DIN 6174 2007 2007–10. Colorimetric evaluation of colour coordinates and colour differences according to the approximate uniform CIELAB Colour Space
[46] Sajed S, Arefi F, Kolahdouz M and Sadeghi M A 2019 *Sens. Actuators* B **298** 126942
[47] Han F, Huang X and Teye E 2019 *J. Food Proc. Engineer.* **42** e12983
[48] Wu J, Li M, Tang H, Su J, He M, Chen G, Guan L and Tian J 2019 *Analyst* **144** 6382–90
[49] Durai W A and Ramu A 2020 *J. Fluoresc.* **30** 215–75
[50] Junaid H M, Batool M, Harun F W, Akhter M S and Shabbir N 2020 *Crit. Rev. Anal. Chem.* 1–18
[51] Uahengo V, Naimhwaka J, Daniel L S, Rahman A, Elzagheid M I, Rhyman L, Ramasami P and Cai P A 2019 *Analyst* **144** 6422–31
[52] Naimhwaka J and Uahengo V 2019 *RSC Adv.* **9** 37926–38
[53] Parchegani F, Orojloo M, Zendehdel M and Amani S 2020 *Spectrochim. Acta* A **229** 117925
[54] Fernandez-Ramos M D, Ogunneye A L, Babarinde N A A, Erenas M M and Capitan-Vallvey L F 2020 *Talanta* **218** 121108
[55] Hosu O, Lettieri M, Papara N, Ravalli A, Sandulescu R, Cristea C and Marrazza G 2019 *Talanta* **204** 525–32
[56] Wei X, Wang Y, Zhao Y and Chen Z 2017 *Biosens. Bioelectron.* **97** 332–7
[57] Yang J, Lu Y, Ao L, Wang F, Jing W, Zhang S and Liu Y 2017 *Sens. Actuators* B **245** 66–73
[58] Sheini A 2020 *Microchim. Acta* **187** 565
[59] Lu W, Dong X, Qiu L, Yan Z, Meng Z, Xue M, He X and Liu X 2017 *J. Hazard. Mater.* **326** 130–37
[60] Liu Y, Minami T, Nishiyabu R, Wang Z and Anzenbacher P 2013 *J. Am. Chem. Soc.* **135** 7705–12
[61] Orlandi G, Calvini R, Pigani L, Foca G, Vasile Simone G, Antonelli A and Ulrici A 2018 *Talanta* **186** 381–8
[62] Mahmoudi M, Lohse S E, Murphy C J and Suslick K S 2016 *ACS Sens.* **1** 17–21

[63] Ranjbar S, Nejad M A F, Parolo C, Shahrokhian S and Merkoçi A 2019 *Anal. Chem.* **91** 14960–66

[64] Sher M, Zhuang R, Demirci U and Asghar W 2017 *Expert Rev. Molecul. Diagnostics* **17** 351–66

[65] Ballard Z S, Joung H A, Goncharov A, Liang J, Nugroho K, Di Carlo D, Garner O B and Ozcan A 2020 *NPJ Digital Med.* **3** 66

[66] Li Z, Fang M, LaGasse M K, Askim J R and Suslick K S 2017 *Ang. Chem. Int. Ed.* **56** 9860–63

[67] Xing C, Liu L, Song S, Feng M, Kuang H and Xu C 2015 *Biosens. Bioelectron.* **66** 445–53

[68] Prosposito P, Burratti L and Venditti I 2020 *Chemosensors* **8** 26

[69] Bigdeli A, Ghasemi F, Golmohammadi H, Abbasi-Moayed S, Nejad M A F, Fahimi-Kashani N, Jafarinejad S, Shahrajabian M and Hormozi-Nezhad M R 2017 *Nanoscale* **9** 16546–63

[70] Ma Y, Li Y, Ma K and Wang Z 2018 *Sci. China Chem.* **61** 643–55

[71] Bordbar M M, Hemmateenejad B, Tashkhourian J and Nami-Ana S F 2018 *Microchim. Acta* **185** 493

[72] Jia M, Pan Y, Zhou J and Zhang M 2021 *Food Chem.* **335** 127566

[73] Ghanem E *et al* 2017 *ACS Sens.* **2** 641–47

[74] Jeningsih, Tan L L, Ulianas A, Heng L Y, Mazlan N-F, Jamaluddin N D, Mohd. Yusof N Y, Khalid B and Ta G C 2020 *Sensors* **20** 1820

[75] Umali A P and Anslyn E V 2010 *Curr. Opin. Chem. Biol.* **14** 685–92

[76] Peveler W J, Yazdani M and Rotello V M 2016 *ACS Sens.* **1** 1282–5

[77] Garau A *et al* 2020 *New J. Chem.* **44** 20834–52

[78] Qiu H, Pu F, Ran X, Ren J and Qu X 2017 *Chem. Eur. J.* **23** 9258–61

[79] Abbasi-Moayed S, Golmohammadi H and Hormozi-Nezhad M R 2018 *Nanoscale* **10** 2492–502

[80] Lin Q, Lu T-T, Zhu X, Sun B, Yang Q-P, Wei T-B and Zhang Y-M 2015 *Chem.Commun.* **51** 1635–8

[81] Lvova L *et al* 2018 *Chem. Commun.* **54** 2747–50

[82] Chen S, Wang Y and Feng L 2020 *Talanta* **210** 120627

[83] Chen L, Tian X, Li Y, Lu L, Nie Y and Wang Y 2020 *J. Hazard. Mater.* **402** 123830

[84] Huang W, SMarsly E, Han J, Bender M, Seehafer K, Wacker I, Schröder R R and Bunz U H F 2017 *ACS Appl. Mater. Interfaces* **9** 3068–74

[85] Huang W, Bender M, Seehafer K, Wacker I, Schröder R R and Bunz U H F 2018 *Macromolecules* **51** 1345–50

[86] Behera P, Mohanty A and De M 2020 *ACS Appl. Nano Mater.* **3** 2846–56

[87] Okada M, Sugai H, Tomita S and Kurita R 2020 *Sensors* **20** 5110

[88] Tomita S, Sugai H, Mimura M, Ishihara S, Shiraki K and Kurita R 2019 *ACS Appl. Mater. Interfaces* **11** 47428–36

[89] Zhang H, Wang B, Seehafer K and Bunz U H F 2020 *Chem. Eur. J.* **26** 1–5

[90] Shen Y, Huang Y, Zhang P, Guo B, Jiang H, Tan C and Jiang Y 2020 *ACS Appl. Bio Mater.* **3** 5639–43

[91] Jafarinejad S, Bigdeli A, Ghazi-Khansari M, Sasanpour P and Hormozi-Nezhad M R 2020 *ACS Chem. Neurosci.* **11** 25–33

[92] Han J, Wang B, Bender M, Kushida S, Seehafer K and Bunz U H F 2017 *ACS Appl. Mater. Interf.* **9** 790–7

[93] Han J, Bender M, Seehafer K and Bunz U H F 2016 *Angew. Chem. Int. Ed.* **55** 7689–92
[94] Han J, Wang B, Bender M, Seehafer K and Bunz U H F 2017 *Analyst* **142** 537–43
[95] Han J, Ma C, Wang B, Bender M, Bojanowski M, Hergert M, Seehafer K, Herrmann A and Bunz U H F 2017 *Chem.* **2** 817–24
[96] Bojanowski N M, Hainer F, Bender M, Seehafer K and Bunz U H F 2018 *Chem. Eur. J.* **24** 4255–58
[97] Ardalan S, Golmohammadi H, Hosseinifard M and Vosough M 2020 *Biosens. Bioelectron.* **168** 112450
[98] Bogomolov A *et al* 2017 *Sensors* **17** 1914
[99] Gade A M, Meadows M K, Ellindton A D and Anslyn E V 2017 *Org. Biomol. Chem.* **15** 9866–74
[100] von Molitor E, Riedel K, Hafner M, Rudolf R and Cesetti T 2020 *Sensors* **20** 1811
[101] Zhang H, Kotlear E A, Kushida S, Maier S, Rominger F, Freudenberg J and Bunz U H F 2020 *Chem. Eur. J.* **26** 8137–43
[102] Xu W, Ren C, Teoh C L, Peng J, Gadre S H, Rhee H-W, Lee C-L K and Chang Y-T 2014 *Anal. Chem.* **86** 8763–9
[103] Lvova L, Galloni P, Floris B, Lundström I, Paolesse R and Di Natale C 2013 *Sensors* **13** 5841–56
[104] Bazzicalupi C *et al* 2013 *Chem. Eur. J.* **19** 14639–53
[105] Lvova L, Guanais Goncalves C, Prodi L, Sgarzi M, Zaccheroni N, Lombardo M, Legin A, Di Natale C and Paolesse R 2017 *Anal. Chim. Acta* **988** 96–103
[106] Lvova L, Di Natale C, Paolesse R, Giorgi L, Fusi V, Garau A and Lippolis V 2016 Proc. Eng. 168 346–50
[107] Lvova L *et al* 2018 *Front. Chem.* **6** 258
[108] Filippini D, Svensson S and Lundström I 2003 *Chem. Commun.* **2** 240–1
[109] Pushina M and Anzenbacher P 2017 *Chem. Commun.* **53** 10074–77
[110] Anzenbacher P Jr, Liu Y, Palacios M A, Minami T, Wang Z and Nishiyabu R 2013 *Chem. Eur. J.* **19** 8497–506
[111] Sun Z, Fan Y Z, Du S Z, Yang Y Z, Ling Y, Li N B and Luo H Q 2020 *Anal. Chem.* **92** 7273–81
[112] Bunz U H F, Seehafer K, Bender M and Porz M 2015 *Chem. Soc. Rev.* **44** 4322–36
[113] Tomita S, Ishihara S and Kurita R 2017 *Sensors* **17** 2194
[114] Behera P and De M 2019 *Chem. Asian J.* **14** 553–60
[115] Naghdi T, Golmohammadi H, Vosough M, Atashi M, Saeedi I and Maghsoudi M T 2019 *Anal. Chim. Acta* **1070** 104–11
[116] Zhang W, Gao N, Cui J, Wang C, Wang S, Zhang G, Dong X, Zhang D and Li G 2017 *Chem. Sci.* **8** 6281–9
[117] Sun Y, Lu F, Yang H, Ding C, Yuan Z and Lu C 2019 *Nanoscale* **11** 12889–97
[118] Ghasemi F, Hormozi-Nezhad M R and Mahmoudi M 2016 *Anal. Chim. Acta* **917** 85–92
[119] Han J, Wang B, Bender M, Kushida S, Seehafer K and Bunz U H F 2016 *ACS Appl. Mater. Interfaces* **9** 790–7
[120] Lvova L, Di Natale C and Paolesse R 2019 Chemical sensors for water potability assessment *Bottled and Packaged Water, the Science of Beverages* ed A M Grumezescu and A M Holban vol 4 (Amsterdam: Elsevier) pp 177–208
[121] Guanais Goncalves C, Dini F, Martinelli E, Catini A, Lundström I, Paolesse R and Di Natale C 2016 *Sens. Actuators* B **236** 997–1004

[122] He Y, Xu B, Li W and Yu H 2015 *J. Agric. Food Chem.* **63** 2930–4

[123] Tropp J, Ihde M, Crater E, Bell N, Bhatta R, Johnson I C, Bonizzoni M and Azoulay J D 2020 *ACS Sens.* **5** 1541–7

[124] Seo D, Oh S, Shin S, Lee M, Hwang Y and Seo S 2019 *Sens. Actuators* B **289** 93–9

[125] Nery E W 2019 *Encyclopedia of Food Security and Sustainability* ed P Ferranti, E M Berry and J R Anderson (Amsterdam: Elsevier) pp 424–32

[126] Guanais Branchini C, Lvova L, Di Natale C and Paolesse R 2016 Wine and combined electronic nose and tongue *Electronic Nose and Tongue in Food Science* ed M L Rodriguez-Mendez (New York: Academic)) pp 291–300

[127] Macias G, Sperling J R, Peveler W J, Burley G A, Neale S L and Clark A W 2019 *Nanoscale* **11** 15216–23

[128] Valli E, Bendini A, Berardinelli A, Ragni L, Riccò B, Grossi M and Gallina Toschi T 2016 *Eur. J. Lipid Sci. Techn.* **118** 1601–19

[129] Tortora L, Stefanelli M, Mastroianni M, Lvova L, Di Natale C, D'Amico A, Filippini D, Lundström I and Paolesse R 2009 *Sens. Actuators* B **142** 457–63

[130] Lvova L 2016 Electronic tongue principles and applications in the food industry *Electronic Nose and Tongue in Food Science* ed M L Rodriguez-Mendez (New York: Academic) pp 149–60

[131] Bagheri S, TermehYousefi A and Mehrmashhadi J 2019 *Rev. Inorg. Chem.* **39** 179–97

[132] Wiskur S L, Floriano P N, Anslyn E V and McDevitt J T 2003 *Angew. Chem. Int. Ed.* **42** 2070–72

[133] Fan J and Ding L 2018 *Analyst* **143** 3775–88

[134] Köstereli Z, Scopelliti R and Severin K 2014 *Chem. Sci.* **5** 2456–60

[135] Hatai J, Motiei L and Margulies D 2017 *J. Am. Chem. Soc.* **139** 2136–9

[136] Zhang L, Huang X, Cao Y, Xin Y and Ding L 2017 *ACS Sens.* **2** 1821–30

[137] Zheng D, Fan J, Huang X, Ding L and Xin Y 2017 *RSC Adv.* **7** 50097

[138] Bo Y, Fan J, Yan S, Ding M, Liu J, Peng J and Ding L 2019 *Sens. Actuators* B. **295** 168–78

[139] Fan J, Zheng D, Huang X, Ding L, Xin Y and Fang Y 2018 *Sens. Actuators* B. **263** 336–46

[140] Fan J, Qi L, Li Y, Tang Q, Ding L and Fang Y 2019 *Sens. Actuators* B. **301** 127144

[141] Qiao M, Cao Y, Liu T, Ding L and Fang Y 2021 *Sens. Actuators* B. **331** 129408

[142] Liu T, Yang L, Zhang J, Liu K, Ding L, Peng H, Belfield K D and Fang Y 2019 *Sens. Actuators* B. **292** 88–93

[143] Lvova L and Nadporozhskaya M 2017 Chemical sensors for soil analysis: principles and applications *New Pesticides and Soil Sensors, Series Nanotechnology in the Agri-Food Industry* ed A M Grumezescu 10 (Amsterdam: Elsevier) pp 637–78

[144] Liu Y, Mettry M, Gill A D, Perez L, Zhong W and Hooley R J 2017 *Anal. Chem.* **89** 11113–21

[145] Yan Q, Ding X-Y, Chen Z-H, Xue S-F, Han X-Y, Lin Z-Y, Yang M, Shi G and Zhang M 2018 *Anal. Chem.* **90** 10536–42

[146] Liu T, Yang L, Feng W, Liu K, Ran Q, Wang W, Liu Q, Peng H, Ding L and Fang Y 2020 *ACS Appl. Mat. Interfaces* **12** 11084–93

[147] Hong J I and Chang B Y 2014 *Lab Chip* **14** 1725–32

[148] Kumar Gunda N S, Naicker S, Shinde S, Kimbahune S, Shrivastava S and Mitra S 2014 *Anal. Methods* **6** 6236–46

[149] Moczko E, Mirkes E M, Cáceres C, Gorban A N and Piletsky S 2016 *Sci. Rep.* **6** 33922

[150] Yang J Y, Jia X D, Wang X Y, Liu M X, Chen M L, Yang T and Wang J H 2020 *Chem. Commun.* **56** 1717–20

[151] Yang J Y, Yang T, Wang X Y, Wang Y T, Liu M X, Chen M L, Yu Y L and Wang J H 2019 *Anal. Chem.* **91** 6012–8

[152] Orlandi G, Calvini R, Foca G, Pigani L, Simone G V and Ulrici A 2018 *Talanta* **195** 181–89

[153] Miao Y, Yang Q, Lv J and Yan G 2017 *New J. Chem.* **41** 14882–89

[154] Cao Z, Lvova L, Paolesse R, Di Natale C, Lundström I and D'Amico A 2014 *Lecture Notes Electr. Eng.* **162** 49–55

[155] Lvova L, Mastroianni M, Di Natale C, Lundström I and Paolesse R 2012 *Electroanalysis* **24** 776–89

[156] Litvinenko S V, Bielobrov D O, Lysenko V and Skryshevsky V A 2016 *Nanoscale Res. Lett.* **11** 374

[157] Jansod S, Cherubini T, Soda Y and Bakker E 2020 *Anal. Chem.* **92** 9138–45

[158] Rho J, Yeon S Y and Chung T D 2020 *Anal. Chem.* **92** 8776–83

[159] Gitlina A Y, Surkova A, Ivonina M V, Sizov V V, Petrovskii S K, Legin A, Starova G L, Koshevoy I O, Grachova E V and Kirsanov D O 2020 *Dyes Pigm.* **180** 108428

[160] Culver H R, Sharma I, Wechsler M E, Anslyn E V and Peppas N A 2017 *Analyst* **142** 3183–93

[161] Anzenbacher P Jr, Lubal P, Buček P, Palacios M A and Kozelkova M E 2010 *Chem. Soc. Rev.* **39** 3954

[162] Massart D L, Vandegiste B G, Deming S N, Michotte Y and Kaufmann L 1988 *Data Handling in Science and Technology Vol.2: Chemometrics: A Textbook* (Amsterdam: Elsevier)

[163] Duda R O, Hart P E and Stork D E 2001 *Pattern Classification.* 2nd Edn (New York: Wiley)

[164] Esbensen K 2004 *Multivariate Data Analysis in Practice* 5th edn (AS, Trondheim: Camo Press)

[165] Fisher R A 1936 *Ann Eugen.* **7** 179–88

[166] Ward J H 1963 *J. Am. Statist. Assoc.* **48** 236–44

[167] Geladi P and Kowlaski B 1986 *Anal. Chim. Acta* **35** 1–17

[168] Wold H 1966 Estimation of principal components and related models by iterative least squares *Multivariate Analysis* ed P R Krishnaiaah (New York: Academic) pp 391–420

Chapter 10

Roadmap for sensorial data analysis

Fernando V Paulovich and Mario Popolin Neto

Sensors and biosensors' data are usually complex entities with patterns hard to understand due to their highly non-linear nature. Computational solutions have been proposed for analyzing sensing data, predominantly using data mining or machine learning techniques and visualization strategies. From machine learning, the most common supported analysis is prediction, either classification or regression. Prediction is commonly used to build calibration models, mapping measures to classes or concentrations of a target analyte. From the visualization, different techniques have been applied for exploratory data analysis. The most popular are dimensionality reduction or multidimensional projection techniques, allowing the investigation of similarity relationships among data samples, and the parallel coordinates, supporting a more detailed view of multidimensional measures. In this chapter, our goal is to present a roadmap for sensing data analysis, discussing the advantages and disadvantages of existing approaches and also the future in combining multiple solutions to support more informative analysis.

10.1 Introduction

The recent years represented a paradigm shift without precedent on sciences regarding the amount of produced information and knowledge dissemination [1]. The idea of collecting as much data as possible to make decisions and understand phenomena has become the norm. Although a positive trend, tasks involving the exploration and interpretation of such data are becoming more costly and complex in their execution, potentially leading to problems in transforming data into valuable information [2]. Aiming to address the inherent difficulties of complex data interpretation, novel computational tools have been deployed. Amongst the solutions, those proposed by machine learning (or data mining) and visualization fields are promising. While machine learning aims to extract useful patterns or create prediction models, visualization seeks to create interactive visual representations enabling the human visual system to augment data analysis.

doi:10.1088/978-0-7503-3687-1ch10

The sensors and biosensors field is not immune to this data deluge effect. In recent years, the analytical process to understand sensing units' particularities became more complex, and better and more sophisticated computational tools are potential solutions. From the machine learning (or data mining) field, classification, regression, and clustering are popular tools for sensorial data analysis. From the visualization, multidimensional projection (or dimensionality reduction) and parallel coordinates have proven their usefulness. In general lines, classification and regression can be used to evaluate the prediction of a sensing unit. Clustering and dimensionality reduction can be employed on distinguishability and reproducibility analysis. Parallel coordinates can bring details on the produced signals. Each different tool has its purpose and can result in important insights into designing or fabricating sensing units.

This chapter is a general roadmap to guide the use of these techniques for sensorial data analysis. Our goal is to serve as a guide, following well-established protocols, on using such techniques to answer essential questions about sensing units. We split this chapter into two major sections, machine learning and visualization, closing each sub-section with a discussion about how to go further and summarizing with a roadmap of utilization. We close this chapter with a general discussion about using machine learning and visualization techniques to perform an analysis covering different perspectives. It is not our goal to be exhaustive but discuss the most common scenarios in the literature and point directions for further exploration.

10.2 Machine learning and data mining

In general, machine learning or data mining techniques can be split into two major groups: supervised and unsupervised techniques (a hybrid approach, called semi-supervised, also exists but is out of our scope) [3]. Supervised techniques aim to create a mathematical model $f(.)$ that learns some property Y related to a data set X. When new data samples X', not contained in the original data set, are received, their property Y' can be approximate, that is $f(X') = Y'$. Therefore, to create such a model, it is necessary to know how Y and X are related, and the model will learn from that. One good example is classification, where, given a set of samples X and their (known) labels Y, a model is created to predict the class of unseen samples.

On the other hand, unsupervised techniques aim to create a model $f(.)$ to discover interesting or unexpected patterns or properties Y in the data X. In this case, no prior knowledge about the relationships between X and Y is necessary or used. The data set is indeed processed to discover those properties. One example is clustering, where $f(.)$ assigns group labels Y to samples X, that is $f(X) = Y$.

Besides clustering and classification, machine learning and data mining literature present other data analysis approaches. Here we focus on them given their prevalence in the sensors and biosensors literature. However, we refer the interested reader to [3] to get more information about other approaches. We start by discussing clustering and how to use it for data analysis in practice.

10.2.1 Clustering

As explained, clustering techniques create a mathematical function $f(.)$ that maps a data set X into group labels Y, considering only information contained in X. In the literature, it is possible to find many techniques, but the usual underlying concept is always the same: to compose groups $C_1, C_2, \ldots, C_k$ of samples, so that the samples inside a group C_i are more related between themselves than the samples in other groups $C_{j\neq i}$. Independently of the employed technique, it is essential to understand that they are stochastic approximations, heavily dependent on parametrization. Multiple runs of the same algorithm or different parametrizations usually produce different results. Also, results are subjective. There is not only one correct solution for the clustering problem. There are multiple depending on the notion of group, that is, how relatedness is defined between samples. For instance, a group can be composed of the most similar samples, or a group can be a dense region in the multidimensional space defined by X.

One example of a similarity-based clustering technique is K-means [4]. Its goal is to produce groups increasing the similarity among samples belonging to the same group. K-means initially selects K samples from X to represent the K groups. Then, each sample x_i in X is assigned to the most similar group considering its representative. After that, the groups' representatives are updated to receive groups' centroids, and the samples are reassigned to the groups. This assignment and representative updating process is then recursively repeated until it reaches a stop criterium. More formally, K-means aims to optimize the sum of squares errors (SSE)

$$SSE = \sum_{i=1}^{K} \sum_{x_j \in C_i} \delta(m_i, x_j)^2 \tag{10.1}$$

where m_i is the representative of group C_i.

Given its simplicity, K-means is one of the most popular techniques for clustering analysis, applied to validate sensors and biosensors units with different applications, from ammonia [5] to food toxin detection [6]. However, results heavily depend on the initial representatives, the number of clusters K, and how similarity is calculated. Usually, initial representatives are randomly defined, giving the technique its stochastic nature. It is highly recommended to execute K-means several times to overcome this issue, using the best result, in terms of SSE or any other quality measure, for the analysis. The number of clusters we will discuss later. However, the most critical factor for understanding K-means results is how distance δ is calculated. Euclidean distance is the usual choice, measuring the squared difference between two samples x_i and x_j or, in other words, the magnitude of the difference. The problem is, if the dynamic range of any variable (frequency in our case) is much larger than the other variables, the analysis will be biased towards that variable. By dynamic range, we refer to the standard deviation of a variable. The data set must be normalized before processing to address this problem. The simplest process is to standardize the data, making the mean of each variable equal to zero and standard deviation equals one. This process is straightforward. Remove from each variable its mean and divide by its standard deviation.

Another problem that Euclidean distance cannot accommodate is if x_i and x_j present noise or shift in phase. Noise can be controlled using some smoothing strategy, for instance, the Savitzky and Golay [7, 8]. The phase shift can be addressed using a distance function that aligns the samples (signals) in the calculation, for instance, the distance time warping (DTW) [9]. The main point is to know what is being measured and use something that semantically makes sense for the analysis. If noise and shift in phase can be ignored, and it makes sense to measure the differences in magnitude between samples, Euclidean is an appropriate choice. In the rest of this chapter, we use the Euclidean distance.

10.2.1.1 Clustering analysis example

So, how to use clustering for the analysis of sensing units? To answer this question, we will use the data presented in [10]. In this paper, different sensing units are used to measure the concentration of phytic acid. The sensing units are made from layer-by-layer (LbL) films with alternating layers of the polyelectrolytes, poly(ally-lamine) hydrochloride (PAH) and poly(vinyl sulfonate) (PVS), or LbL films of PAH alternated with layers of the enzyme phytase, all adsorbed on interdigitated gold electrodes. The sensors were used in impedance spectroscopy measurements in the frequency range from 1 to 10^7 Hz, as it was exposed to different concentrations of phytic acid (10^{-2}, 10^{-3}, 10^{-4}, 10^{-5}, and 10^{-6} M). The measurements were performed using three gold interdigitated electrodes: a bare electrode, one coated with seven PAH/PVS bilayers, and another with two PAH/PVS bilayers covered with five PAH/phytase bilayers. Each solution was measured ten times, including the buffer, thereby resulting in 60 measures (data instances) in 71 different frequencies per sensor. In our analysis, the data comprising the real (z_r) and imaginary (z_i) parts of the impedance at the various frequencies were transformed into capacitance using:

$$c_f = \frac{1}{2\pi f}\left(\frac{-z_i}{(z_r)^2 + (z_i)^2}\right) \tag{10.2}$$

where f is the frequency [11]. Capacitance is here just for illustration, and the same procedure can be applied in potentiometry or voltammetry, considering the due differences.

Considering this type of data, a typical clustering analysis involves measuring how well a sensor can distinguish the concentrations. In our example, if a sensor can separate the five different concentrations of phytic acid and the buffer. In this process, for each sensor, we run the K-means algorithm varying the number of clusters K, measuring the quality of the resulting clusters. What we expect is to have the best result when $K = 6$. If the best result is smaller than that, it means that some concentrations are mixed. Conversely, if it is larger, concentrations are split. Notice, as mentioned before, to handle the stochastic nature of K-means, for each K, we run the algorithm several times and get the best value.

Although the straightforward choice for measuring the quality of clusters produced by the K-means technique would be the SSE measure, it is usually not the right choice. SSE tends to improve as K increases since as clusters get smaller or more compact, the distances to the centroids reduce. To avoid this problem, the separation between clusters also needs to be considered, so the best clustering result is the one in which the clusters are compact and well-separated. This is measured using the silhouette coefficient [12], computed as

$$s = \frac{1}{N}\sum_{i}^{N}\frac{b_i - a_i}{\max(a_i, b_i)} \tag{10.3}$$

where a_i is the average distance between x_i and the samples belonging to the same cluster of x_i, and b_i is the average distance between x_i and all other samples in the closest cluster. The closest cluster is the one with the smallest average distance to x_i. The silhouette varies in the range [−1, 1]. Values close to zero indicate overlapping clusters, negative values indicate that instances were assigned to wrong clusters, and values close to one indicate well-composed clusters [12].

Figure 10.1 presents the clustering analysis of the three sensing units using the silhouette coefficient. The PAH/phytase and PAH/PVS units' best results are with $K = 7$, showing that they detect seven groups of similar samples, splitting some of the five concentrations or the buffer (we will come back to this soon) into two groups. The bare electrode detects only two groups, so mixing the existing concentrations. Notice that the PAH/phytase presents the larger values for the silhouette, indicating a better distinguishability power than the other sensors. Although the best result for it is for $K = 7$, the result for $K = 6$, which is the optimal result, is not much different. However, the $K = 7$ for the PAH/PVS and $K = 2$ for the bare electrode are more prominent, indicating that these sensing units are 'more confident' on the wrong answer.

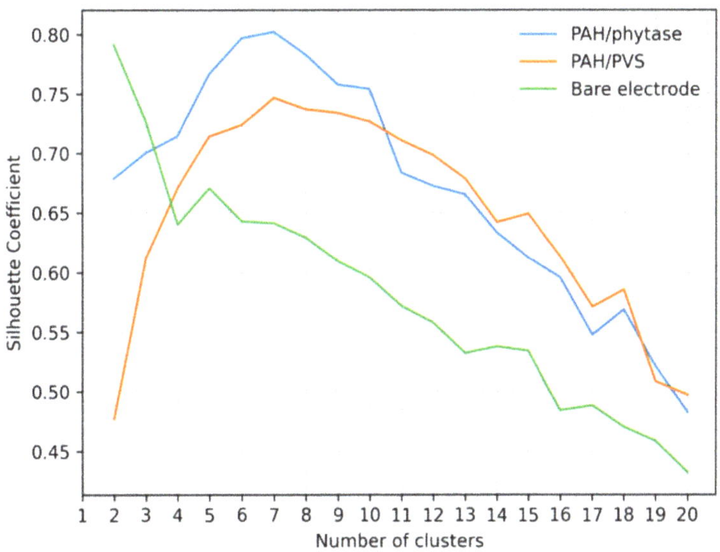

Figure 10.1. Silhouette coefficient analysis of the three sensing units varying the number of clusters. The bare electrode distinguishes only two groups. The PAH/phytase and the PAH/PVS detect seven, with a better distinguishability power associate with the PAH/phytase (larger value of silhouette).

The silhouette analysis limitation ignores the real classes or the ground truth. It only uses the calculated clusters and distance. Hence, we do not know if the created clusters are composed of instances of the same real class or concentration. The silhouette is more useful when the ground truth is unknown, but in the specific case of sensing units, such information is available. We have the real class (concentration) a sample belongs to. In this scenario, it is preferred to use two different measures, homogeneity and completeness [13]. We say a clustering result is homogeneous if the computed clusters only contain instances of the same real class, in our case, concentration. A clustering is complete when all the same real class instances are assigned to the same cluster. The best clustering result is the one having high homogeneity and completeness. Typically, these two measures are combined giving rise to the V-measure score, computed as:

$$\text{V-measure} = \frac{(1+\beta)_{*}\text{homogeneity}_{*}\text{completeness}}{\text{beta}_{*}\text{homogeneity} + \text{completeness}} \tag{10.4}$$

where β is the weight attribute to homogeneity versus completeness, usually set to one. The V-measure score ranges in [0,1], with one indicating a complete and homogeneous clustering [13].

Figure 10.2 presents the results of the V-measure score for the three sensing units. For the PAH/phytase unit, the best results happen when $K = 6$ attaining the best value possible equals 1. It means that the generated clusters only contain instances of the same real class, and the instances of the same class are assigned to the same cluster. The best value is attained for the bare electrode and the PAH/PVS units

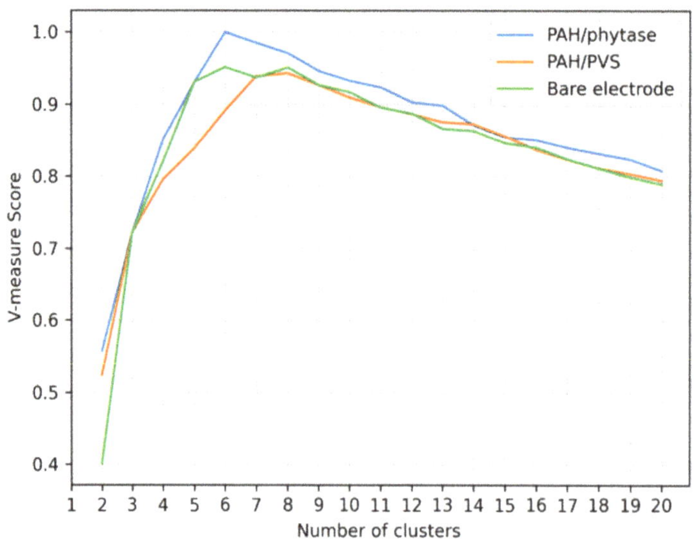

Figure 10.2. V-measure score analysis of the three sensing units varying the number of clusters. The PAH/phytase results in six pure and complete clusters, indicating a perfect result in distinguishing the five concentrations and buffer. The bare electrode and the PAH/PVS present better results with seven clusters, showing that at least one concentration is split.

when $K = 8$, and it is below 1. This indicates that the clusters are not pure nor complete, with an interesting result indicating that the bare electrode is slightly better than the PAH/PVS. It is difficult to understand why this happens, but we will use visualization to help in interpreting the results. Visualization is presented in section 10.3.

10.2.1.2 Discussion and roadmap

In this section, we presented a simple protocol to analyze sensing units using clustering. In summary, the idea is to run a clustering algorithm several times, varying the number of clusters, and getting the best result for each number. Here we present the silhouette coefficient and the V-measure score as alternatives to measure a clustering quality. However, many other different metrics can be found in the literature [3]. Despite that, the core idea is the same: to check if the detected clusters match with the real classes, for the sensing units, concentrations. In our example with five concentrations and a buffer, the best sensing unit should present six pure and complete clusters (V-measure) compact and well separated (silhouette).

Throughout the analysis, we have used the K-means technique. Notice that K-means is one option among other possible techniques. K-means defines globular clusters and may not be the best option. For clusters with varying shapes, DBScan [14, 15] is a better option. DBScan creates clusters based on density, so segmenting the data sets into non-hyper-spherical groups. Hierarchical clustering [16] is also another option. Independently of the technique being used, the analytical protocol is similar. Check if the technique's best results match the real classes (concentrations) —notice that DBScan cannot be evaluated using silhouette. However, the most important point is interpreting the results from the perspective of the type of generated clusters. In this case, K-means is usually one of the most appropriate for analyzing the distinguishability power of sensing units. Its process follows the idea of creating compact and different groups, that is, to find groups presenting the best separability.

To summarize this section, figure 10.3 presents the roadmap for clustering analysis of sensing units. In the first step, the data needs to be normalized and the distance defined. The normalization is necessary to reduce disparities in the dynamic range of each variable/frequency. Each variable/frequency must be normalized individually, for instance, making the average of each frequency equal zero and the standard deviation equal one. In terms of distance, typically, the Euclidean distance is a good choice since it is easy to understand. After that, the clustering technique is executed, varying the number of clusters. If the technique presents any random behavior, it must be executed several times per each number of clusters, getting the best result. Finally, in the last step, quality metrics evaluate the resulting clustering for each different number of clusters. It is expected that the best number of clusters be equal (or close) to the number of concentrations (or real classes). There are different ways to measure the quality of clustering. We suggest using the silhouette coefficient to measure the groups/concentrations' distinguishability and the V-measure score to measure their purity and completeness. Good sensing units will present high values for both the silhouette and V-measure.

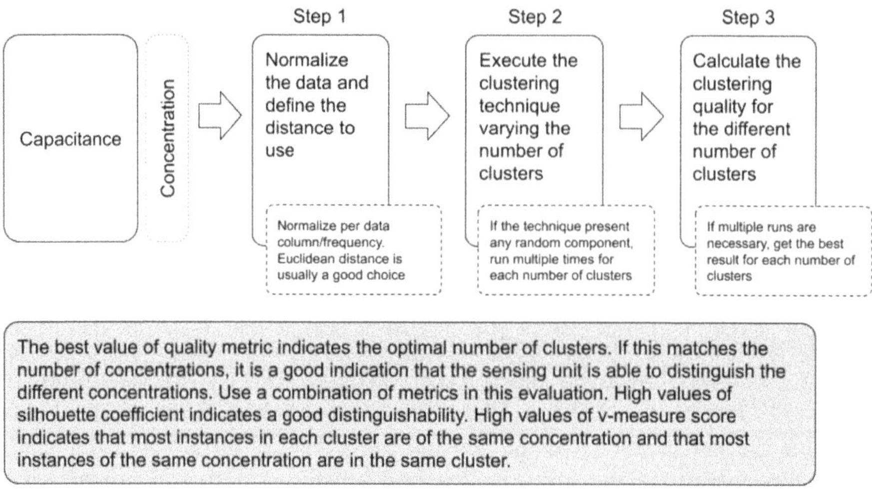

Figure 10.3. Roadmap for clustering analysis of sensing units. First, normalize the data and define the distance to use. Then execute the clustering varying the number of clusters. A good sensing unit will present the detected number of clusters equal to the number of real classes (or concentrations).

10.2.2 Classification and regression

Clustering is very useful to check the distinguishability of sensing units, but it cannot be used for prediction. In the phytic acid example, clustering cannot be used to discover the concentration of such acid in an unknown sample. In this case, classification or regression are the choices. Unlike clustering, both classification and regression models require a 'target' variable in the data set that will be predicted or approximated by the technique. For regression, this target variable is continuous, whereas, for classification, it is categorical ('yes' and 'no' or 'Class A,' 'Class B,' and 'Class C,' etc).

In the process of creating a prediction model, one of the most critical elements is its evaluation. As for clustering, different quality metrics can be used. However, this process needs to follow a particular protocol to ensure that the produced model is also generic and unbiased. That is, it is accurate when applied to unknown samples, not only for the ones used to create the model. To create unbiased models, usually, a data set is split into two sets [17]: the TRAIN for model inference and TEST for model evaluation using the performance measures. It is worth mentioning that these two sets must be completely disjoint. The TEST set plays the role of unknown instances, whereas the model learns how to approximate the target variable using the TRAIN set. Figure 10.4 depicts the phytic acid data set split into TRAIN and TEST sets following the standard approach of 70% for training and 30% for testing, stratifying the instances between the sets according to the target column. In this way, the whole dataset with 50 instances, 10 per each target (phytic acid concentration, 10^{-2}, 10^{-3}, 10^{-4}, 10^{-5}, and 10^{-6} M) results in a TRAIN set with 35 (70%) instances (seven per each target) and a TEST set with 15 (30%) instances (three per each target). In this section, we removed the buffer from the prediction since the target variable is concentration.

Instance	C (nF) @ F10000000 (Hz)	C (nF) @ F7943282 (Hz)		C (nF) @ F2 (Hz)	C (nF) @ F1 (Hz)	Target				
1	-2.43235380e-01	-3.43870020e-01		1.10277420e+02	1.14119274e+02	10^-2	K1	KFOLD CROSS-VALIDATION	MODEL SELECTION	TRAIN
2	-2.38352120e-01	-3.26316800e-01		9.97104760e+01	1.02495740e+02	10^-3				
3	-2.37502800e-01	-3.22933740e-01		9.91222500e+01	1.00881260e+02	10^-4				
4	-2.37578760e-01	-3.23414050e-01		1.02162815e+02	1.04538710e+02	10^5				
5	-2.38809170e-01	-3.25497380e-01		1.03510840e+02	1.06443906e+02	10^-6				
⋮			…	⋮			⋮			
31	-2.43142920e-01	-3.43695270e-01		1.10213854e+02	1.13470390e+02	10^-2	K7			
32	-2.38115830e-01	-3.25918850e-01		9.88327300e+01	1.00781705e+02	10^-3				
33	-2.37768500e-01	-3.23708920e-01		1.01613644e+02	1.03935710e+02	10^-4				
34	-2.37770660e-01	-3.23897000e-01		1.01689260e+02	1.03848800e+02	10^5				
35	-2.38669300e-01	-3.25857480e-01		1.03570690e+02	1.06992890e+02	10^-6				
36	-2.43229740e-01	-3.43932400e-01		1.10739435e+02	1.13990610e+02	10^-2	R²	RECALL	PERFORMANCE MEASURES	TEST
37	-2.38101950e-01	-3.25865360e-01		9.89150800e+01	1.00496390e+02	10^-3				
38	-2.37514650e-01	-3.23095080e-01		9.93530240e+01	1.02341400e+02	10^-4				
39	-2.37635940e-01	-3.23503340e-01		1.01456530e+02	1.04707300e+02	10^5	MSE	PRECISION		
40	-2.38562670e-01	-3.25526700e-01		1.04374095e+02	1.04374095e+02	10^-6				
⋮			…	⋮						
46	-2.43121400e-01	-3.43757360e-01		1.10283466e+02	1.13643510e+02	10^-2				
47	-2.38273620e-01	-3.26353440e-01		1.01200050e+02	1.03495740e+02	10^-3	RMSE	ACCURACY		
48	-2.37955500e-01	-3.23761130e-01		1.00463490e+02	1.03591205e+02	10^-4				
49	-2.37470020e-01	-3.23166660e-01		1.01418960e+02	1.03611356e+02	10^5				
50	-2.38785700e-01	-3.26078100e-01		1.04085540e+02	1.06622730e+02	10^-6				

Figure 10.4. Phytic acid data set split model for inference and evaluation of prediction models. TRAIN model selection (KFold Cross-Validation, seven folds) and creation (whole TRAIN set), and TEST for model evaluation using performance measures, such as accuracy, precision, and recall for classification models. RMSE, MSE, and R^2 for regression models.

Typically, classification and regression techniques receive different hyperparameters as input that influences the final results (as the number of clusters for the K-means technique). To find the best parametrization, a model selection experiment can be performed using the KFold Cross-Validation technique [17, 18]. In the KFold Cross-Validation, a data set is split into k folds (stratified by the target). A particular fold is left out to be used to evaluate, via performance measures (validation), the model created with the remaining folds. This process is repeated k times until each fold is used for validation, and the performance average is taken as the final measure. In the model selection experiment, each combination of values (from a finite range) for the hyperparameters is assessed by a KFold Cross-Validation. The best parametrization is the one that results in the best KFold Cross-Validation performance. Finally, a model is created using the whole TRAIN set using the best combination of hyperparameters. Figure 10.4 depicts the KFold Cross-Validation for model inference splitting the TRAIN set. A total of seven folds are randomly created, stratifying the instances between the folds according to the target column. In this way, each fold has five instances (one per target).

The settings presented in Figure 10.4 are used to select, create and evaluate regression and classification models in the following sections. It is worth noting that the classification and regression models reported here are only for didactic purposes. They must not be taken as a benchmark for the domain of phytic acid concentrations prediction.

10.2.2.1 Regression analysis example

As a regression model technique, we have chosen partial least squares (PLS) to exemplify this section. PLS is well adopted in the field of chemometrics [17, 19], used

Table 10.1. PLS model selection varying the number of components considering different performance measures. Even the bare electrode can predict the concentration of phytic acid with high precision.

Unit	Number of components	RMSE	MSE	R^2
Bare electrode	4	0.0001	1.865E-08	0.999
PAH/phytase	5	0.0001	8.864E-09	0.999
PAH/PVS	5	0.0007	4.938E-07	0.968

as a calibration model for sensing data, and multivariate calibration when more than one variable is used [19]. For PLS models, the data must be normalized (or scaled). In this way, the TRAIN and TEST sets are standardized individually (after the split), removing the mean of each frequency and dividing by the standard deviation. It is worth mentioning that standardization must be done after the split between TRAIN and TEST. Executing this process before the split results in data leakage, generating super estimated performance measures. The reason is simple. In real-world applications, the TEST data is not available when creating the prediction model, so the TRAIN normalization should not use any TEST information.

The number of components used to create a PLS regression model is a hyper-parameter. In this way, ranging from 1 to 5 components, we perform a model selection (five possible components, five possible PLS models, built using seven folds, figure 10.4) to choose the number of components that result in the best KFold Cross-Validation performance.

Table 10.1 presents the best number of components for the three different sensing units with the performance measures RMSE (root mean square error), MSE (mean square error), and R^2 (coefficient of determination) obtained on the TEST. Even for the bare electrode, the best PLS models provided excellent results, with low values for RMSE and MSE errors and high R^2 (between 0 and 1), where 1 indicates that the PLS model predicted the target (phytic acid concentration) perfectly.

The PLS regression model analysis can be performed using a bar chart of the model coefficients. This is useful to indicate how each frequency contributes to the final prediction. Figure 10.5 presents the PLS model visual representation for the PAH/PVS sensing unit (third row in table 10.1). Frequencies from 398 107 (Hz) to 5011 873 (Hz) mostly contribute to predicting low concentration instances (red bars). Frequencies from 39 810 (Hz) to 316 227 (Hz) mostly contribute to the prediction of high concentrations (blue bars). Despite being fast and determinist, the PLS technique fits linear models and may fail to map non-linear relationships between variables and target values [17].

10.2.2.2 Classification analysis example

We have chosen the random forest (RF) [20] as the technique to exemplify classification on sensing units. RF is an ensemble approach of decision trees (DTs), commonly used as a classification or regression method, that can provide

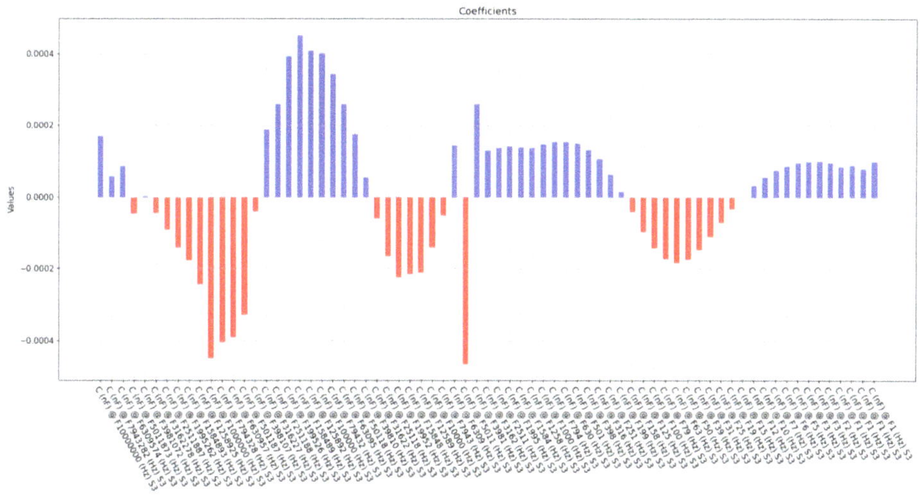

Figure 10.5. Visual analysis of PLS coefficients for the PAH/PVS sensing unit. Different frequency ranges contribute to predicting high (in blue) and low concentrations (in red).

good performance measures [20, 21]. Although more complex models, such as deep neural networks, are also good options for achieving high performance in classification tasks, they can suffer overfitting. Overfitting is the phenomenon where a model learns the TRAIN set correctly but predicts poorly the TEST set [17]. In other words, it is suitable for the known instances but does not generalize well for the unknown cases. RF models use the bagging strategy (bootstrapped training samples) and are less susceptible to overfitting [17, 20].

Regression problems, that is, continuous numbers as target values (e.g., phytic acid concentrations), can be transformed into classification problems (multiclass) by the discretization of the target values [22]. In this way, the phytic acid concentrations 10^{-2}, 10^{-3}, 10^{-4}, 10^{-5}, and 10^{-6} M can be seen as five concentration classes. DT and RF models do not require special data preprocessing (e.g., normalization). It can be used as produced by the sensing unit or its transformation to capacitance. As the PLS, the RF technique presents different hyperparameters, such as number of trees and split criterion. Therefore, we use the same KFold Cross-Validation process to select a good prediction model: create different models varying the hyperparameters and validate using the seven folds; using the best parametrization to create the RF model using the entire TRAIN set. It is worth mentioning that there are other less time-consuming strategies to select good RF models [21, 23]. It is out of our scope to discuss them.

When running this classification protocol for the three sensing units, all of them resulted in perfect accuracy, precision, and recall (the usual quality metrics for classification). From an analytical perspective, there is not much to discuss beyond that the three sensing units can effectively classify phytic acid concentrations. For the sake of illustration, we select the worse model for each sensing unit for the rest of this section. Table 10.2 presents different performance measures considering the

Table 10.2. Evaluation of one RF model using accuracy, precision, and recall used to classify phytic acid concentration considering three distinct sensing units.

		Class 10^{-2}		Class 10^{-3}		Class 10^{-4}		Class 10^{-5}		Class 10^{-6}	
Unit	**Accur.**	**Prec.**	**Rec.**	**Prec.**	**Rec.**	**Prec.**	**Rec.**	**Prec.**	**Rec.**	**Prec.**	**Rec.**
Bare electrode	100%	100%	100%	100%	100%	100%	100%	100%	100%	100%	100%
PAH/phytase	93%	100%	67%	75%	100%	100%	100%	100%	100%	100%	100%
PAH/PVS	60%	0%	0%	38%	100%	75%	100%	0%	0%	100%	100%

TEST set of these models. The RF model using the PAH/phytase unit's data presents 93% accuracy for the various concentrations. Specifically, for the concentration 10^{-2}, it presents 100% of precision and 67% of recall. So, all instances classified as class 10^{-2} are correct (100% precision), but only two out of three instances of class 10^{-2} (TEST set contains five instances per class) are classified as 10^{-2} (67% recall). A similar analysis can be performed for the concentration 10^{-3}. The model provided 75% precision and 100% recall, meaning that 75% of the instances classified as 10^{-3} are correct, and all 10^{-3} instances are classified correctly (100% recall).

In classification tasks, it is recommended a combination of quality metrics to report results. Accuracy is an aggregate measure of the quality considering all the classes together. However, it hides information about specific classes. For instance, in our example, the RF model produced for the PAH/phytase sensing unit performs poorly to predict instances with low concentration while it excels when processing instances with large concentrations. This information is revealed when precision and recall are used.

In these examples, we have shown how to create and compare classification models considering different sensing units. However, it is also possible to use the signal of the three units together as an e-tongue and create a single classification. In general, if we have N sensing units producing the signals X_1, X_2, ...X_N considering the same analyte in the different concentrations and the same number of samples, this can be done by concatenating the signals into a single dataset X and using the above classification procedure [24]. In the phytic acid example, the number of samples remains the same, 50, but the number of frequencies is the summation of the original number of frequencies produced by each sensing unit, $71 \times 3 = 213$. Although we are discussing the use of RF, this process can be applied to any classification technique. Just be observant of problems with overfitting, next discussed.

10.2.3 Discussion and roadmap

In this section, we present the steps involved in prediction tasks, regression or classification. The central idea of such tasks is to create an accurate model that can be further applied to make predictions of unknown instances. This is called model generality, and sometimes it is neglected in the literature. The opposite of generality is model overfitting, and it may be caused by different aspects, for instance, wrong parametrizations. A classic example of classification technique overfitting due to parametrization is to set the number of neighbors in a k-nearest neighbor classifier to one [3].

In general, a predictor (classifier or regressor) can be viewed as an approximation function that combines the input data, resulting in a prediction. This combination process presents parameters (these are combination 'coefficients,' not the technique hyperparameters) that are inferred during the training process (from the TRAIN set). If the model is complex, e.g., artificial neural networks, the number of parameters is large, and large TRAIN sets are required. It is possible to train a complex with small TRAIN sets. However, the result is typically an overfitted model [25]. Therefore, as a rule of thumb, always try the simplest classification technique first (k-nearest neighbors, decision trees, etc) before testing the complex ones.

The key concept to avoid the overfitting issue is evaluation. This chapter presents the strategy of splitting the data set into TRAIN and TEST, using Kfold Cross-Validation. It is usually a good choice. However, more rigorous methods can be found in the literature, such as the KFold Nested Cross-Validation [26, 27]. We invite the interested reader to check [22]. Notice that if the prediction is not the goal, if the idea is to evaluate groups' existence, clustering analysis is the way to go.

To summarize this section, Figure 10.6 presents the roadmap for prediction analysis. The first step is to divide the data set into TRAIN and TEST sets. The TRAIN is used to construct the model, and the TEST is for validation. Usually, a 70/30 splitting is a good choice (70% TRAIN and 30% TEST), but 80/20 is also reported in the literature. After the splitting, normalize the data if required. Do not normalize the data before splitting to avoid problems with data leakage.

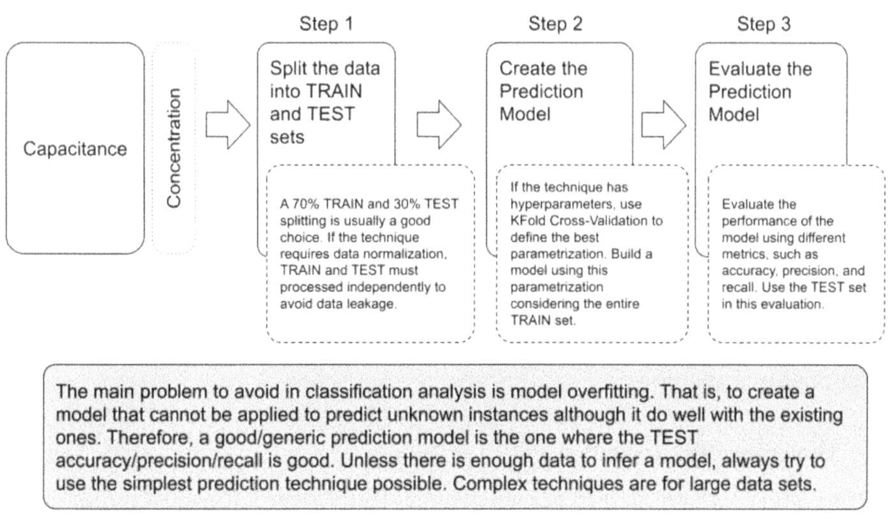

Figure 10.6. Roadmap for prediction analysis of sensing units. First, split the data into TRAIN and TEST sets. Then execute the prediction technique with different hyperparametrizations to define the best model. Evaluate the final model using different metrics. A good sensing unit will result in a generic prediction model and can be used to predict unknown instances.

The TRAIN and TEST sets must be independently normalized. After the splitting, the prediction model is created. Usually, those models have hyperparameters that need to be set. To test different parametrizations, use a KFold Cross-Validation approach, where the TRAIN data is split into subsets, and those are used to create different models varying the parameters. The best parametrization is then used to create the final model considering the entire TRAIN set. In the last step, the model performance is evaluated using different metrics using the TEST set. The error to be reported is the TEST error. If it is low (high accuracy, precision, and recall), the model is probably generic enough to be used to predict unknown instances. It is worth mentioning that the described approach for data set splits (TRAIN, TEST, and KFold Cross-Validation on TRAIN) requires a certain number of samples. There are more suitable approaches for small data sets, such as KFold Nested Cross-Validation [26, 27].

10.3 Visualization

It is undeniable that machine learning and data mining have become essential tools for any person interested in analyzing and better understanding data. However, despite their popularity, most of the existing techniques lack fine-grain support for understanding certain data particularities. For some, even the provided results may be difficult to understand (if possible). One example is clustering. Although the silhouette coefficient can be used, up to an extent, to express how compact and separate a group (concentration) is, it is difficult to have an idea about the relationships between groups. For instance, is it difficult to conclude if there is a particular group more dissimilar to others or an instance with unexpected behavior. To help answering those questions, visualization techniques have been devised to allows users to extract more detailed insights from data using the human visual ability [28].

The idea of visualization is to communicate information using graphical representations encoding different data elements into visual variables (position, color, shape, texture, etc). There are several different approaches to visualize data depending on how the visual variables are combined. It is out of our scope to discuss the existing possibilities. Here we focus on the more popular techniques for sensing unit analysis [29], viz., the dimensionality reduction or multidimensional projections [30, 31], and the parallel coordinates [32]. Both techniques serve different analytical application tasks and are typically used together to support a detailed analysis of sensing units' distinguishability and reproducibility. We start our discussion with dimensionality reduction.

10.3.1 Dimensionality reduction

Dimensionality reduction or multidimensional projection is the most popular visualization technique for sensing units' analysis. In general terms, multidimensional projection techniques aim at mapping data instances into points on the visual space, usually, a two-dimensional plane, preserving as much as possible among the points, the distance or neighborhood relationships among the instances. More formally, given a data set $X = \{x_1, x_2, \ldots, x_N\}$ and $Y = \{y_1, y_2, \ldots, y_N\} \in R^2$ the respective points, a projection technique is a injective function $f{:}X \rightarrow Y$ that maps each x_i instance into a point y_i so that the pairwise distance $d(y_i, y_j)$ between points approaches the pairwise distance $\delta(x_i, x_j)$ between instances.

Currently, there is a multitude of techniques devoted to producing multidimensional projection layouts. The difference is how the process of approaching distances is performed. Some methods seek to preserve all the pairwise distances (as much as possible); they are called global techniques. Others seek to maintain only the distances to nearest neighbors; they are called local techniques. The choice between the two groups depends on the type of analysis performed. For sensing units' analysis, where the goal is to analyze the distinguishability power and reproducibility of a unit, global techniques are preferred since they are tailored to preserve the distances between groups of instances. Typically, local techniques are not reliable for group analysis (later, we discuss that).

In the literature, dimensionality reduction has been used for sensing units' analysis considering different applications. For example, to detect acid, salty, and umami flavors [33], to distinguish different types of soil [34], to diagnose Hepatitis C [35], Leishmaniosis, and Chagas diseases [36], to detect gluten [37], glucose and triglycerides [38], to mention a few. The most popular in this context is principal component analysis (PCA) [39]. PCA is a global technique that seeks to find orthogonal linear combinations, the principal components, which best represent the data variability. In this process, the first component presents the linear combination representing the largest variance, the second, the linear combination, orthogonal to the first component, representing the second largest variance, and so on. There are as many principal components as the number of original dimensions. However, typically, the first ones capture most of the data variance, so most of the principal components can be discarded with a small loss in the data variance information. One attractive property of PCA is that the subspace composed by the first k principal components has the smallest average deviation from X among all subspaces of dimension k. That is, among all matrices A with maximum rank equal to k, PCA produces the matrix that minimizes the norm of the reconstruction error $||X - A||$ [40]. Consequently, PCA preserves the original space's Euclidean distances into the reduced one [41].

Despite its popularity, PCA reduces the dimensions of the data set through linear combinations. So, if a dataset presents non-linear relationships among the attributes, in our case, the frequencies, the produced layout will present large errors considering the distance preservation. To circumvent this problem, non-linear techniques can be used. For sensing units' analysis, the two most prevalent methods are the Interactive Document Map (IDMAP) [42] or Force Scheme [43] and the Sammon's Mapping [44]. Both are similar. They define a cost function and optimize it using a gradient descent approach. We will focus on the IDMAP technique.

In the original IDMAP, the cost function is defined by:

$$\frac{1}{N^2} \sum_{x_i \in X} \sum_{x_j \in X} \left| \frac{\delta(x_i, x_j) - \delta_{\min}}{\delta_{\max} - \delta_{\min}} - d(y_i, y_j) \right| \tag{10.5}$$

where δ_{max} and δ_{min} are the maximum and minimum distances among samples. The problem with this formulation is that the two-dimensional layout will be normalized, not reflecting the original distances. We suggest not using such a normalization, optimizing only the module of differences between the distances. In this chapter, we do that and incorporate other elements into the gradient descent process. On the original paper, the optimization step size is fixed. Here, use an adaptative strategy to reduce the step as the optimization is executed to stabilize the optimization. Also, we randomly initialize the position of the two-dimensional points. So, as in the clustering analysis, IDMAP must be run multiple times, getting the best result for analysis.

Independent of the employed technique, a multidimensional projection always produces a layout that approximates the input data set. Unless the original data set already lies in a two-dimensional plane, some information is lost in the process. In other words, it is impossible to make $d(y_i, y_j) = \delta(x_i, x_j), \forall\, x_i, x_j \in X$. The layout always presents some degree of error. There are different ways to evaluate a projection's quality, but, typically, for global projections, stress is the standard metric. The most popular definition for stress is given by Kruskal [45], computed as:

$$\text{Stress} = \sqrt{\frac{\sum_{i,j}(\delta(x_i, x_j) - d(y_i, y_j))^2}{\sum \delta(y_i, y_j)^2}} \tag{10.6}$$

Stress uses the square of the differences between distances to emphasize bigger errors and normalize the result by a factor to put it in the range [0,1] so comparisons among distinct projections can be made. A projection with stress equal to zero is a perfect projection. Larger values indicate worse projections. As a general guide, based on empirical evidence, Kruskal established that a projection with stress in the range [0, 0.025] is perfect, in the range [0.025, 0.05] is excellent, in the range [0.05, 0.1] is good, in the range [0.1, 0.2] is fair, and in the range [0.2, 1] is poor [45].

As in the clustering analysis, the most critical element is how the distance between the instances ($\delta(x_i, x_j)$) is calculated. By definition, PCA always preserves the Euclidean distances between the instances (there is no option in this case). In contrast, IDMAP can be used to preserve other metrics. Again, the most important point is to understand what the distance metric is measuring and interpret the produced layout from that perspective. We recommend using the Euclidean distance for the IDMAP as well to ease the interpretation. Also, it is highly recommended to normalize the data before processing, for instance, making the mean of the variables/ frequencies equals zero, and the standard deviation equals one.

10.3.1.1 Dimensionality reduction analysis example

So, how to use a projection for sensing units' analysis? Basically, a projection layout can visually express two different features of a sensing unit, the distinguishability power and the reproducibility degree [46]. A sensing unit with high distinguishability

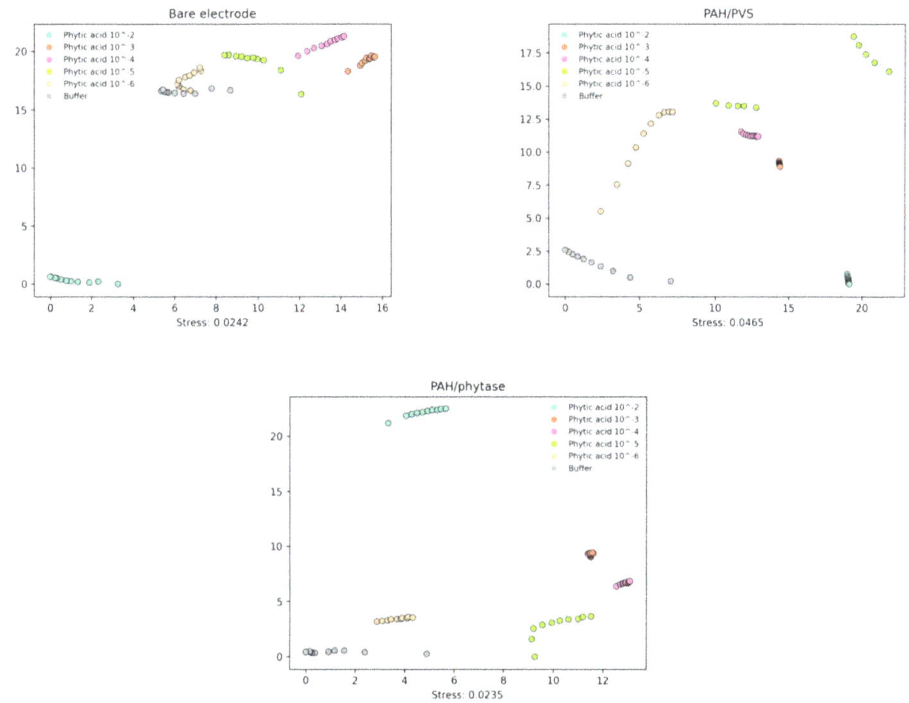

Figure 10.7. Projections of three different sensing units used to detect phytic acid concentration. A good sensing unit presents well-separated groups (distinguishability) with the points overlapped (reproducibility). In this example, the PAH/phytase is the best.

presents well-composed groups (cohesive and separate) representing the different concentrations. A unit with a high degree of reproducibility presents the points (or circles) representing the same concentration all overlapped. Highly reproducible nominally equal sensing units should produce the same signal when measuring the same concentration; this, in theory and practice may not be true.

Figure 10.7 shows IDMAP projections for the three sensing units crafted to detect phytic acid concentrations previously described. We can observe that the bare electrode mixes the smallest concentration 10^{-6} (yellow) and the buffer (gray), and it looks like two major groups are composed, one with the highest concentration 10^{-2} (teal) and another with the rest. Also, the reproducibility is low since the points are spread. The PAH/PVS separates all the concentrations but, interestingly, splits the concentration 10^{-5} (light green) into two groups. In terms of reproducibility, the lowest concentrations are spread, while for the highest 10^{-3} and 10^{-2} (orange and teal) there is a good overlap. The PAH/phytase presents the best results, with all concentrations well separated and less spread. In this sensing unit, the concentrations with good overlap are different, 10^{-4} and 10^{-3} (orange and purple). According to Kruskal classification, the bare electrode (stress: 0.0242) and PAH/phytase (stress: 0.0235) projections are perfect, while the PAH/PVS is excellent (stress: 0.0465). This indicates that the data produced by the PAH/PVS

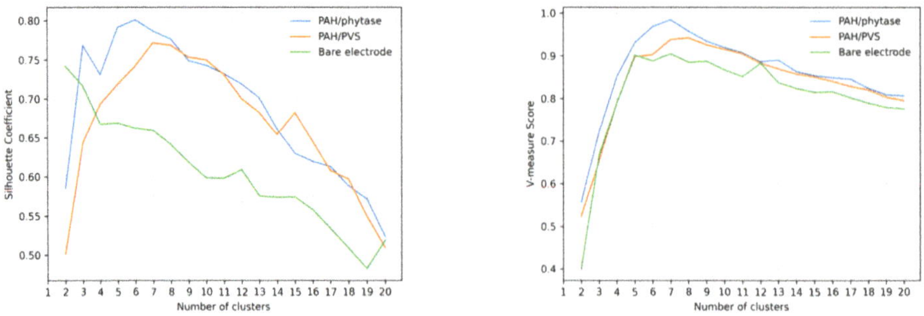

Figure 10.8. Clustering analysis combined with dimensionality reduction. The clustering analysis previously presented is applied using the projections that are input. This type of analysis can hide important information given the fact that the dimensionality reduction is an approximation. Whenever possible, it is recommended to use clustering and dimensionality reduction together.

sensor is more complex than the other two sensors, needing more than two dimensions to be well represented. Notice that the stress does not reflect the quality of a sensing unit. It is only a numeric evaluation of the projection's quality and reliability.

At this point, the careful reader may have noticed that the clustering and dimensionality reduction strategies have similar analytical goals. Both aim to evaluate the cohesion and separation of groups, in our case, concentrations. Although they may result in similar conclusions, for instance, the previous analysis of figure 10.7 aligns quite well with the clustering analysis presented in figure 10.2, where the bare electrode detects two major groups, the PAH/PVS seven and the PAH-Phytase six, with more cohesive and separate groups (figure 10.1), they are complementary and should be used together. Clustering analysis is a quantitative evaluation in which the clustering validity metrics can be used to measure and compare different sensing units in terms of the existing groups. On the other hand, dimensionality reduction is a qualitative analysis, allowing one to visually check groups cohesion and separation in detail, instance per instance.

Notice that the combination of clustering and dimensionality reduction has already been proposed in the literature, for instance, to analyze sensing units' discrimination power of black tea [47] and Portuguese honey [48]. A PCA plot is first created in both cases. Then the resulting layout is evaluated using different clustering quality metrics. We only recommend using this type of analysis if the projection is very reliable, if its stress is relatively low. Remember that a projection is an approximation of the original data and may lose important information. Figure 10.8 presents a clustering analysis using as input the projections of figure 10.7. It is possible to notice that, compared to figures 10.1 and 10.2, most of the peaks of the best number of groups change for the silhouette coefficient and V-measure score metrics. Even though the projections are of perfect and excellent quality, they are still approximations. We suggest using both analyses together whenever possible.

10.3.1.2 Discussion and roadmap

The use of dimensionality reduction to analyze data is quite widespread, not only for sensing units but also in other domains, such as biology [49], political sciences [50], and sociology [51]. As mentioned, there is a multitude of available techniques that can be used. The choice depends on the type of analysis being performed. For sensing units, usually, it makes more sense to use global techniques. Local techniques typically are not good choices. One example is the well-known t-SNE [52]. t-SNE is very good to create groups. However, it presents some significant limitations. We suggest the interested reader check [53] for more details, but, in summary: (1) its parametrizations may create groups that do not exist or may hide existing groups; (2) the sizes of the clusters do not reflect how cohesive a group is; (3) distance between groups might not reflect the separation between groups; (4) and the most important, it does not preserve distances. Thereby, when reading a layout, the distances between points do not reflect the similarity among instances. They reflect the probabilities of the instances to be neighbors in the original space. If the analysis being performed is not based on similarity, and group cohesion and separation are not part of the conclusions, it is acceptable to use t-SNE. Otherwise, we recommend PCA or IDMAP.

As mentioned, the dimensionality reduction always incurs errors, and, usually, this is not equally spread across the points. To check that, one possibility is to map on top of each point its error. Figure 10.9 presents examples of mapping the stress

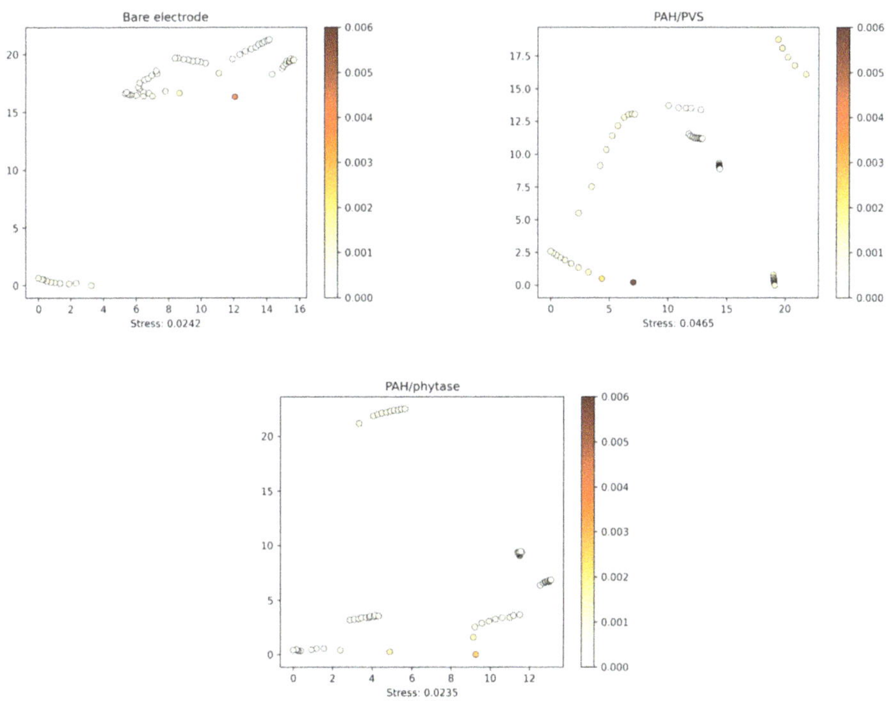

Figure 10.9. Projections of figure 10.7 colored according to the error. Most of the points are projected with high accuracy. Only a few present some errors.

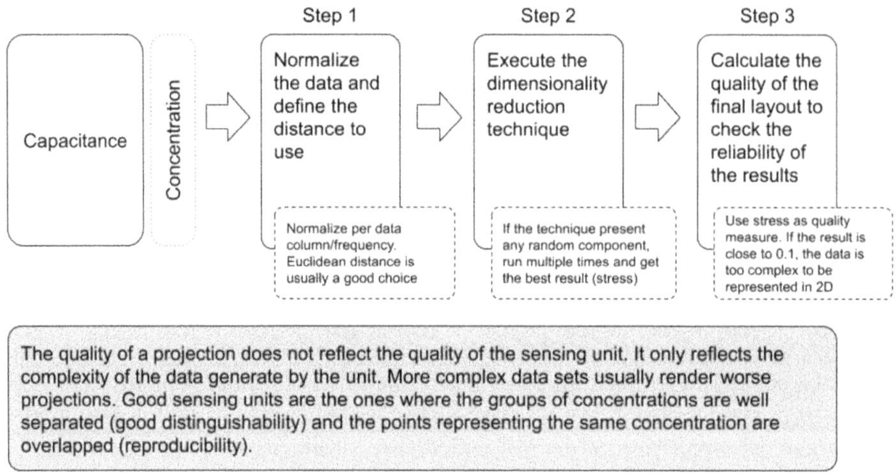

Figure 10.10. A roadmap for sensing unit analysis using dimensionality reduction techniques. The first step is to normalize the data to avoid problems with differences between the frequencies. After that, the technique can be executed multiple times, getting the best layout for analysis. Finally, the layout's overall quality must be computed to verify its reliability.

per point of each projection of figure 10.7. For most points, the error is very low. Only one or a few points per projection present some error. So, as indicated by the Kruskal classification, most of these projections' points are of perfect and excellent quality. Notice that other metrics can be used to measure the quality of a projection, such as neighborhood preservation [54] or trustworthiness [55]. However, it is not our goal to show how to analyze the quality of a projection, and from the perspective of distance and global structure (distance between groups), stress is the best choice.

To summarize this section, Figure 10.10 presents the roadmap for dimensionality reduction analysis of sensing units. In the first step, the data needs to be normalized and the distance defined. As clustering, dimensionality reduction is based on distance, so normalization is necessary to reduce disparities in the dynamic range of each individual variable/frequency. In terms of distance, typically, the Euclidean distance is a good choice since it matches the human perception, and, as mentioned before, it is easy to understand. After the normalization, the dimensionality reduction technique is executed. If it presents any random components (e.g., initialization), run it multiple times and get the best result. In the last step, the quality of a projection is measured. There are different ways to do that. For sensing unit analysis, where the goal is to evaluate group (concentrations) formation, a global metric is preferred, like the Kruskal stress. Notice that the idea is to evaluate the projection to check if it is reliable. Stress, or any other projection quality metric, cannot be used to evaluate the sensing unit. Stress is not about data quality; it only reflects projection faithfulness. If fair or poor projections are generated, maybe the data is too complex to be embedded into two dimensions, and projections may not be a reliable option for analysis. If the projection presents good quality, a good sensing unit will be the one where the groups of concentrations are well separated (good distinguishability), and points representing the same concentration are overlapped (reproducibility). The interesting property of a projection is that this type of analysis can be visually executed per group, evaluating for which concentrations the sensing unit presents a good distinguishability power and a high degree of reproducibility— something not possible to be performed using clustering analysis.

10.3.2 Parallel coordinates

The limitation of dimensionality reduction techniques is the lack of ability to answer why groups are composed or why a specific instance differs from the others. To address this issue, parallel coordinates [32] can be used, shedding some light on how the data attributes (frequencies) influence the similarity among instances. On parallel coordinates, each data dimension, in our case, frequency, is represented as a vertical axis that is parallel to the other axes. An instance is represented as a polyline connecting the axes at positions proportional to the values the instance present in each dimension. Therefore, it is possible to check and compare the instances or groups of instances considering distinct frequencies or frequency ranges.

In the literature, parallel coordinates have been used for the analysis of sensing units' applied to different scenarios, for instance, to distinguishing soil samples enriched with plant macronutrients [34], to detect catechol with high sensitivity [56], and to detect phytic acid [57]. In all these examples, the idea is the same, to use parallel coordinates to understand a dimensionality reduction layout. Figure 10.11 presents

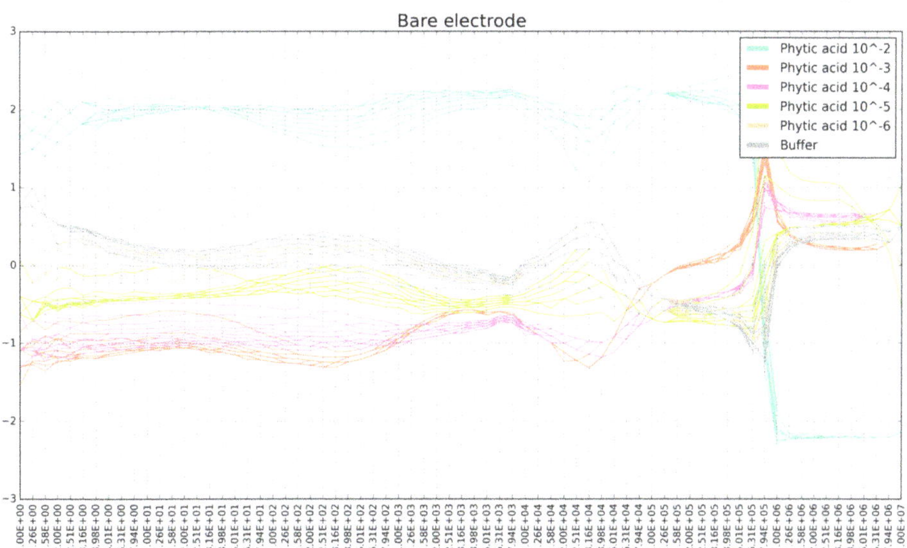

Figure 10.11. Parallel coordinates for the three sensing units built to detect phytic acid concentrations. Used in conjunction with the projection of figure 10.7, these visual representations can help understand the patterns observed in terms of group formation and separation.

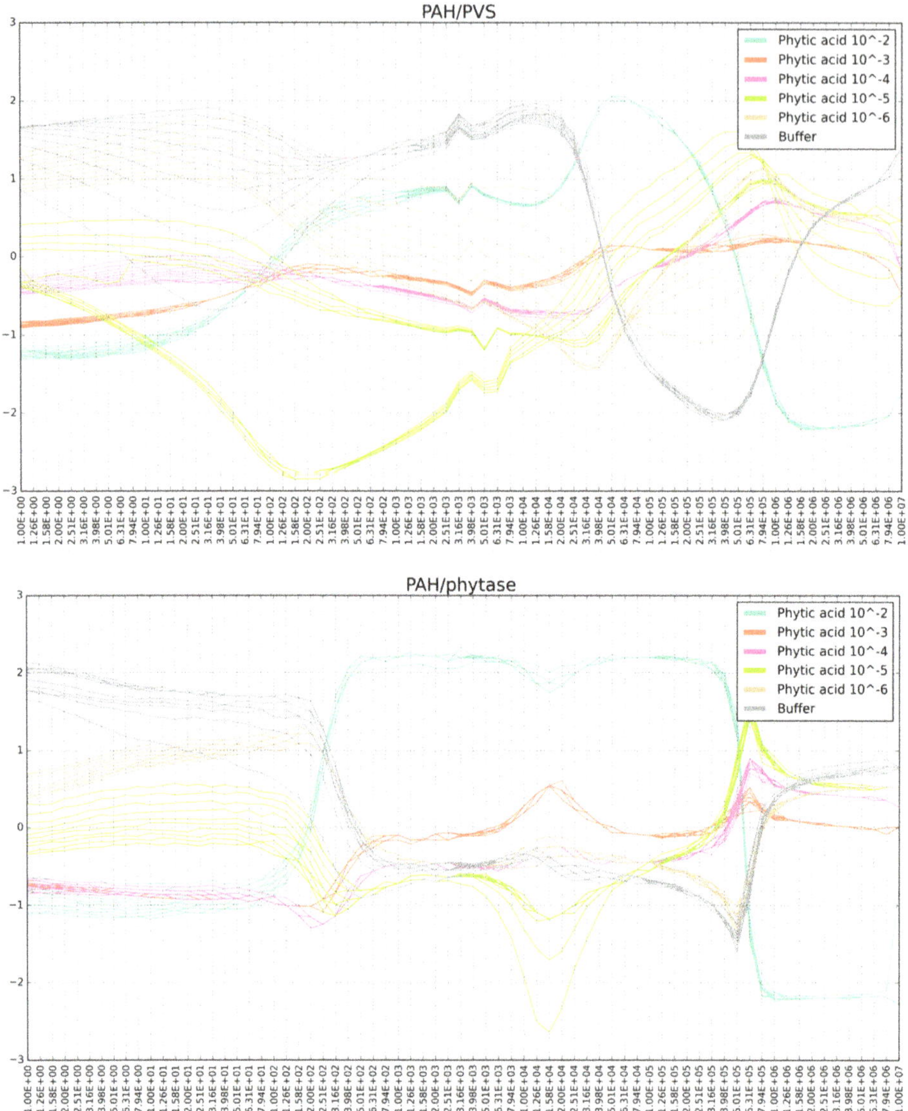

Figure 10.11. (Continued.)

parallel coordinates layouts for the three sensing units previously described, designed to detect phytic acid concentrations. As discussed, lines represent the instances, color is the concentration, and the (dashed) vertical axes represent the frequencies. The lower frequency is on the left, the largest on the right. Notice that, in this example, the data was normalized following the dimensionality reduction normalization, removing the average of each column and dividing by the standard deviation.

Since our goal is to help interpret a projection, it is crucial to use the same (normalized) data. Also, using such a normalization, some data patterns are visible that are obfuscated by the differences in dynamic ranges of signals captured in different frequencies (which is usually the norm since the magnitude of the signal varies considerably among the frequencies).

The bare electrode's parallel coordinates representation indicates that the largest concentration 10^{-2} (teal) is very different from the others. The buffer (gray) and the lowest concentration 10^{-6} (yellow) instances are mixed, and instances of the same concentration are spread, indicating a low degree of reproducibility for this sensing unit. This explains the patterns observed in the projection of figure 10.7. The PAH/PVS parallel coordinates reveal that large concentrations 10^{-2} (teal), 10^{-3} (orange), and 10^{-4} (pink) instances are very similar (almost the same lines), with a little spread for the 10^{-4} instances considering the low frequencies. Buffer instances are much more spread on the lower-mid frequencies but are concentrated for large frequencies. The 10^{-6} instances are spread, and the 10^{-5} is divided into two groups on almost all frequency spectra. This explains the observations made for the PAH/PVS projection presented in figure 10.7. The PAH/PVS works quite well for the larger concentrations, especially for mid-high frequencies. The instances of each different concentration present very similar values while these are different among concentrations. However, this sensing unit presents problems as the frequency decreases.

Finally, the parallel coordinates representation for the PAH/phytase sensing unit shows that the instances are spread in the lower frequencies for most of the concentrations. In contrast, most are concentrated for mid-high frequencies, especially for high frequencies. Again, the patterns presented on the projections of figure 10.7 can be observed here. The instances of concentrations 10^{-3} and 10^{-4} are very compacted, including in the low frequencies. The 10^{-2} instances are quite different from the others. Notice in the mid-frequencies for the concentration 10^{-5}, one instance is very different from the others. This can also be noticed on the projection. In general, in this sensing unit, the measures are more concentrated and separated than the bare electrode and PAH/PVS units. The parallel coordinates confirm the conclusions attained from the clustering and dimensionality reduction but in a more detailed level.

10.3.2.1 Discussion and roadmap

Parallel coordinates is a well-established technique for multidimensional data visualization, but it is not the only one [58]. Many other techniques could be explored for the analysis of sensing units, such as pixel-oriented techniques [59], grand tour [60], and RadViz [61]. Visualization is a rich field that is growing fast given the necessity to involve humans in the data analysis cycle. Whenever a more refined grain level analysis of data is needed, visualization techniques, or a combination of techniques, for instance, dimensionality reduction layouts with parallel coordinates, are very helpful resources.

One crucial point is that for data where the dimensions are continuous values and related, parallel coordinates are reduced to line graphs (y-coordinates are capacitance and the x-coordinates are frequencies). The difference is the normalization. Due to the

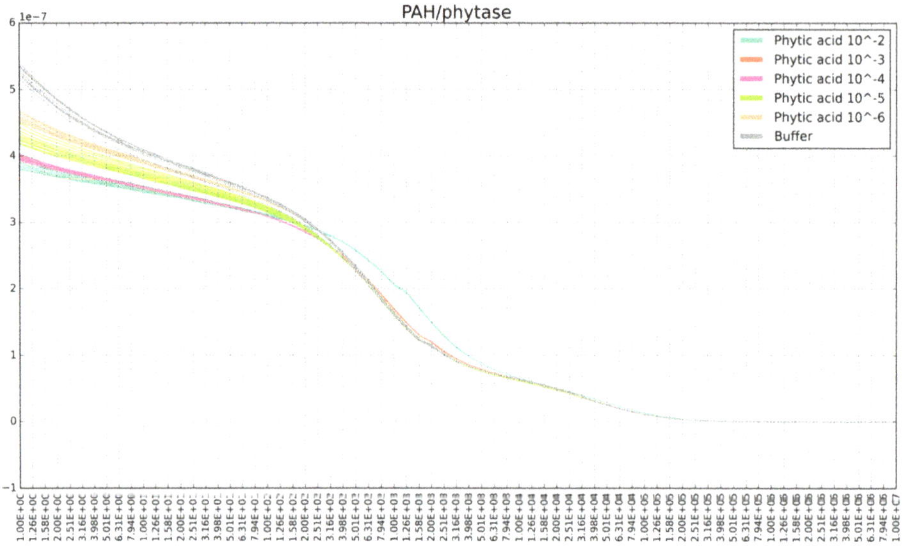

Figure 10.12. Parallel coordinates without normalization. Without normalization, for sensing units, the parallel coordinates are reduced to line plots. However, this hides important information since, for data analysis, usually magnitude is not the most relevant factor. The separation between the groups is.

normalization, the magnitude of the signal is ignored in the visual representation, and the focus of the analysis is on the separability and reproducibility of the signal. For instance, in figure 10.12, we remove the normalization from the data, and it is clear the response in the lower frequencies is much stronger than on the high frequencies. However, compared to figure 10.11, it is clear that the instance of the same concentration is much more similar between themselves and separate from the other concentrations on the high frequencies. Notice that magnitude is not important for data analysis tasks involving the distinguishability power and reproducibility degree. The difference between groups is.

To summarize this section, Figure 10.13 presents the roadmap for parallel coordinates analysis of sensing units. In the first step, the data is normalized. If the analytical goal is to check frequency ranges with high separability and reproducibility, normalize the data per frequency to eliminate signal magnitude effects, we recommend normalizing removing the average of the frequencies, dividing by the standard deviation. If the parallel coordinates are used to understand dimensionality reduction layouts, use the same normalization employed to create the projection layouts. A good sensing unit will present the lines representing the same concentration very similar (overlapped) and separated from the other concentrations. From a data analysis perspective, this does not need to happen in all frequency. The selection of the frequencies to be used, for instance, for prediction, can be made using some external algorithm for frequency selection [57] or can be performed by hand, assisted by the parallel coordinates visual representation.

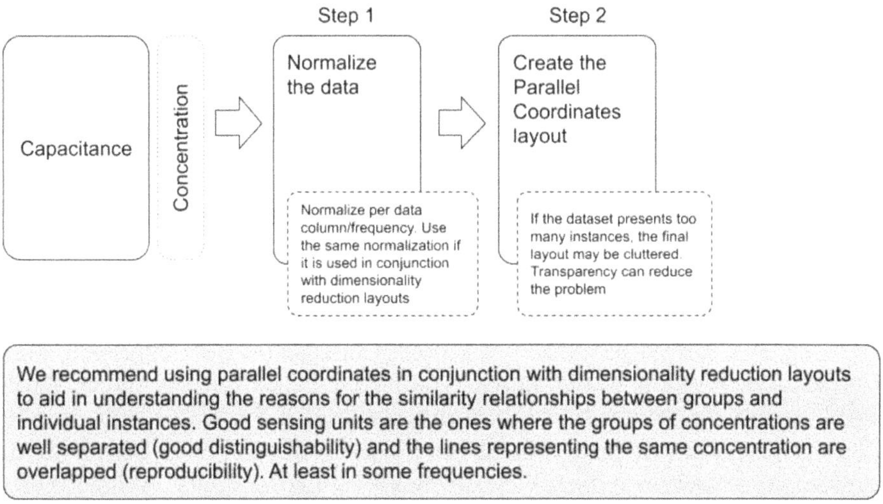

Figure 10.13. Roadmap of sensing units' analysis using parallel coordinates. If the parallel coordinates representation is intended to understand projection layouts, use the same normalization employed on the dimensionality reduction task. For large data sets, cluttered layouts may be produced. Transparency can be used to mitigate the problem.

10.4 Conclusions and the complete analytical roadmap

In this chapter, we presented how to use the most popular machine learning, data mining, and visualization approaches to analyze sensing units. Our goal was to show some technical details and how to use different techniques properly. It was not our goal to present a comprehensive survey of the existing literature in sensing analysis or discuss in-depth the elements of machine learning, data mining, and visualization. Indeed, these fields are very active, and entire books have been dedicated to the points discussed here (you can find books on classification [62], clustering [16], dimensionality reduction [41], and parallel coordinates [63]). Our idea was to show roadmaps to guide an analyst, discussing important points and considerations that sometimes are overlooked. For instance, most of these methods are heavily parametrized and/or present a stochastic behavior, so that multiple runs, combining different parameters sets, are necessary to produce reliable results. Or that a combination of metrics is usually required to understand the results produced by these techniques. Using a single metric may hide important information.

Among the tips and tricks we discussed, we think the most important is the notion that the techniques have clear different purposes and can (or should) be used in conjunction. Classification and regression are for prediction. They do not say much about the data characteristics. They are black-boxes analytical tools (especially the more complex methods). The conclusion will orbit around the distinguishability of a sensing unit, but anything further may be speculation. Recently, some attempts have been made to interpreting classification models, helping in answering questions involving the reasons for a particular prediction [24, 64–66]. Although fascinating, those techniques are out of our scope. In contrast, clustering is for the analysis of group formation, helping answer questions about distinguishability and reproducibility of a sensing unit. It cannot be used for prediction. Dimensionality reduction has similar goals but allows for a more detailed discussion of the differences between groups (concentrations) or individual instances. Finally, the parallel coordinates are typically used to explain the patterns discovered in a dimensionality reduction layout, allowing the analysis of separability considering the different frequency ranges.

A complete analytical roadmap, therefore, should be based on what is planned to be investigated or the sensing unit's final intended application. Is the goal to examine if a sensing unit can be used for prediction? Classification or regression analysis should be used. Is the objective understanding the behavior of a unit in terms of separability and reproducibility? Clustering analysis is the way to go. Is a more detailed view about where a unit fails or excels necessary? A combination of dimensionality reduction and parallel coordinates is a good choice. A dimensionality layout will help infer the property of individual groups of instances, and the parallel coordinates will help explain that considering the instances' values per frequency. All in all, a good analysis may involve all these techniques.

We have tried to cover the most critical points when using the analytical techniques presented here. It is safe to say that the outlined process is correct for similar sensing units, but different units or scenarios may demand additional considerations. For instance, if the data presents class imbalance, some class (concentration) has many more elements than others, the analysis is different. Although the proliferation of libraries and tools has substantially increased the popularity of machine learning, data mining, and visualization techniques, bad choices in technique usage (parametrization, evaluation, etc) may lead to incorrect or misleading conclusions. We highly recommend studying in-depth these analytical techniques before applying if the application deviates from what was outlined here.

References

[1] Butler D 2006 2020 computing: everything, everywhere *Nature* **440** 402–5

[2] De Mauro A, Greco M and Grimaldi M 2015 What is big data? A consensual definition and a review of key research topics *AIP Conf. Proc.* **1644** 97–104

[3] Tan P-N, Steinbach M and Kumar V 2005 *Introduction to Data Mining* 1st ed. (Upper Saddle River, NJ: Pearson)

[4] MacQueen J 1967 Some methods for classification and analysis of multivariate observations Proc. of the Fifth Berkeley Symp. on Mathematical Statistics and Probability 1 pp 281–96 Accessed: Jan. 27, 2021. [Online]. Available: https://projecteuclid.org/euclid.bsmsp/1200512992

[5] Santos M C, Bianchi A G C, Ushizima D M, Pavinatto F J and Bianchi R F 2017 Ammonia gas sensor based on the frequency-dependent impedance characteristics of ultrathin polyaniline films *Sens. Actuators A Phys.* **253** 156–64

[6] Ghosh H and RoyChaudhuri C 2015 Noise spectroscopy as an efficient tool for impedance based sub-femtomolar toxin detection in complex mixture using nanoporous silicon oxide *Biosens. Bioelectron.* **67** 757–62
[7] Savitzky A and Golay M J E 1964 Smoothing and differentiation of data by simplified least squares procedures *Anal. Chem.* **36** 1627–39
[8] Steinier J, Termonia Y and Deltour J 1972 Comments on smoothing and differentiation of data by simplified least square procedure *Anal. Chem.* **44** 1906–9
[9] Tormene P, Giorgino T, Quaglini S and Stefanelli M 2009 Matching incomplete time series with dynamic time warping: an algorithm and an application to post-stroke rehabilitation *Artif. Intell. Med.* **45** 11–34
[10] Moraes M L *et al* 2010 Strategies to optimize biosensors based on impedance spectroscopy to detect phytic acid using layer-by-layer films *Anal. Chem.* **82** 3239–46
[11] Macdonald J R and Evgenij B 2005 *Impedance Spectroscopy; Theory, Experiment, and Applications* 2nd edn. (Hoboken, NJ: Wiley)
[12] Rousseeuw P J 1987 Silhouettes: a graphical aid to the interpretation and validation of cluster analysis *J. Comput. Appl. Math.* **20** 53–65
[13] Rosenberg A and Hirschberg J 2007 V-measure: a conditional entropy-based external cluster evaluation measure EMNLP-CoNLL 2007 – Proc. of the 2007 Joint Conf. on Empirical Methods in Natural Language Processing and Computational Natural Language Learning pp 410–20 Accessed: Jan. 27, 2021. [Online]. Available: https://www.aclweb.org/anthology/D07-1043
[14] Ester M, Kriegel H-P, Sander J and Xu X 1996 A density-based algorithm for discovering clusters in large spatial databases with noise Proc. of the 2nd Int. Conf. on Knowledge Discovery and Data Mining pp 226–31 Accessed: Jan. 27, 2021. [Online]. Available: www.aaai.org
[15] Schubert E, Sander J, Ester M, Kriegel H P and Xu X 2017 DBSCAN revisited, revisited: why and how you should (still) use DBSCAN *ACM Trans. Database Syst.* **42** 1–21
[16] Rokach L and Maimon O 2006 Clustering methods *Data Mining and Knowledge Discovery Handbook* (Berlin: Springer) 321–52 pp
[17] James G, Witten D, Hastie T and Tibshirani R 2013 *An Introduction to Statistical Learning with Applications in R.* (New York: Springer)
[18] Kohavi R 1995 A study of cross-validation and bootstrap for accuracy estimation and model selection Proc. of the 14th Int. Joint Conf. on Artificial Intelligence 2 1137–43
[19] Kirsanov D, Mednova O, Vietoris V, Kilmartin P A and Legin A 2012 Towards reliable estimation of an 'electronic tongue' predictive ability from PLS regression models in wine analysis *Talanta* **90** 109–16
[20] Breiman L 2001 Random forests *Mach. Learn.* **45** 5–32
[21] Biau G and Scornet E 2016 A random forest guided tour *TEST An Off. J. Spanish Soc. Stat. Oper. Res.* **25** 197–227
[22] Salman R and Kecman V 2012 Regression as classification 2012 Proc. of IEEE Southeastcon pp 1–6
[23] Kruppa J, Schwarz A, Arminger G and Ziegler A 2013 Consumer credit risk: individual probability estimates using machine learning *Expert Syst. Appl.* **40** 5125–31
[24] Popolin Neto M, Soares A C, Oliveira O N and Paulovich F V 2021 Machine learning used to create a multidimensional calibration space for sensing and biosensing data *Bull. Chem. Soc. Jpn.* **94** 1553–62

[25] Everitt A and Skrondal B S 2010 *Cambridge Dictionary of Statistics* 4th edn (Cambridge: Cambridge University Press)

[26] Varma S and Simon R 2006 Bias in error estimation when using cross-validation for model selection *BMC Bioinf.* **7** 91

[27] Tsamardinos I, Rakhshani A and Lagani V 2014 Performance-estimation properties of cross-validation-based protocols with simultaneous hyper-parameter optimization *Artificial Intelligence: Methods and ApplicationsLecture Notes on Computer Science* vol 8445 (Berlin: Springer) pp 1–14

[28] Ward M, Grinstein G and Keim D 2010 *Interactive Data Visualization: Foundations, Techniques, and Applications.* (Nattick, MA: A. K. Peters, Ltd)

[29] Oliveira O N, Pavinatto F J, Constantino C J L, Paulovich F V and de Oliveira M C F 2012 Information visualization to enhance sensitivity and selectivity in biosensing *Biointerphases* **7** 1–15

[30] Nonato L G and Aupetit M 2019 Multidimensional projection for visual analytics: linking techniques with distortions, tasks, and layout enrichment *IEEE Trans. Vis. Comput. Graph.* **25** 2650 73

[31] Espadoto M, Martins R M, Kerren A, Hirata N S T and Telea A C 2019 Towards a quantitative survey of dimension reduction techniques *IEEE Trans. Vis. Comput. Graph.* **27** 2153–73

[32] Inselberg A 1985 The plane with parallel coordinates *Vis. Comput.* **1** 69–91

[33] Manzoli A *et al* 2014 Layer-by-layer fabrication of AgCl-PANI hybrid nanocomposite films for electronic tongues *Phys. Chem. Chem. Phys.* **16** 24275–81

[34] Braunger M L *et al* 2017 Microfluidic electronic tongue applied to soil analysis *Chemosensors* **5**

[35] Lima L R, Gonçalves A-M B, Paulovich F V, Oliveira O N Jr, Ribeiro S J L and Moraes M L 2018 Electrical immunosensor made with antigenic peptide NS5A-1 immobilized onto silk fibroin for diagnosing hepatitis C *J. Braz. Chem. Soc.* **29** 2054–9

[36] Perinoto Â C *et al* 2010 Biosensors for efficient diagnosis of leishmaniasis: innovations in bioanalytics for a neglected disease *Anal. Chem.* **82** 9763–8

[37] Daikuzono C M *et al* 2017 Information visualization and feature selection methods applied to detect gliadin in gluten-containing foodstuff with a microfluidic electronic tongue *ACS Appl. Mater. Interfaces* **9** 19646–52

[38] Moraes M L *et al* 2012 Detection of glucose and triglycerides using information visualization methods to process impedance spectroscopy data *Sens. Actuators, B Chem.* **166–167** 231–8

[39] Jolliffe I T 2002 *Principal Component Analysis.* (New York: Springer)

[40] Eckart C and Young G 1936 The approximation of one matrix by another of lower rank *Psychometrika* **1** 211–8

[41] Cox T and Cox M 2000 *Multidimensional Scaling* 2nd edn (London: Chapman and Hall/CRC)

[42] Minghim R, Paulovich F V and de Andrade Lopes A 2006 Content-based text mapping using multi-dimensional projections for exploration of document collections *Proc. SPIE* **6060** 60600S–12

[43] Tejada E, Minghim R and Nonato L G 2003 On improved projection techniques to support visual exploration of multidimensional data sets *Inf. Vis.* **2** 218–31

[44] Sammon J W 1969 A nonlinear mapping for data structure analysis *IEEE Trans. Comput.* **C–18** 401–9

[45] Kruskal J B 1964 Multidimensional scaling by optimizing goodness of fit to a nonmetric hypothesis *Psychometrika* **29** 1–27

[46] Paulovich F V *et al* 2011 Using multidimensional projection techniques for reaching a high distinguishing ability in biosensing *Anal. Bioanal. Chem.* **400** 1153–9

[47] Bhondekar A P, Vig R, Gulati A, Singla M L and Kapur P 2011 Performance evaluation of a novel iTongue for Indian black tea discrimination *IEEE Sens. J.* **11** 3462–8

[48] Ulloa P A, Guerra R, Cavaco A M, da Costa A M R, Figueira A C and Brigas A F 2013 Determination of the botanical origin of honey by sensor fusion of impedance e-tongue and optical spectroscopy *Comput. Electron. Agric.* **94** 1–11

[49] Kobak D and Berens P 2019 The art of using t-SNE for single-cell transcriptomics *Nat. Commun.* **10** 1–14

[50] Rabinowitz G B 1975 An introduction to nonmetric multidimensional scaling *Am. J. Pol. Sci.* **19** 343

[51] Amato P R 1990 Dimensions of the family environment as perceived by children: a multidimensional scaling analysis *J. Marriage Fam.* **52** 613

[52] Van Der Maaten L and Hinton G 2008 Visualizing data using t-SNE *J. Mach. Learn. Res.* **9** 2579–625

[53] Wattenberg M, Viégas F and Johnson I 2017 How to use t-SNE effectively *Distill* **1** e2

[54] Fadel S G, Fatore F M, Duarte F S L G and Paulovich F V 2015 LoCH: a neighborhood-based multidimensional projection technique for high-dimensional sparse spaces *Neurocomputing* **150** 546–56

[55] Venna J and Kaski S 2006 Visualizing gene interaction graphs with local multidimensional scaling ESANN 2006 Proc. – European Symp. on Artificial Neural Networks pp 557–62 Accessed: Jan. 27, 2021. [Online]. Available http://bioinformatics.icmb.utexas.edu/lgl/

[56] Aoki P H B *et al* 2013 Molecularly designed layer-by-layer (LbL) films to detect catechol using information visualization methods *Langmuir* **29** 7542–50

[57] Paulovich F V, Moraes M L, Maki R M, Ferreira M, Oliveira O N and De Oliveira M C F 2011 Information visualization techniques for sensing and biosensing *Analyst* **136** 1344–50

[58] Ferreira de Oliveira M C, Levkowitz H, de Oliveira M C F and Levkowitz H 2003 From visual data exploration to visual data mining: a survey *IEEE Trans. Vis. Comput. Graph.* **9** 378–94

[59] Keim D A 2000 Designing pixel-oriented visualization techniques: theory and applications *IEEE Trans. Vis. Comput. Graph.* **6** 59–78

[60] Asimov D 1985 The grand tour: a tool for viewing multidimensional data *SIAM J. Sci. Stat. Comput.* **6** 128–43

[61] Hoffman P, Grinstein G and Pinkney D 1999 Dimensional anchors: a graphic primitive for multidimensional multivariate information visualizations Proc. of the 1999 Workshop on New Paradigms in Information Visualization and Manipulation in conjunction with the 8th ACM Internation Conf. on Information and Knowledge Management *(Nov. 1999) (NPIVM)* pp 9–16

[62] Aggarwal C C 2015 *Data Classification: Algorithms and Applications* 1st edn (London: Chapman and Hall/CRC)

[63] Inselberg A 2009 *Parallel Coordinates: Visual Multidimensional Geometry and Its Applications.* (New York: Springer)

[64] Carvalho D V, Pereira E M and Cardoso J S 2019 Machine learning interpretability: a survey on methods and metrics *Electronics* **8**

[65] Guidotti R, Monreale A, Ruggieri S, Turini F, Giannotti F and Pedreschi D 2018 A survey of methods for explaining black box models *ACM Comput. Surv.* **51**

[66] Popolin Neto M and Paulovich F V 2021 Explainable matrix – visualization for global and local interpretability of random forest classification ensembles *IEEE Trans. Vis. Comput. Graph.* **27** 1427–37

www.ingramcontent.com/pod-product-compliance
Ingram Content Group UK Ltd.
Pitfield, Milton Keynes, MK11 3LW, UK
UKHW051901100926
8556IPUK00005B/716

9 780750 336857